Modeling and Management of Stochastic Systems

Modeling and Management of Stochastic Systems

Edited by **William Taylor**

CLANRYE INTERNATIONAL

New Jersey

Published by Clanrye International,
55 Van Reypen Street,
Jersey City, NJ 07306, USA
www.clanryeinternational.com

Modeling and Management of Stochastic Systems
Edited by William Taylor

International Standard Book Number: 978-1-63240-354-4 (Hardback)

Printed in the United States of America.

Contents

Preface

This book has been a concerted effort by a group of academicians, researchers and scientists, who have contributed their research works for the realization of the book. This book has materialized in the wake of emerging advancements and innovations in this field. Therefore, the need of the hour was to compile all the required researches and disseminate the knowledge to a broad spectrum of people comprising of students, researchers and specialists of the field.

Stochastic control deals with the uncertainties in data observation playing a crucial role in data evolution. Stochastic control plays a crucial role in a number of scientific and applied disciplines including engineering, finance, communications and medicine. Stochastic modeling is one of the most useful techniques for formulation of optimal decision-making strategies in applications. This book provides a compilation of exceptional investigations in different aspects of stochastic systems and their behavior. It presents a distinct analysis on practical aspects of calculus and stochastic modeling including applications derived from computer science, engineering and statistics. This book will be of great utility to readers with knowledge about stochastic calculus and basic probability theory. It will specifically serve as a useful resource for PhD students and researchers in stochastic control.

At the end of the preface, I would like to thank the authors for their brilliant chapters and the publisher for guiding us all-through the making of the book till its final stage. Also, I would like to thank my family for providing the support and encouragement throughout my academic career and research projects.

Editor

Chapter 1

Design of Estimation Algorithms from an Innovation Approach in Linear Discrete-Time Stochastic Systems with Uncertain Observations

J. Linares-Pérez, R. Caballero-Águila and I. García-Garrido

Additional information is available at the end of the chapter

1. Introduction

The least-squares estimation problem in linear discrete-time stochastic systems in which the signal to be estimated is always present in the observations has been widely treated; as is well known, the Kalman filter [12] provides the least-squares estimator when the additive noises and the initial state are Gaussian and mutually independent.

Nevertheless, in many real situations, usually the measurement device or the transmission mechanism can be subject to random failures, generating observations in which the state appears randomly or which may consist of noise only due, for example, to component or interconnection failures, intermittent failures in the observation mechanism, fading phenomena in propagation channels, accidental loss of some measurements or data inaccessibility at certain times. In these situations where it is possible that information concerning the system state vector may or may not be contained in the observations, at each sampling time, there is a positive probability (called *false alarm probability*) that only noise is observed and, hence, that the observation does not contain the transmitted signal, but it is not generally known whether the observation used for estimation contains the signal or it is only noise. To describe this interrupted observation mechanism (*uncertain observations*), the observation equation, with the usual additive measurement noise, is formulated by multiplying the signal function at each sampling time by a binary random variable taking the values one and zero (Bernoulli random variable); the value one indicates that the measurement at that time contains the signal, whereas the value zero reflects the fact that the signal is missing and, hence, the corresponding observation is only noise. So, the observation equation involves both an additive and a multiplicative noise, the latter modeling the uncertainty about the signal being present or missing at each observation.

Linear discrete-time systems with uncertain observations have been widely used in estimation problems related to the above practical situations (which commonly appear, for example, in

Communication Theory). Due to the multiplicative noise component, even if the additive noises are Gaussian, systems with uncertain observations are always non-Gaussian and hence, as occurs in other kinds of non-Gaussian linear systems, the least-squares estimator is not a linear function of the observations and, generally, it is not easily obtainable by a recursive algorithm; for this reason, research in this kind of systems has focused special attention on the search of suboptimal estimators for the signal (mainly linear ones).

In some cases, the variables modeling the uncertainty in the observations can be assumed to be independent and, then, the distribution of the multiplicative noise is fully determined by the probability that each particular observation contains the signal. As it was shown by Nahi [17] (who was the first who analyzed the least-squares linear filtering problem in this kind of systems assuming that the state and observation additive noises are uncorrelated) the knowledge of the aforementioned probabilities allows to derive estimation algorithms with a recursive structure similar to the Kalman filter. Later on, Monzingo [16] completed these results by analyzing the least-squares smoothing problem and, subsequently, [3] and [4] generalized the least-squares linear filtering and smoothing algorithms considering that the additive noises of the state and the observation are correlated.

However, there exist many real situations where this independence assumption of the Bernoulli variables modeling the uncertainty is not satisfied; for example, in signal transmission models with stand-by sensors in which any failure in the transmission is detected immediately and the old sensor is then replaced, thus avoiding the possibility of the signal being missing in two successive observations. This different situation was considered by [9] by assuming that the variables modeling the uncertainty are correlated at consecutive time instants, and the proposed least-squares linear filtering algorithm provides the signal estimator at any time from those in the two previous instants. Later on, the state estimation problem in discrete-time systems with uncertain observations, has been widely studied under different hypotheses on the additive noises involved in the state and observation equations and, also, under several hypotheses on the multiplicative noise modeling the uncertainty in the observations (see e.g. [22] - [13], among others).

On the other hand, there are many engineering application fields (for example, in communication systems) where sensor networks are used to obtain all the available information on the system state and its estimation must be carried out from the observations provided by all the sensors (see [6] and references therein). Most papers concerning systems with uncertain observations transmitted by multiple sensors assume that all the sensors have the same uncertainty characteristics. In the last years, this situation has been generalized by several authors considering uncertain observations whose statistical properties are assumed not to be the same for all the sensors. This is a realistic assumption in several application fields, for instance, in networked communication systems involving heterogeneous measurement devices (see e.g. [14] and [8], among others). In [7] it is assumed that the uncertainty in each sensor is modeled by a sequence of independent Bernoulli random variables, whose statistical properties are not necessarily the same for all the sensors. Later on, in [10] and [1] the independence restriction is weakened; specifically, different sequences of Bernoulli random variables correlated at consecutive sampling times are considered to model the uncertainty at each sensor. This form of correlation covers practical situations where the signal cannot be missing in two successive observations. In [2] the least-squares linear and quadratic problems are addressed when the Bernoulli variables describing the uncertainty in the observations are correlated at instants that differ two units of time. This study covers more general practical situations, for example, in sensor networks where sensor failures may happen and a failed sensor is replaced not immediately, but two sampling times after having failed. However, even if it is assumed that any failure in the transmission results from sensor failures, usually the failed sensor may not be replaced immediately but after m instants of time; in such situations, correlation among the random variables modeling the uncertainty in the observations at times k and $k+m$ must be considered and new algorithms must be deduced.

The current chapter is concerned with the state estimation problem for linear discrete-time systems with uncertain observations when the uncertainty at any sampling time k depends only on the uncertainty at the previous time $k-m$; this form of correlation allows us to consider certain models in which the signal cannot be missing in $m+1$ consecutive observations.

The random interruptions in the observation process are modeled by a sequence of Bernoulli variables (at each time, the value one of the variable indicates that the measurement is the current system output, whereas the value zero reflects that only noise is available), which are correlated only at the sampling times $k-m$ and k. Recursive algorithms for the filtering and fixed-point smoothing problems are proposed by using an innovation approach; this approach, based on the fact that the innovation process can be obtained by a causal and invertible operation on the observation process, consists of obtaining the estimators as a linear combination of the innovations and simplifies considerably the derivation of the estimators due to the fact that the innovations constitute a white process.

The chapter is organized as follows: in Section 2 the system model is described; more specifically, we introduce the linear state transition model perturbed by a white noise, and the observation model affected by an additive white noise and a multiplicative noise describing the uncertainty. Also, the pertinent hypotheses to address the least-squares linear estimation problem are established. In Section 3 this estimation problem is formulated using an innovation approach. Next, in Section 4, recursive algorithms for the filter and fixed-point smoother are derived, including recursive formulas for the estimation error covariance matrices. Finally, the performance of the proposed estimators is illustrated in Section 5 by a numerical simulation example, where a two-dimensional signal is estimated and the estimation accuracy is analyzed for different values of the uncertainty probability and several values of the time period m.

2. Model description

Consider linear discrete-time stochastic systems with uncertain observations coming from multiple sensors, whose mathematical modeling is accomplished by the following equations.

The state equation is given by

$$x_k = F_{k-1}x_{k-1} + w_{k-1}, \quad k \geq 1, \tag{1}$$

where $\{x_k;\ k \geq 0\}$ is an n-dimensional stochastic process representing the system state, $\{w_k;\ k \geq 0\}$ is a white noise process and F_k, for $k \geq 0$, are known deterministic matrices.

We consider scalar uncertain observations $\{y_k^i;\ k \geq 1\}$, $i = 1, \ldots, r$, coming from r sensors and perturbed by noises whose statistical properties are not necessarily the same for all the sensors. Specifically, we assume that, in each sensor and at any time k, the observation y_k^i, perturbed by an additive noise, can have no information about the state (thus being only noise) with a known probability. That is,

$$y_k^i = \begin{cases} H_k^i x_k + v_k^i, \text{ with probability } \overline{\theta}_k^i \\ v_k^i, \text{ with probability } 1 - \overline{\theta}_k^i \end{cases}$$

where, for $i = 1, \ldots, r$, $\{v_k^i;\ k \geq 1\}$ is the observation additive noise process of the i-th sensor and H_k^i, for $k \geq 1$, are known deterministic matrices of compatible dimensions. If we introduce $\{\theta_k^i;\ k \geq 1\}$, $i = 1, \ldots, r$, sequences of Bernoulli random variables with $P[\theta_k^i = 1] = \overline{\theta}_k^i$, the observations of the state can be rewritten as

$$y_k^i = \theta_k^i H_k^i x_k + v_k^i, \quad k \geq 1, \quad i = 1, \ldots, r. \tag{2}$$

Remark 1. If $\theta_k^i = 1$, which occurs with known probability $\overline{\theta}_k^i$, the state x_k is present in the observation y_k^i coming from the i-th sensor at time k, whereas if $\theta_k^i = 0$ such observation only contains additive noise, v_k^i, with probability $1 - \overline{\theta}_k^i$. This probability is called *false alarm probability* and it represents the probability that only noise is observed or, equivalently, that y_k^i does not contain the state.

The aim is to address the state estimation problem considering all the available observations coming from the r sensors. For convenience, denoting $y_k = (y_k^1, \ldots, y_k^r)^T$, $v_k = (v_k^1, \ldots, v_k^r)^T$, $H_k = (H_k^{1T}, \ldots, H_k^{rT})^T$ and $\Theta_k = Diag(\theta_k^1, \ldots, \theta_k^r)$, Equation (2) is equivalent to the following stacked observation equation

$$y_k = \Theta_k H_k x_k + v_k, \quad k \geq 1. \tag{3}$$

2.1. Model hypotheses

In order to analyze the least-squares linear estimation problem of the state x_k from the observations $y_1, \ldots, y_L$, with $L \geq k$, some considerations must be taken into account. On the one hand, it is known that the linear estimator of x_k, is the orthogonal projection of x_k onto the space of n-dimensional random variables obtained as linear transformations of the observations $y_1, \ldots, y_L$, which requires the existence of the second-order moments of such observations. On the other hand, we consider that the variables describing the uncertainty in the observations are correlated in instants that differ m units of time to cover many practical situations where the independence assumption on such variables is not realistic. Specifically, the following hypotheses are assumed:

Hypothesis 1. The initial state x_0 is a random vector with $E[x_0] = \overline{x}_0$ and $Cov[x_0] = P_0$.

Hypothesis 2. The state noise $\{w_k;\ k \geq 0\}$ is a zero-mean white sequence with $Cov[w_k] = Q_k$, $\forall k \geq 0$.

Hypothesis 3. The observation additive noise $\{v_k;\ k \geq 1\}$ is a zero-mean white process with $Cov[v_k] = R_k$, $\forall k \geq 1$.

Hypothesis 4. For $i = 1, \dots, r$, $\{\theta_k^i;\ k \geq 1\}$ is a sequence of Bernoulli random variables with $P[\theta_k^i = 1] = \overline{\theta}_k^i$. For $i, j = 1, \dots, r$, the variables θ_k^i and θ_s^j are independent for $|k - s| \neq 0, m$ and $Cov[\theta_k^i, \theta_s^j]$ are known for $|k - s| = 0, m$. Defining $\theta_k = (\theta_k^1, \dots, \theta_k^r)^T$ the covariance matrices of θ_k and θ_s will be denoted by $K_{k,s}^\theta$.

Finally, we assume the following hypothesis on the independence of the initial state and noises:

Hypothesis 5. The initial state x_0 and the noise processes $\{w_k;\ k \geq 0\}$, $\{v_k;\ k \geq 1\}$ and $\{\theta_k;\ k \geq 1\}$ are mutually independent.

Remark 2. For the derivation of the estimation algorithms a matrix product, called *Hadamard product*, which is simpler than the conventional product, will be considered. Let $A, B \in M_{mn}$, the Hadamard product (denoted by $\circ$) of A and B is defined as $[A \circ B]_{ij} = A_{ij} B_{ij}$. From this definition it is easily deduced (see [7]) the next property that will be needed later.

For any random matrix $G_{m \times m}$ independent of $\{\Theta_k;\ k \geq 1\}$, the following equality is satisfied

$$E[\Theta_k G_{m \times m} \Theta_s] = E[\theta_k \theta_s^T] \circ E[G_{m \times m}].$$

Particularly, denoting $\overline{\Theta}_k = E[\Theta_k]$, it is immediately clear that

$$E[(\Theta_k - \overline{\Theta}_k) G_{m \times m} (\Theta_s - \overline{\Theta}_s)] = K_{k,s}^\theta \circ E[G_{m \times m}]. \tag{4}$$

Remark 3. Several authors assume that the observations available for the estimation come either from multiple sensors with identical uncertainty characteristics or from a single sensor (see [20] for the case when the uncertainty is modeled by independent variables, and [19] for the case when such variables are correlated at consecutive sampling times). Nevertheless, in the last years, this situation has been generalized by some authors considering multiple sensors featuring different uncertainty characteristics (see e.g. [7] for the case of independent uncertainty, and [1] for situations where the uncertainty in each sensor is modeled by variables correlated at consecutive sampling times). We analyze the state estimation problem for the class of linear discrete-time systems with uncertain observation (3), which, as established in Hypothesis 4, are characterized by the fact that the uncertainty at any sampling time k depends only on the uncertainty at the previous time $k - m$; this form of correlation allows us to consider certain models in which the signal cannot be absent in $m + 1$ consecutive observations.

3. Least-squares linear estimation problem

As mentioned above, our aim in this chapter is to obtain the least-squares linear estimator, $\widehat{x}_{k/L}$, of the signal x_k based on the observations $\{y_1, \dots, y_L\}$, with $L \geq k$, by recursive formulas. Specifically, the problem is to derive recursive algorithms for the least-squares linear filter ($L = k$) and fixed-point smoother (fixed k and $L > k$) of the state using uncertain observations (3). For this purpose, we use an innovation approach as described in [11].

Since the observations are generally nonorthogonal vectors, we use the *Gram-Schmidt orthogonalization procedure* to transform the set of observations $\{y_1, \dots, y_L\}$ into an equivalent set of orthogonal vectors $\{\nu_1, \dots, \nu_L\}$; equivalent in the sense that they both generate the same linear subspace; that is,

$$\mathcal{L}(y_1, \dots, y_L) = \mathcal{L}(\nu_1, \dots, \nu_L) = \mathcal{L}_L.$$

Let $\{\nu_1, \dots, \nu_{k-1}\}$ be the set of orthogonal vectors satisfying $\mathcal{L}(\nu_1, \dots, \nu_{k-1}) = \mathcal{L}(y_1, \dots, y_{k-1})$, the next orthogonal vector, ν_k, corresponding to the new observation y_k, is obtained by projecting y_k onto $\mathcal{L}_{k-1}$; specifically

$$\nu_k = y_k - Proj\{y_k \text{ onto } \mathcal{L}_{k-1}\},$$

and, because of the orthogonality of $\{\nu_1, \dots, \nu_{k-1}\}$ the above projection can be found by projecting y_k along each of the previously found orthogonal vectors ν_i, for $i \leq k-1$,

$$Proj\{y_k \text{ onto } \mathcal{L}_{k-1}\} = \sum_{i=1}^{k-1} Proj\{y_k \text{ along } \nu_i\} = \sum_{i=1}^{k-1} E[y_k \nu_i^T] \left(E[\nu_i \nu_i^T] \right)^{-1} \nu_i.$$

Since the projection of y_k onto $\mathcal{L}_{k-1}$ is $\widehat{y}_{k/k-1}$, the one-stage least-squares linear predictor of y_k, we have that

$$\widehat{y}_{k/k-1} = \sum_{i=1}^{k-1} T_{k,i} \Pi_i^{-1} \nu_i, \quad k \geq 2 \tag{5}$$

where $T_{k,i} = E[y_k \nu_i^T]$ and $\Pi_i = E[\nu_i \nu_i^T]$ is the covariance of ν_i.

Consequently, by starting with $\nu_1 = y_1 - E[y_1]$, the orthogonal vectors ν_k are determined by $\nu_k = y_k - \widehat{y}_{k/k-1}$, for $k \geq 2$. Hence, ν_k can be considered as the "new information" or the "innovation" in y_k given $\{y_1, \dots, y_{k-1}\}$.

In summary, the observation process $\{y_k;\ k \geq 1\}$ has been transformed into an equivalent white noise $\{\nu_k;\ k \geq 1\}$ known as innovation process. Taking into account that both processes satisfy that

$$\nu_i \in \mathcal{L}(y_1, \dots, y_i) \quad \text{and} \quad y_i \in \mathcal{L}(\nu_1, \dots, \nu_i), \quad \forall i \geq 1,$$

we conclude that such processes are related to each other by a causal and causally invertible linear transformation, thus making the innovation process be uniquely determined by the observations.

This consideration allows us to state that the least-squares linear estimator of the state based on the observations, $\widehat{x}_{k/L}$, is equal to the least-squares linear estimator of the state based on the innovations $\{\nu_1, \dots, \nu_L\}$. Thus, projecting x_k separately onto each ν_i, $i \leq L$, the following general expression for the estimator $\widehat{x}_{k/L}$ is obtained

$$\widehat{x}_{k/L} = \sum_{i=1}^{L} S_{k,i} \Pi_i^{-1} \nu_i, \quad k \geq 1, \tag{6}$$

where $S_{k,i} = E[x_k \nu_i^T]$. This expression is the starting point to derive the recursive filtering and fixed-point smoothing algorithms in the next section.

4. Least-squares linear estimation recursive algorithms

In this section, using an innovation approach, recursive algorithms are proposed for the filter, $\widehat{x}_{k/k}$, and the fixed-point smoother, $\widehat{x}_{k/L}$, for fixed k and $L > k$.

4.1. Linear filtering algorithm

In view of the general expression (6) for $L = k$, it is clear that the state filter, $\widehat{x}_{k/k}$, is obtained from the one-stage state predictor, $\widehat{x}_{k/k-1}$, by

$$\widehat{x}_{k/k} = \widehat{x}_{k/k-1} + S_{k,k}\Pi_k^{-1}\nu_k, \quad k \geq 1; \quad \widehat{x}_{0/0} = \overline{x}_0. \tag{7}$$

Hence, an equation for the predictor $\widehat{x}_{k/k-1}$ in terms of the filter $\widehat{x}_{k-1/k-1}$ and expressions for the innovation ν_k, its covariance matrix Π_k and the matrix $S_{k,k}$ are required.

State predictor $\widehat{x}_{k/k-1}$. From hypotheses 2 and 5, it is immediately clear that the filter of the noise w_{k-1} is $\widehat{w}_{k-1/k-1} = E[w_{k-1}] = 0$ and hence, taking into account Equation (1), we have

$$\widehat{x}_{k/k-1} = F_{k-1}\widehat{x}_{k-1/k-1}, \quad k \geq 1. \tag{8}$$

Innovation ν_k. We will now get an explicit formula for the innovation, $\nu_k = y_k - \widehat{y}_{k/k-1}$, or equivalently for the one-stage predictor of the observation, $\widehat{y}_{k/k-1}$. For this purpose, taking into account (5), we start by calculating $T_{k,i} = E[y_k\nu_i^T]$, for $i \leq k-1$.

From the observation equation (3) and hypotheses 3 and 5, it is clear that

$$T_{k,i} = E\left[\Theta_k H_k x_k \nu_i^T\right], \quad i \leq k-1. \tag{9}$$

Now, for $k \leq m$ or $k > m$ and $i < k-m$, hypotheses 4 and 5 guarantee that Θ_k is independent of the innovations ν_i, and then we have that $T_{k,i} = \overline{\Theta}_k H_k E[x_k\nu_i^T] = \overline{\Theta}_k H_k S_{k,i}$. So, after some manipulations, we obtain

I. For $k \leq m$, it is satisfied that $\widehat{y}_{k/k-1} = \overline{\Theta}_k H_k \sum_{i=1}^{k-1} S_{k,i}\Pi_i^{-1}\nu_i$, and using (6) for $L = k-1$ it is obvious that

$$\widehat{y}_{k/k-1} = \overline{\Theta}_k H_k \widehat{x}_{k/k-1}, \quad k \leq m. \tag{10}$$

II. For $k > m$, we have that $\widehat{y}_{k/k-1} = \overline{\Theta}_k H_k \sum_{i=1}^{k-(m+1)} S_{k,i}\Pi_i^{-1}\nu_i + \sum_{i=1}^{m} T_{k,k-i}\Pi_{k-i}^{-1}\nu_{k-i}$ and adding and subtracting $\sum_{i=1}^{m} \overline{\Theta}_k H_k S_{k,k-i}\Pi_{k-i}^{-1}\nu_{k-i}$, the following equality holds

$$\widehat{y}_{k/k-1} = \overline{\Theta}_k H_k \sum_{i=1}^{k-1} S_{k,i}\Pi_i^{-1}\nu_i + \sum_{i=1}^{m}(T_{k,k-i} - \overline{\Theta}_k H_k S_{k,k-i})\Pi_{k-i}^{-1}\nu_{k-i}, \quad k > m. \tag{11}$$

Next, we determine an expression for $T_{k,k-i} - \overline{\Theta}_k H_k S_{k,k-i}$, for $1 \leq i \leq m$.

Taking into account (9), it follows that

$$T_{k,k-i} - \overline{\Theta}_k H_k S_{k,k-i} = E\left[(\Theta_k - \overline{\Theta}_k) H_k x_k \nu_{k-i}^T\right], \quad 1 \le i \le m, \tag{12}$$

or equivalently,

$$T_{k,k-i} - \overline{\Theta}_k H_k S_{k,k-i} = E\left[(\Theta_k - \overline{\Theta}_k) H_k x_k y_{k-i}^T\right] - E\left[(\Theta_k - \overline{\Theta}_k) H_k x_k \widehat{y}_{k-i/k-(i+1)}^T\right].$$

To calculate the first expectation, we use again (3) for y_{k-i} and from hypotheses 3 and 5, we have that

$$E\left[(\Theta_k - \overline{\Theta}_k) H_k x_k y_{k-i}^T\right] = E\left[(\Theta_k - \overline{\Theta}_k)\, H_k x_k x_{k-i}^T H_{k-i}^T \Theta_{k-i}\right]$$

which, using Property (4), yields

$$E\left[(\Theta_k - \overline{\Theta}_k) H_k x_k y_{k-i}^T\right] = K_{k,k-i}^{\theta} \circ \left(H_k E[x_k x_{k-i}^T] H_{k-i}^T\right).$$

Now, denoting $D_k = E[x_k x_k^T]$ and $\mathbb{F}_{k,i} = F_{k-1} \cdots F_i$, from Equation (1) it is clear that $E[x_k x_{k-i}^T] = \mathbb{F}_{k,k-i} D_{k-i}$, and hence

$$E\left[(\Theta_k - \overline{\Theta}_k) H_k x_k y_{k-i}^T\right] = K_{k,k-i}^{\theta} \circ \left(H_k \mathbb{F}_{k,k-i} D_{k-i} H_{k-i}^T\right)$$

where D_k can be recursively obtained by

$$\begin{aligned} D_k &= F_{k-1} D_{k-1} F_{k-1}^T + Q_{k-1}, \quad k \ge 1; \\ D_0 &= P_0 + \overline{x}_0 \overline{x}_0^T. \end{aligned} \tag{13}$$

Summarizing, we have that

$$\begin{aligned} T_{k,k-i} - \overline{\Theta}_k H_k S_{k,k-i} = {} & K_{k,k-i}^{\theta} \circ \left(H_k \mathbb{F}_{k,k-i} D_{k-i} H_{k-i}^T\right) \\ & - E\left[(\Theta_k - \overline{\Theta}_k) H_k x_k \widehat{y}_{k-i/k-(i+1)}^T\right], \quad 1 \le i \le m. \end{aligned} \tag{14}$$

Taking into account the correlation hypothesis of the variables describing the uncertainty, the right-hand side of this equation is calculated differently for $i = m$ or $i < m$, as shown below.

(a) For $i = m$, since Θ_k is independent of the innovations ν_i, for $i < k - m$, we have that $E\left[(\Theta_k - \overline{\Theta}_k) H_k x_k \widehat{y}_{k-m/k-(m+1)}^T\right] = 0$, and from (14)

$$T_{k,k-m} - \overline{\Theta}_k H_k S_{k,k-m} = K_{k,k-m}^{\theta} \circ \left(H_k \mathbb{F}_{k,k-m} D_{k-m} H_{k-m}^T\right). \tag{15}$$

(b) For $i < m$, from Hypothesis 4, $K_{k,k-i}^{\theta} = 0$ and, hence, from (14)

$$T_{k,k-i} - \overline{\Theta}_k H_k S_{k,k-i} = -E\left[(\Theta_k - \overline{\Theta}_k) H_k x_k \widehat{y}_{k-i/k-(i+1)}^T\right].$$

Now, from expression (5),

$$\widehat{y}_{k-i/k-(i+1)} = \sum_{j=1}^{k-(i+1)} T_{k-i,j}\Pi_j^{-1}\nu_j,$$

and using again that Θ_k is independent of ν_i, for $i \neq k-m$, it is deduced that

$$T_{k,k-i} - \overline{\Theta}_k H_k S_{k,k-i} = -E[(\Theta_k - \overline{\Theta}_k) H_k x_k \nu_{k-m}^T]\Pi_{k-m}^{-1} T_{k-i,k-m}^T$$

or, equivalently, from (12) for $i = m$, (15) and noting

$$\Psi_{k,k-m} = K_{k,k-m}^{\theta} \circ \left(H_k \mathbb{F}_{k,k-m} D_{k-m} H_{k-m}^T\right) \Pi_{k-m}^{-1},$$

we have that

$$T_{k,k-i} - \overline{\Theta}_k H_k S_{k,k-i} = -\Psi_{k,k-m} T_{k-i,k-m}^T, \quad i < m. \tag{16}$$

Next, substituting (15) and (16) into (11) and using (6) for $\widehat{x}_{k/k-1}$, it is concluded that

$$\widehat{y}_{k/k-1} = \overline{\Theta}_k H_k \widehat{x}_{k/k-1} + \Psi_{k,k-m}\left[\nu_{k-m} - \sum_{i=1}^{m-1} T_{k-i,k-m}^T \Pi_{k-i}^{-1}\nu_{k-i}\right], \quad k > m. \tag{17}$$

Finally, using (3) and (16) and taking into account that, from (1), $S_{k,k-i} = \mathbb{F}_{k,k-i} S_{k-i,k-i}$, the matrices $T_{k,k-i}$ in (17) are obtained by

$$T_{k,k-i} = \overline{\Theta}_k H_k \mathbb{F}_{k,k-i} S_{k-i,k-i}, \quad 2 \leq k \leq m, \quad 1 \leq i \leq k-1,$$

$$T_{k,k-i} = \overline{\Theta}_k H_k \mathbb{F}_{k,k-i} S_{k-i,k-i} - \Psi_{k,k-m} T_{k-i,k-m}^T, \quad k > m, \quad 1 \leq i \leq m-1.$$

Matrix $S_{k,k}$. Since $\nu_k = y_k - \widehat{y}_{k/k-1}$, we have that $S_{k,k} = E[x_k \nu_k^T] = E[x_k y_k^T] - E[x_k \widehat{y}_{k/k-1}^T]$. Next, we calculate these expectations.

I. From Equation (3) and the independence hypothesis, it is clear that $E[x_k y_k^T] = D_k H_k^T \overline{\Theta}_k$, $\forall k \geq 1$, where $D_k = E[x_k x_k^T]$ is given by (13).

II. To calculate $E[x_k \widehat{y}_{k/k-1}^T]$, the correlation hypothesis of the random variables θ_k must be taken into account and two cases must be considered:

(a) For $k \leq m$, from (10) we obtain

$$E[x_k \widehat{y}_{k/k-1}^T] = E[x_k \widehat{x}_{k/k-1}^T] H_k^T \overline{\Theta}_k.$$

By using the orthogonal projection lemma, which assures that $E[x_k \widehat{x}_{k/k-1}^T] = D_k - P_{k/k-1}$, where $P_{k/k-1} = E[(x_k - \widehat{x}_{k/k-1})(x_k - \widehat{x}_{k/k-1})^T]$ is the prediction error covariance matrix, we get

$$E[x_k \widehat{y}_{k/k-1}^T] = (D_k - P_{k/k-1}) H_k^T \overline{\Theta}_k, \quad k \leq m.$$

(b) For $k > m$, from (17) it follows that

$$E[x_k\widehat{y}_{k/k-1}^T] = E[x_k\widehat{x}_{k/k-1}^T]H_k^T\overline{\Theta}_k + E[x_k\nu_{k-m}^T]\Psi_{k,k-m}^T$$
$$- E\left[x_k\left(\sum_{i=1}^{m-1} T_{k-i,k-m}^T\Pi_{k-i}^{-1}\nu_{k-i}\right)^T\right]\Psi_{k,k-m}^T,$$

hence, using again the orthogonal projection lemma and taking into account that $S_{k,k-i} = E[x_k\nu_{k-i}^T]$, for $1 \le i \le m$, it follows that

$$E[x_k\widehat{y}_{k/k-1}^T] = (D_k - P_{k/k-1})\,H_k^T\overline{\Theta}_k + S_{k,k-m}\Psi_{k,k-m}^T$$
$$- \sum_{i=1}^{m-1} S_{k,k-i}\Pi_{k-i}^{-1}T_{k-i,k-m}\Psi_{k,k-m}^T, \quad k > m.$$

Then, substituting these expectations in the expression of $S_{k,k}$ and simplifying, it is clear that

$$S_{k,k} = P_{k/k-1}H_k^T\overline{\Theta}_k, \quad 1 \le k \le m,$$
$$S_{k,k} = P_{k/k-1}H_k^T\overline{\Theta}_k - \left(S_{k,k-m} - \sum_{i=1}^{m-1} S_{k,k-i}\Pi_{k-i}^{-1}T_{k-i,k-m}\right)\Psi_{k,k-m}^T, \quad k > m. \tag{18}$$

Now, an expression for the prediction error covariance matrix, $P_{k/k-1}$, is necessary. From Equation (1), it is immediately clear that

$$P_{k/k-1} = F_{k-1}P_{k-1/k-1}F_{k-1}^T + Q_{k-1}, \quad k \ge 1,$$

where $P_{k/k} = E[(x_k - \widehat{x}_{k/k})(x_k - \widehat{x}_{k/k})^T]$ is the filtering error covariance matrix. From Equation (7), it is concluded that

$$P_{k/k} = P_{k/k-1} - S_{k,k}\Pi_k^{-1}S_{k,k}^T, \quad k \ge 1; \quad P_{0/0} = P_0.$$

Covariance matrix of the innovation $\boldsymbol{\Pi_k = E[\nu_k\nu_k^T]}$. From the orthogonal projection lemma, the covariance matrix of the innovation is obtained as $\Pi_k = E[y_ky_k^T] - E[\widehat{y}_{k/k-1}\widehat{y}_{k/k-1}^T]$.

From (3) and using Property (4), we have that

$$E[y_ky_k^T] = E[\theta_k\theta_k^T] \circ \left(H_kD_kH_k^T\right) + R_k, \quad k \ge 1.$$

To obtain $E[\widehat{y}_{k/k-1}\widehat{y}_{k/k-1}^T]$ two cases must be distinguished again, due to the correlation hypothesis of the Bernoulli variables θ_k:

I. For $k \le m$, Equation (10) and Property (4) yield

$$E[\widehat{y}_{k/k-1}\widehat{y}_{k/k-1}^T] = \left(\overline{\theta}_k\overline{\theta}_k^T\right) \circ \left(H_kE[\widehat{x}_{k/k-1}\widehat{x}_{k/k-1}^T]H_k^T\right)$$

and in view of the orthogonal projection lemma,

$$E[\widehat{y}_{k/k-1}\widehat{y}_{k/k-1}^T] = \left(\overline{\theta}_k\overline{\theta}_k^T\right) \circ \left(H_k(D_k - P_{k/k-1})H_k^T\right), \quad k \le m.$$

II. For $k > m$, an analogous reasoning, but using now Equation (17), yields

$$E[\widehat{y}_{k/k-1}\widehat{y}^T_{k/k-1}] = \left(\overline{\theta}_k\overline{\theta}^T_k\right) \circ \left(H_k(D_k - P_{k/k-1})H^T_k\right) + \Psi_{k,k-m}\Pi_{k-m}\Psi^T_{k,k-m}$$
$$+ \Psi_{k,k-m}\sum_{i=1}^{m-1} T^T_{k-i,k-m}\Pi^{-1}_{k-i}\Pi_{k-i}\Pi^{-1}_{k-i}T_{k-i,k-m}\Psi^T_{k,k-m}$$
$$+ \overline{\Theta}_k H_k E[\widehat{x}_{k/k-1}\nu^T_{k-m}]\Psi^T_{k,k-m} + \Psi_{k,k-m}E[\nu_{k-m}\widehat{x}^T_{k/k-1}]H^T_k\overline{\Theta}_k$$
$$- \overline{\Theta}_k H_k E[\widehat{x}_{k/k-1}\sum_{i=1}^{m-1}\nu^T_{k-i}\Pi^{-1}_{k-i}T_{k-i,k-m}]\Psi^T_{k,k-m}$$
$$- \Psi_{k,k-m}\sum_{i=1}^{m-1} T^T_{k-i,k-m}\Pi^{-1}_{k-i}E[\nu_{k-i}\widehat{x}^T_{k/k-1}]H^T_k\overline{\Theta}_k.$$

Next, again from the orthogonal projection lemma, $E[\widehat{x}_{k/k-1}\nu^T_{k-i}] = E[x_k\nu^T_{k-i}] = S_{k,k-i}$, for $1 \le i \le m$, and therefore

$$E[\widehat{y}_{k/k-1}\widehat{y}^T_{k/k-1}] = \left(\overline{\theta}_k\overline{\theta}^T_k\right) \circ \left(H_k(D_k - P_{k/k-1})H^T_k\right) + \Psi_{k,k-m}\Pi_{k-m}\Psi^T_{k,k-m}$$
$$+ \Psi_{k,k-m}\sum_{i=1}^{m-1} T^T_{k-i,k-m}\Pi^{-1}_{k-i}T_{k-i,k-m}\Psi^T_{k,k-m}$$
$$+ \overline{\Theta}_k H_k\left(S_{k,k-m}\Psi^T_{k,k-m} - \sum_{i=1}^{m-1} S_{k,k-i}\Pi^{-1}_{k-i}T_{k-i,k-m}\Psi^T_{k,k-m}\right)$$
$$+ \left(\Psi_{k,k-m}S^T_{k,k-m} - \Psi_{k,k-m}\sum_{i=1}^{m-1} T^T_{k-i,k-m}\Pi^{-1}_{k-i}S^T_{k,k-i}\right)H^T_k\overline{\Theta}_k.$$

Finally, from Equation (18), we have

$$S_{k,k-m}\Psi^T_{k,k-m} - \sum_{i=1}^{m-1} S_{k,k-i}\Pi^{-1}_{k-i}T_{k-i,k-m}\Psi^T_{k,k-m} = -(S_{k,k} - P_{k/k-1}H^T_k\overline{\Theta}_k),$$

and hence,

$$E[\widehat{y}_{k/k-1}\widehat{y}^T_{k/k-1}] = \left(\overline{\theta}_k\overline{\theta}^T_k\right) \circ \left(H_k(D_k - P_{k/k-1})H^T_k\right) + \Psi_{k,k-m}\Pi_{k-m}\Psi^T_{k,k-m}$$
$$+ \Psi_{k,k-m}\sum_{i=1}^{m-1} T^T_{k-i,k-m}\Pi^{-1}_{k-i}T_{k-i,k-m}\Psi^T_{k,k-m}$$
$$- \overline{\Theta}_k H_k\left(S_{k,k} - P_{k/k-1}H^T_k\overline{\Theta}_k\right) - \left(S^T_{k,k} - \overline{\Theta}_k H_k P_{k/k-1}\right)H^T_k\overline{\Theta}_k, \ k > m.$$

Finally, since $K^{\theta}_{k,k} = E[\theta_k\theta^T_k] - \overline{\theta}_k\overline{\theta}^T_k$, the above expectations lead to the following expression for the innovation covariance matrices

$$\Pi_k = K^{\theta}_{k,k} \circ \left(H_k D_k H^T_k\right) + R_k + \overline{\Theta}_k H_k S_{k,k}, \quad k \le m,$$

$$\Pi_k = K^{\theta}_{k,k} \circ \left(H_k D_k H_k^T\right) + R_k - \Psi_{k,k-m}\left(\Pi_{k-m} + \sum_{i=1}^{m-1} T^T_{k-i,k-m}\Pi^{-1}_{k-i}T_{k-i,k-m}\right)\Psi^T_{k,k-m}$$

$$+\overline{\Theta}_k H_k S_{k,k} + S^T_{k,k}H_k^T\overline{\Theta}_k - \overline{\Theta}_k H_k P_{k/k-1}H_k^T\overline{\Theta}_k, \quad k > m.$$

All these results are summarized in the following theorem.

Theorem 1. The linear filter, $\widehat{x}_{k/k}$, of the state x_k is obtained as

$$\widehat{x}_{k/k} = \widehat{x}_{k/k-1} + S_{k,k}\Pi_k^{-1}\nu_k, \quad k \geq 1; \quad \widehat{x}_{0/0} = \overline{x}_0,$$

where the state predictor, $\widehat{x}_{k/k-1}$, is given by

$$\widehat{x}_{k/k-1} = F_{k-1}\widehat{x}_{k-1/k-1}, \quad k \geq 1.$$

The innovation process satisfies

$$\nu_k = y_k - \overline{\Theta}_k H_k \widehat{x}_{k/k-1}, \quad k \leq m,$$

$$\nu_k = y_k - \overline{\Theta}_k H_k \widehat{x}_{k/k-1} + \Psi_{k,k-m}\left[\nu_{k-m} - \sum_{i=1}^{m-1} T^T_{k-i,k-m}\Pi^{-1}_{k-i}\nu_{k-i}\right], \quad k > m,$$

where $\overline{\Theta}_k = E[\Theta_k]$ and $\Psi_{k,k-m} = K^{\theta}_{k,k-m} \circ \left(H_k \mathbb{F}_{k,k-m} D_{k-m} H^T_{k-m}\Pi^{-1}_{k-m}\right)$, with $\circ$ the Hadamard product, $\mathbb{F}_{k,i} = F_{k-1}\cdots F_i$ and $D_k = E[x_k x_k^T]$ recursively obtained by

$$D_k = F_{k-1}D_{k-1}F^T_{k-1} + Q_{k-1}, \quad k \geq 1; \qquad D_0 = P_0 + \overline{x}_0\overline{x}_0^T.$$

The matrices $T_{k,k-i}$ are given by

$$T_{k,k-i} = \overline{\Theta}_k H_k \mathbb{F}_{k,k-i} S_{k-i,k-i}, \quad 2 \leq k \leq m, \quad 1 \leq i \leq k-1,$$

$$T_{k,k-i} = \overline{\Theta}_k H_k \mathbb{F}_{k,k-i} S_{k-i,k-i} - \Psi_{k,k-m} T^T_{k-i,k-m}, \quad k > m, \quad 1 \leq i \leq m-1.$$

The covariance matrix of the innovation, $\Pi_k = E[\nu_k \nu_k^T]$, satisfies

$$\Pi_k = K^{\theta}_{k,k} \circ \left(H_k D_k H_k^T\right) + R_k + \overline{\Theta}_k H_k S_{k,k}, \quad k \leq m,$$

$$\Pi_k = K^{\theta}_{k,k} \circ \left(H_k D_k H_k^T\right) + R_k - \Psi_{k,k-m}\left(\Pi_{k-m} + \sum_{i=1}^{m-1} T^T_{k-i,k-m}\Pi^{-1}_{k-i}T_{k-i,k-m}\right)\Psi^T_{k,k-m}$$

$$+\overline{\Theta}_k H_k S_{k,k} + S^T_{k,k}H_k^T\overline{\Theta}_k - \overline{\Theta}_k H_k P_{k/k-1}H_k^T\overline{\Theta}_k, \quad k > m.$$

The matrix $S_{k,k}$ is determined by the following expression

$$S_{k,k} = P_{k/k-1}H_k^T\overline{\Theta}_k, \quad k \leq m,$$

$$S_{k,k} = P_{k/k-1}H_k^T\overline{\Theta}_k - \left(S_{k,k-m} - \sum_{i=1}^{m-1} S_{k,k-i}\Pi^{-1}_{k-i}T_{k-i,k-m}\right)\Psi^T_{k,k-m}, \quad k > m,$$

where $P_{k/k-1}$, the prediction error covariance matrix, is obtained by

$$P_{k/k-1} = F_{k-1}P_{k-1/k-1}F_{k-1}^T + Q_{k-1}, \quad k \geq 1,$$

with $P_{k/k}$, the filtering error covariance matrix, satisfying

$$P_{k/k} = P_{k/k-1} - S_{k,k}\Pi_k^{-1}S_{k,k}^T, \quad k \geq 1; \qquad P_{0/0} = P_0. \tag{19}$$

4.2. Linear fixed-point smoothing algorithm

The following theorem provides a recursive fixed-point smoothing algorithm to obtain the least-squares linear estimator, $\widehat{x}_{k/k+N}$, of the state x_k based on the observations $\{y_1, \ldots, y_{k+N}\}$, for $k \geq 1$ fixed and $N \geq 1$. Moreover, to measure of the estimation accuracy, a recursive formula for the error covariance matrices, $P_{k/k+N} = E\left[(x_k - \widehat{x}_{k/k+N})(x_k - \widehat{x}_{k/k+N})^T\right]$, is derived.

Theorem 2. For each fixed $k \geq 1$, the fixed-point smoothers, $\widehat{x}_{k/k+N}$, $N \geq 1$ are calculated by

$$\widehat{x}_{k/k+N} = \widehat{x}_{k/k+N-1} + S_{k,k+N}\Pi_{k+N}^{-1}\nu_{k+N}, \quad N \geq 1, \tag{20}$$

whose initial condition is the filter, $\widehat{x}_{k/k}$, given in (7).

The matrices $S_{k,k+N}$ are calculated from

$$\begin{aligned} S_{k,k+N} &= \left(D_k\mathbf{F}_{k+N,k}^T - M_{k,k+N-1}F_{k+N-1}^T\right)H_{k+N}^T\overline{\Theta}_{k+N}, \quad k \leq m-N, \quad N \geq 1, \\ S_{k,k+N} &= \left(D_k\mathbf{F}_{k+N,k}^T - M_{k,k+N-1}F_{k+N-1}^T\right)H_{k+N}^T\overline{\Theta}_{k+N} \\ &\quad - \left(S_{k,k+N-m} - \sum_{i=1}^{m-1} S_{k,k+N-i}\Pi_{k+N-i}^{-1}T_{k+N-i,k+N-m}\right) \\ &\quad \times \Psi_{k+N,k+N-m}^T, \quad k > m-N, \quad N \geq 1. \end{aligned} \tag{21}$$

where the matrices $M_{k,k+N}$ satisfy the following recursive formula:

$$\begin{aligned} M_{k,k+N} &= M_{k,k+N-1}F_{k+N-1}^T + S_{k,k+N}\Pi_{k+N}^{-1}S_{k+N,k+N}^T, \quad N \geq 1, \\ M_{k,k} &= D_k - P_{k/k}. \end{aligned} \tag{22}$$

The innovations ν_{k+N}, their covariance matrices Π_{k+N}, the matrices $T_{k+N,k+N-i}$, $\Psi_{k+N,k+N-m}$, D_k and $P_{k/k}$ are given in Theorem 1.

Finally, the fixed-point smoothing error covariance matrix, $P_{k/k+N}$, verifies

$$P_{k/k+N} = P_{k/k+N-1} - S_{k,k+N}\Pi_{k+N}^{-1}S_{k,k+N}^T, \quad N \geq 1, \tag{23}$$

with initial condition the filtering error covariance matrix, $P_{k/k}$, given by (19).

Proof. From the general expression (6), for each fixed $k \geq 1$, the recursive relation (20) is immediately clear.

Now, we need to prove (21) for $S_{k,k+N} = E[x_k v_{k+N}^T] = E[x_k y_{k+N}^T] - E[x_k \widehat{y}_{k+N/k+N-1}^T]$, thus being necessary to calculate both expectations.

I. From Equation (3), taking into account that $E[x_k x_{k+N}^T] = D_k \mathbb{F}_{k+N,k}^T$ and using that Θ_{k+N} and v_{k+N} are independent of x_k, we obtain

$$E[x_k y_{k+N}^T] = D_k \mathbb{F}_{k+N,k}^T H_{k+N}^T \overline{\Theta}_{k+N}, \quad N \geq 1.$$

II. Based on expressions (10) and (17) for $\widehat{y}_{k+N/k+N-1}$, which are different depending on $k+N \leq m$ or $k+N > m$, two options must be considered:

(a) For $k \leq m-N$, using (10) for $\widehat{y}_{k+N/k+N-1}$ with (8) for $\widehat{x}_{k+N/k+N-1}$, we have that

$$E\left[x_k \widehat{y}_{k+N/k+N-1}^T\right] = M_{k,k+N-1} F_{k+N-1}^T H_{k+N}^T \overline{\Theta}_{k+N},$$

where $M_{k,k+N-1} = E\left[x_k \widehat{x}_{k+N-1/k+N-1}^T\right]$.

(b) For $k > m-N$, by following a similar reasoning to the previous one but starting from (17), we get

$$E[x_k \widehat{y}_{k+N/k+N-1}^T] = M_{k,k+N-1} F_{k+N-1}^T H_{k+N}^T \overline{\Theta}_{k+N} + \left(S_{k,k+N-m} - \sum_{i=1}^{m-1} S_{k,k+N-i} \Pi_{k+N-i}^{-1} T_{k+N-i,k+N-m}\right) \Psi_{k+N,k+N-m}^T.$$

Then, the replacement of the above expectations in $S_{k,k+N}$ leads to expression (21).

The recursive relation (22) for $M_{k,k+N} = E\left[x_k \widehat{x}_{k+N/k+N}^T\right]$ is immediately clear from (7) for $\widehat{x}_{k+N/k+N}$ and its initial condition $M_{k,k} = E[x_k \widehat{x}_{k/k}]$ is calculated taking into account that, from the orthogonality, $E[x_k \widehat{x}_{k/k}^T] = E[\widehat{x}_{k/k} \widehat{x}_{k/k}^T] = D_k - P_{k/k}$.

Finally, since $P_{k/k+N} = E\left[x_k x_k^T\right] - E\left[\widehat{x}_{k/k+N} \widehat{x}_{k/k+N}^T\right]$, using (20) and taking into account that $\widehat{x}_{k/k+N-1}$ is uncorrelated with ν_{k+N}, we have

$$P_{k/k+N} = E\left[x_k x_k^T\right] - E\left[\widehat{x}_{k/k+N-1} \widehat{x}_{k/k+N-1}^T\right] - S_{k,k+N} \Pi_{k+N}^{-1} S_{k,k+N}^T, \quad N \geq 1$$

and, consequently, expression (23) holds.

□

5. Numerical simulation example

In this section, we present a numerical example to show the performance of the recursive algorithms proposed in this chapter. To illustrate the effectiveness of the proposed estimators, we ran a program in MATLAB which, at each iteration, simulates the state and the observed values and provides the filtering and fixed-point smoothing estimates, as well as the corresponding error covariance matrices, which provide a measure of the estimators accuracy.

Consider a two-dimensional state process, $\{x_k;\ k \geq 0\}$, generated by the following first-order autoregressive model

$$x_k = \left(1 + 0.2\sin\left(\frac{(k-1)\pi}{50}\right)\right)\begin{pmatrix}0.8 & 0\\ 0.9 & 0.2\end{pmatrix} x_{k-1} + w_{k-1}, \quad k \geq 1$$

with the following hypotheses:

- The initial state, x_0, is a zero-mean Gaussian vector with covariance matrix given by $P_0 = \begin{pmatrix}0.1 & 0\\ 0 & 0.1\end{pmatrix}$.
- The process $\{w_k; k \geq 0\}$ is a zero-mean white Gaussian noise with covariance matrices $Q_k = \begin{pmatrix}0.36 & 0.3\\ 0.3 & 0.25\end{pmatrix}$, $\forall k \geq 0$.

Suppose that the scalar observations come from two sensors according to the following observation equations:

$$y_k^i = \theta_k^i x_k + v_k^i, \quad k \geq 1, \quad i = 1, 2.$$

where $\{v_k^i;\ k \geq 1\}$, $i = 1, 2$, are zero-mean independent white Gaussian processes with variances $R_k^1 = 0.5$ and $R_k^2 = 0.9$, $\forall k \geq 1$, respectively.

According to our theoretical model, it is assumed that, for each sensor, the uncertainty at time k depends only on the uncertainty at the previous time $k - m$. The variables θ_k^i, $i = 1, 2$, modeling this type of uncertainty correlation in the observation process are modeled by two independent sequences of independent Bernoulli random variables, $\{\gamma_k^i;\ k \geq 1\}, i = 1, 2$, with constant probabilities $P[\gamma_k^i = 1] = \gamma_i$. Specifically, the variables θ_k^i are defined as follows

$$\theta_k^i = 1 - \gamma_{k+m}^i(1 - \gamma_k^i), \quad i = 1, 2.$$

So, if $\theta_k^i = 0$, then $\gamma_{k+m}^i = 1$ and $\gamma_k^i = 0$, and hence, $\theta_{k+m}^i = 1$; this fact guarantees that, if the state is absent at time k, after $k + m$ instants of time the observation necessarily contains the state. Therefore, there cannot be more than m consecutive observations consisting of noise only.

Moreover, since the variables γ_k^i and γ_s^i are independent, θ_k^i and θ_s^i also are independent for $|k - s| \neq 0, m$. The common mean of these variables is $\overline{\theta}^i = 1 - \gamma_i(1 - \gamma_i)$ and its covariance function is given by

$$K_{k,s}^{\theta} = E[(\theta_k^i - \overline{\theta}^i)(\theta_s^i - \overline{\theta}^i)] = \begin{cases} 0 & \text{if} \quad |k-s| \neq 0, m, \\ -(1-\overline{\theta}^i)^2 & \text{if} \quad |k-s| = m, \\ \overline{\theta}^i(1-\overline{\theta}^i) & \text{if} \quad |k-s| = 0. \end{cases}$$

To illustrate the effectiveness of the respective estimators, two hundred iterations of the proposed algorithms have been performed and the results obtained for different values of the uncertainty probability and several values of m have been analyzed.

Let us observe that the mean function of the variables θ_k^i, for $i = 1,2$ are the same if $1 - \gamma_i$ is used instead of γ_i; for this reason, only the case $\gamma_i \leq 0.5$ will be considered here. Note that, in such case, the false alarme probability at the i-th sensor, $1 - \overline{\theta}^i$, is an increasing function of γ_i.

Firstly, the values of the first component of a simulated state together with the filtering and the fixed-point smoothing estimates for $N = 2$, obtained from simulated observations of the state for $m = 3$ and $\gamma^1 = \gamma^2 = 0.5$ are displayed in Fig. 1. This graph shows that the fixed-point smoothing estimates follow the state evolution better than the filtering ones.

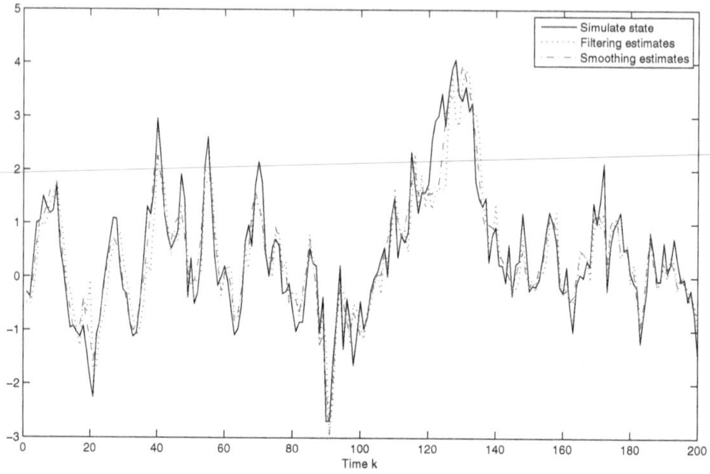

Figure 1. First component of the simulate state, filtering and fixed-point smoothing estimates for $N = 2$, when $m = 3$ and $\gamma^1 = \gamma^2 = 0.5$.

Next, assuming again that the Bernoulli variables of the observations are correlated at sampling times that differ three units of time ($m = 3$), we compare the effectiveness of the proposed filtering and fixed-point smoothing estimators considering different values of the probabilities γ_1 and γ_2, which provides different values of the false alarm probabilities $1 - \overline{\theta}^i$, $i = 1,2$; specifically, $\gamma_1 = 0.2$, $\gamma_2 = 0.4$ and $\gamma_1 = 0.1$, $\gamma_2 = 0.3$. For these values, Fig. 2 shows the filtering and fixed-point smoothing error variances, when $N = 2$ and $N = 5$, for the first state component. From this figure it is observed that:

i) As both γ_1 and γ_2 decrease (which means that the false alarm probability decreases), the error variances are smaller and, consequently, better estimations are obtained.

ii) The error variances corresponding to the fixed-point smoothers are less than those of the filters and, consequently, agreeing with the comments on the previous figure, the fixed-point smoothing estimates are more accurate.

iii) The accuracy of the smoothers at each fixed-point k is better as the number of available observations increases.

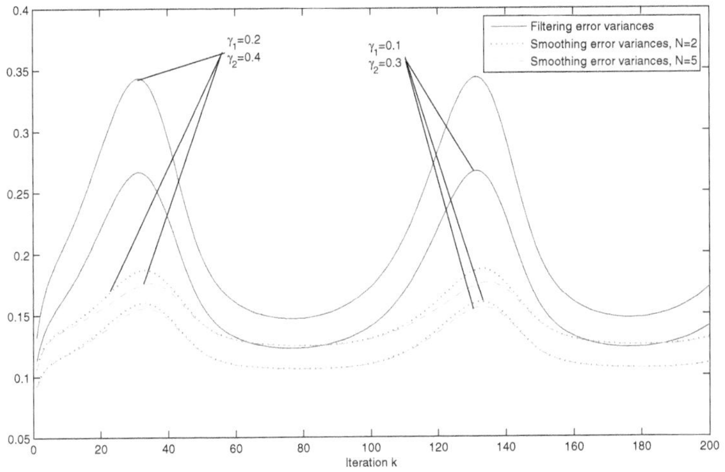

Figure 2. Filtering and smoothing error variances for the first state component for $\gamma_1 = 0.2, \gamma_2 = 0.4$ and $\gamma_1 = 0.1, \gamma_2 = 0.3$, when $m = 3$.

On the other hand, in order to show more precisely the dependence of the error variance on the values γ_1 and γ_2, Fig. 3 displays the filtering and fixed-point smoothing error variances of the first state component, at a fixed iteration (namely, $k = 200$) for $m = 3$, when both γ_1 and γ_2 are varied from 0.1 to 0.5, which provide different values of the probabilities $\overline{\theta}^1$ and $\overline{\theta}^2$. More specifically, we have considered the values $\gamma_i = 0.1, 0.2, 0.3, 0.4, 0.5$, which lead to the false alarm probabilities $1 - \overline{\theta}^i = 0.09, 0.16, 0.22, 0.24, 0.25$, respectively.

In this figure, both graphs (corresponding to the filtering and fixed-point smoothing error variances) corroborate the previous results, showing again that, as the false alarm probability increases, the filtering and fixed-point smoothing error variances ($N = 2$) become greater and consequently, worse estimations are obtained. Also, it is concluded that the smoothing error variances are better than the filtering ones.

Analogous results to those of Fig. 1-3 are obtained for the second component of the state. As example, Fig. 4 shows the filtering and fixed-point smoothing error variances of the second state component, at $k = 200$, versus γ_1 for constant values of γ_2, when $m = 3$ and similar comments to those made from Fig. 3 are deduced.

Finally, for $\gamma_1 = 0.2, \gamma_2 = 0.4$ the performance of the estimators is compared for different values of m; specifically, for $m = 1, 3, 6$, the filtering error variances of the first state component are displayed in Fig. 5. From this figure it is gathered that the estimators are more accurate as the values of m are lower. In other words, a greater distance between the instants at which the variables are correlated (which means that more consecutive observations may not contain state information) yields worse estimations.

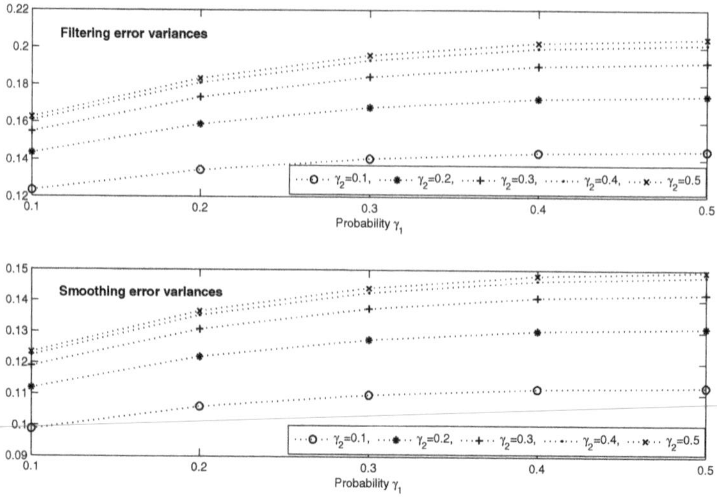

Figure 3. Filtering error variances and smoothing error variances for $N = 2$ of the first state component at $k = 200$ versus γ_1 with γ_2 varying from 0.1 to 0.5 when $m = 3$.

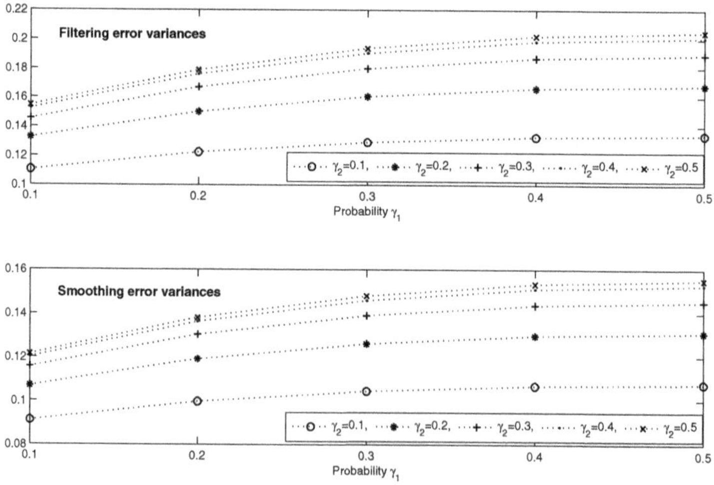

Figure 4. Filtering error variances and smoothing error variances for $N = 2$ of the second state component at $k = 200$ versus γ_1 with γ_2 varying from 0.1 to 0.5 when $m = 3$.

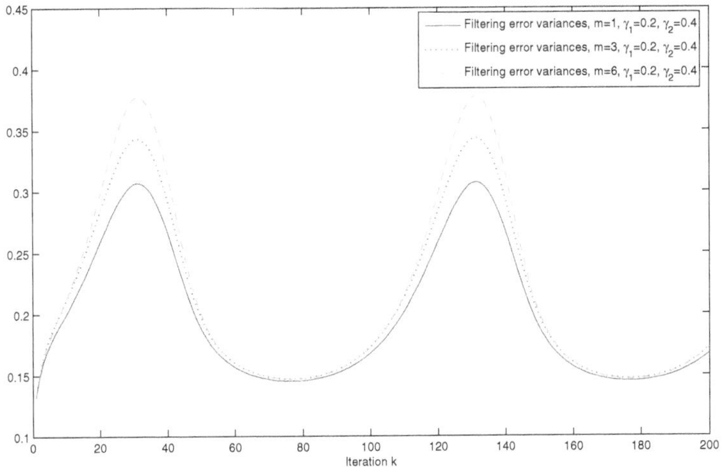

Figure 5. Filtering error variances for $\gamma_1 = 0.2$, $\gamma_2 = 0.4$ and $m = 1, 3, 6$.

6. Conclusions and future research

In this chapter, the least-squares linear filtering and fixed-point smoothing problems have been addressed for linear discrete-time stochastic systems with uncertain observations coming from multiple sensors. The uncertainty in the observations is modeled by a binary variable taking the values one or zero (Bernoulli variable), depending on whether the signal is present or absent in the corresponding observation, and it has been supposed that the uncertainty at any sampling time k depends only on the uncertainty at the previous time $k-m$. This situation covers, in particular, those signal transmission models in which any failure in the transmission is detected and the old sensor is replaced after m instants of time, thus avoiding the possibility of missing signal in $m+1$ consecutive observations.

By applying an innovation technique, recursive algorithms for the linear filtering and fixed-point smoothing estimators have been obtained. This technique consists of obtaining the estimators as a linear combination of the innovations, simplifying the derivation of these estimators, due to the fact that the innovations constitute a white process.

Finally, the feasibility of the theoretical results has been illustrated by the estimation of a two-dimensional signal from uncertain observations coming from two sensors, for different uncertainty probabilities and different values of m. The results obtained confirm the greater effectiveness of the fixed-point smoothing estimators in contrast to the filtering ones and conclude that more accurate estimations are obtained as the values of m are lower.

In recent years, several problems of signal processing, such as signal prediction, detection and control, as well as image restoration problems, have been treated using quadratic estimators and, generally, polynomial estimators of arbitrary degree. Hence, it must be noticed that the current chapter can be extended by considering the least-squares polynomial estimation problems of arbitrary degree for such linear systems with uncertain observations correlated in instants that differ m units of time. On the other hand, in practical engineering, some recent progress on the filtering and control problems for nonlinear stochastic systems with uncertain observations is being achieved. Nonlinearity and stochasticity are two important sources that are receiving special attention in research and, therefore, filtering and smoothing problems for nonlinear systems with uncertain observations would be relevant topics on which further investigation would be interesting.

Acknowledgements

This research is supported by Ministerio de Educación y Ciencia (Programa FPU and grant No. MTM2011-24718) and Junta de Andalucía (grant No. P07-FQM-02701).

Author details

J. Linares-Pérez
Dpto. de Estadística e I.O. Universidad de Granada. Avda. Fuentenueva. 18071. Granada, Spain

R. Caballero-Águila
Dpto. de Estadística e I.O. Universidad de Jaén. Paraje Las Lagunillas. 23071. Jaén, Spain

I. García-Garrido
Dpto. de Estadística e I.O. Universidad de Granada. Avda. Fuentenueva. 18071. Granada, Spain

7. References

[1] Caballero-Águila, R., Hermoso-Carazo, A. & Linares-Pérez, J. [2011]. Linear and quadratic estimation using uncertain observations from multiple sensors with correlated uncertainty, *Signal Processing* 91(No. 2): 330–337.

[2] García-Garrido, I., Linares-Pérez, J., Caballero-Águila, R. & Hermoso-Carazo, A. [2012]. A solution to the filtering problem for stochastic systems with multi-sensor uncertain observations, *International Mathematical Forum* 7(No. 18): 887–903.

[3] Hermoso-Carazo, A. & Linares-Pérez, J. [1994]. Linear estimation for discrete-time systems in the presence of time-correlated disturbances and uncertain observations, *IEEE Transactions on Automatic Control* 39(No. 8): 1636 –1638.

[4] Hermoso-Carazo, A. & Linares-Pérez, J. [1995]. Linear smoothing for discrete-time systems in the presence of correlated disturbances and uncertain observations, *IEEE Transactions on Automatic Control* 40(No. 8): 243–251.

[5] Hermoso-Carazo, A., Linares-Pérez, J., Jiménez-López, J., Caballero-Águila, R. & Nakamori, S. [2008]. Recursive fixed-point smoothing algorithm from covariances based on uncertain observations with correlation in the uncertainty, *Applied Mathematics and Computation* 203(No. 1): 243–251.

[6] Hespanha, J., Naghshtabrizi, P. & Xu, Y. [2007]. A survey of recent results in networked control systems, *Proceedings of the IEEE* 95(No. 1): 138 –172.

[7] Hounkpevi, F. & Yaz, E. [2007]. Robust minimum variance linear state estimators for multiple sensors with different failure rates, *Automatica* 43(No. 7): 1274–1280.

[8] Huang, R. & Záruba, G. [2007]. Incorporating data from multiple sensors for localizing nodes in mobile ad hoc networks, *IEEE Transactions on Mobile Computing* 6(No. 9): 1090–1104.
[9] Jackson, R. & Murthy, D. [1976]. Optimal linear estimation with uncertain observations, *IEEE Transactions on Information Theory* 22(No. 3): 376–378.
[10] Jiménez-López, J., Linares-Pérez, J., Nakamori, S., Caballero-Águila, R. & Hermoso-Carazo, A. [2008]. Signal estimation based on covariance information from observations featuring correlated uncertainty and coming from multiple sensors, *Signal Processing* 88(No. 12): 2998–3006.
[11] Kailath, T., Sayed, A. & Hassibi, B. [2000]. *Linear Estimation*, Prentice Hall, New Jersey.
[12] Kalman, R. [1960]. A new approach to linear filtering and prediction problems, *Transactions of the ASME. Journal of Basic Engineering* (No. 82 (Series D)): 35–45.
[13] Ma, J. & Sun, S. [2011]. Optimal linear estimators for systems with random sensor delays, multiple packet dropouts and uncertain observations, *IEEE Transactions on Signal Processing* 59(No. 11): 5181–5192.
[14] Malyavej, V., Manchester, I. & Savkin, A. [2006]. Precision missile guidance using radar/multiple-video sensor fusion via communication channels with bit-rate constraints, *Automatica* 42(No. 5): 763–769.
[15] Moayedi, M., Foo, Y. & Soh, Y. [2010]. Adaptive kalman filtering in networked systems with random sensor delays, multiple packet dropouts and missing measurements, *IEEE Transactions on Signal Processing* 58(No. 3): 1577–1578.
[16] Monzingo, R. [1981]. Discrete linear recursive smoothing for systems with uncertain observations, *IEEE Transactions on Automatic Control* AC-26(No. 3): 754–757.
[17] Nahi, N. [1969]. Optimal recursive estimation with uncertain observation, *IEEE Transactions on Information Theory* 15(No. 4): 457–462.
[18] Nakamori, S., Caballero-Águila, R., Hermoso-Carazo, A., Jiménez-López, J. & Linares-Pérez, J. [2006]. Least-squares vth-order polynomial estimation of signals from observations affected by non-independent uncertainty, *Applied Mathematics and Computation* Vol. 176(No. 2): 642–653.
[19] Nakamori, S., Caballero-Águila, R., Hermoso-Carazo, A., Jiménez-López, J. & Linares-Pérez, J. [2007]. Signal polynomial smoothing from correlated interrupted observations based on covariances, *Mathematical Methods in the Applied Sciences* Vol. 30(No. 14): 1645–1665.
[20] NaNacara, W. & Yaz, E. [1997]. Recursive estimator for linear and nonlinear systems with uncertain observations, *Signal Processing* Vol. 62(No. 2): 215–228.
[21] Sahebsara, M., Chen, T. & Shah, S. [2007]. Optimal $\mathcal{H}_2$ filtering with random sensor delay, multiple packet dropout and uncertain observations, *International Journal of Control* 80(No. 2): 292–301.
[22] Sinopoli, B., Schenato, L., Franceschetti, M., Poolla, K., Jordan, M. & Sastry, S. [2004]. Kalman filtering with intermittent observations, *IEEE Transactions on Automatic Control* 49(No. 9): 1453–1464.
[23] Wang, Z., Yang, F., Ho, D. & Liu, X. [2005]. Robust finite-horizon filtering for stochastic systems with missing measurements, *IEEE Signal Processing Letters* 12(No. 6): 437–440.

[24] Wang, Z., Yang, F., Ho, D. & Liu, X. [2006]. Robust H_∞ filtering for stochastic time-delay systems with missing measurements, *IEEE Transactions on Signal Processing* 54(No. 7): 2579–2587.

Chapter 2

Stochastic Modelling of Structural Elements

David Opeyemi

Additional information is available at the end of the chapter

1. Introduction

One of the primary goals of structural engineers is to assure proper levels of safety for the structures they design. This seemingly simple task is complicated by uncertainties associated with the materials with which the structure was designed and the loads they must resist, as well as our inaccuracies in analysis and design. Structural reliability and probabilistic analysis/design are tools which can be employed to quantify these uncertainties and inaccuracies, and produce designs and design procedures meeting acceptable levels of safety. Recent researches in the area of structural reliability and probabilistic analysis have centred on the development of probability-based design procedures. These include load modelling, ultimate and service load performance and evaluation of current levels of safety/reliability in design (Farid Uddim, 2000; Afolayan, 1999; Afolayan, 2003; Afolayan and Opeyemi, 2008; Opeyemi, 2009).

Deterministic methods are very subjective and are generally not based on a systematic assessment of reliability, especially when we consider their use in the entire structure. These methods can produce structures with some "over designed" components and perhaps some "under designed" components. The additional expense incurred in constructing the over designed components probably does not contribute to the overall reliability of the structure, so this is not a very cost-effective way to produce reliable structures. In other words, it would be better to redirect some of the resources used to build the over designed components toward strengthening the under designed ones. Therefore, there is increasing interest in adopting reliability-based design methods in civil engineering. These methods are intended to quantify reliability, and thus may be used to develop balance designs that are both more reliable and less expensive. Also, according to Coduto (2001), the methods can be used to better evaluate the various sources of failure and use this information to develop design and construction methods that are both more reliable and more robust - one that is insensitive to variations in materials, construction techniques, and environment.

The reliability of an engineering system can be defined as its ability to fulfil its design purpose for some time period. The theory of probability provides the fundamental basis to measure this ability. The reliability of a structural element can be viewed as the probability of its satisfactory performance according to some performance functions under specific service and extreme conditions within a stated time period. In estimating this probability, system uncertainties are modelled using random variables with mean values, variances, and probability distribution functions. Many methods have been proposed for structural reliability assessment purposes, such as First Order Second Moment (FOSM) method, Advanced Second Moment (ASM) method, and Computer-based Monte Carlo Simulation (MCS) (e.g., Ang and Tang,1990;Ayyub and Haldar,1984; White and Ayyub,1985; Ayyub and McCuen,1997) as reported by Ayyub and Patev (1998).

The concept of the First-Order Reliability Method (FORM) is a powerful tool for estimating nominal probability level of failure associated with uncertainties and it is the method adopted for the reliability estimations in this text. The general problem to which FORM provides an approximate solution is as follows. The state of a system is a function of many variables some of which are uncertain. These uncertain variables are random with joint distribution function $F_x(x) = P(\bigcap_{i=1}^{n}\{X_i \le x_i\})$ defining the stochastic model. For FORM, it is required that F_X (x), is at least locally continuously differentiable, i. e., that probability densities exist. The random variables $X = (X_1,...X_n)^T$ are called basic variables. The locally sufficiently smooth (at least once differentiable) state function is denoted by g(X). It is defined such that g(X)>0 corresponds to favourable (safe, intact, acceptable) state. g(X) =0 denotes the so-called limit state or the failure boundary. Therefore, g(X) <0 (sometimes also g(X) ≤ 0) defines the failure (unacceptable, adverse) domain, F. The function g(X) can be defined as an analytic function or an algorithm (e.g., a finite element code). In the context of FORM it is convenient but necessary only locally that g(X) is a monotonic function in each component of X. Among other useful information FORM produces an approximation to:

$$P_f = P(X \in F) = P(g(X) \le 0) = \int_{g(x) \le 0} dF_X(x) = \phi(-\beta_R) \qquad (1)$$

In which β_R = the reliability or safety index (Melchers, 2002).

Mathematical models, be they deterministic or stochastic, are intended to mimic real world systems. In particular, they can be used to predict how systems will behave under specified conditions. In scientific work, we may be able to conduct experiments to see if model predictions agree with what actually happens in practice. But in many situations, experimentation is impossible. Even if experimentation is conceivable in principle, it may be impractical for ethical or financial reasons. In these circumstances, the model can only be tested less formally, for example by seeking expert opinion on the predictions of the model.

The examples based on the author's research work introduce a different collection of stochastic models on analysis of the carrying capacities of piles based on static and dynamic approaches with special consideration to steel and precast concrete as pile types in cohesive, cohesionless, and intermediate soils between sand and clay and layered soils. Steel and precast concrete piles were analysed based on static pile capacity equations under various soils such as cohesive, cohesionless, and intermediate soils between sand and clay and layered soils. Likewise, three (3) dynamic formulae, namely: Hiley, Janbu and Gates were closely examined and analysed using the first-order reliability concept (Afolayan and Opeyemi, 2008; Opeyemi, 2009).

2. General methods and procedures for modelling structural elements

Structural reliability is concerned with the calculation and prediction of the probability of limit state violation for an engineered structural system at any stage during its life. In particular, the study of structural safety is concerned with violation of the ultimate or safety limit states for the structure. In probabilistic assessments any uncertainty about a variable expressed in terms of its probability density function is taken into account explicitly. This is not the case in traditional ways of measuring safety, such as the "factor of safety" or "load factor". These are "deterministic" measures since the variables describing the structure; its strength and the applied loads are assumed to take on known (if conservative) values about which there are assumed to be no uncertainty. The loads which are applied to a structure fluctuate with time and are of uncertain value at any one point in time. This is carried over directly to the load effects (or internal actions) S. Somewhat similarly the structural resistance R will be a function of time (but not a fluctuating one) owing to deterioration and similar actions. Loads have a tendency to increase, and resistances to decrease, with time. It is usual also for the uncertainty in both these quantities to increase with time. This means that the probability density functions fS ( ) and fR ( ) become wider and flatter with time and that the mean values of S and R also change with time. The safety limit state will be violated whenever, at any time t.

$$R(t) - S(t) < 0 \text{ or } \frac{R(t)}{S(t)} < 1 \tag{2}$$

The probability that this occurs for any one load application (or load cycle) is the probability of limit state violation, or simply the probability of failure P_f.

2.1. Reliability analysis

For basic structural reliability problem only one load effect S resisted by one resistance R is considered. Each described by a known probability density function FS( ) and FR( ) respectively. S is obtained from the applied loading Q through a structural analysis, but R and S are expressed in the same units.

For convenience, but without loss of generality, only the safety of a structural element is considered and as usual, that structural element will be considered to have failed if its resistance R is less than the stress resultant S acting on it. The probability of failure, P_f of the structural element will be stated as follows:

$$P_f = P\ (G\ (R, S)\ 0) \tag{3}$$

Where G ( ) is termed the "limit state function" and the probability of failure is identical with the probability of limit state violation.

In general, R is a function of material properties and element or structure dimensions, while S is a function of applied loads Q, material densities and perhaps dimensions of the structure, each of which may be a random variable. Also, R and S may not be independent, such as when some loads act to oppose failure (e.g. overturning) or when

$$P_f = P(R - S \leq 0) = \int_{-\infty}^{\infty} F_R(x) f_S(x)\, dx \tag{4}$$

2.1.1. Modeling procedures

The first step is to define the variables involved in the generalized reliability problem. The fundamental variables which define and characterize the behaviour and safety of a structure, termed "basic" variables, usually are variables employed in conventional structure analysis and design. Examples are dimensions densities or unit weights, materials, loads, material strengths, etc. It is very convenient to choose the basic variables such that they are independent, since dependence between basic variables usually adds some complexity to a reliability analysis.

The probability distributions will then be assigned to the basic variables, depending on the knowledge that is available. If it can be assumed that past observations and experience for similar structure can be used, validly, for the structure under consideration the probability distributions might be inferred directly from such observed data. More generally, subjective information may be employed or some combination of techniques is required. Sometimes physical reasoning may be used to suggest an appropriate probability distribution.

The parameters of the distribution are estimated from the data by using one of the usual methods viz: methods of moments, maximum likelihood, or order statistics.

Finally, when the model parameters have been selected, the model would be compared with the data. A graphical plot on appropriate probability paper is often very revealing, but analytical "goodness of fit" tests could be used also.

When or after the basic variables and their probability distributions have been established, the next step is to replace the simple R – S form of limit state function with a generalized version expressed directly in terms of basic variables.

The vector X will be used to represent all the basic variables involved in the problem. Then the resistance R will be expressed as R – GR(X) and the loading or load effect as S = GS(X). The limit state function G (R, S) can be generalized. When the functions GR(X) and GS(X) are used in G (R,S), the resulting limit state function will be written simply as G(X), where X is the vector of all relevant basic variables and G( ) is some function expressing the relationship between the limit state and the basic variables.

The limit state equation G(x) = 0 now defines the boundary between the satisfactory or 'safe' domain G>0 and the unsatisfactory or 'unsafe' domain G≤O in n-dimensional basic variable space.

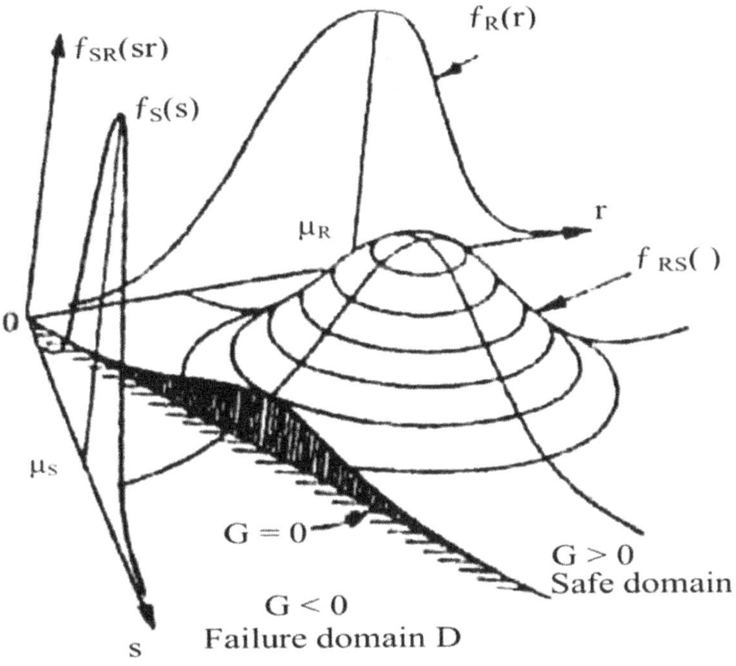

Figure 1. The n-dimensional basic variable space.

With the limit state function expressed as G(x), the generalization of

$$P_f = P(R - S \leq 0) = \iint_D f_{RS}(r,s)\,dr\,ds \tag{5}$$

Becomes

$$P_f = P\left(G(x) \le 0\right) = \int \ldots \int_{G(x)\le 0} f_x(x)\,dx \tag{6}$$

Sidestepping the integration process completely by transforming fx(x) in (6) to a multi-normal probability density function and using some remarkable properties which may then be used to determine, approximately, the probability of failure – the so-called 'First Order Second Moment' methods.

Rather than use approximate (and numerical) methods to perform the integration required in the reliability integral; the probability density function fx ( ) in the integrand is simplified. In this case the reliability estimation with its each variable is represented only by its first two moments i.e. by its mean and standard deviation. This is known as the "Second – Moment" level of representation. A convenient way in which the second-moment representation might be interpreted is that each random variable is represented by the normal distribution. This continuous probability distribution is described completely by its first two moments.

Because of their inherent simplicity, the so-called "Second-moment" methods have become very popular. Early works by Mayer (1926), Freudenthal (1956), Rzhanitzyn (1957) and Basler (1961) contained second moment concepts. Not until the late 1960s, however, was the time ripe for the ideas to gain a measure of acceptance, prompted by the work of Cornell (1969) as reported by Melchers (1999).

When resistance R and load effects are each Second-moment random variables (i.e. such as having normal distributions), the limit state equation is the "safety margin" Z = R – R and the probability of failure Pf is

$$P_f = \phi(-\beta) = \beta = \frac{U_z}{\sigma_z} \tag{7}$$

Where β is the (simple) 'safety' (or 'reliability') index and ϕ( ) the standard normal distribution function.

2.1.2. *First Order Reliability Method (FORM)*

FORM has been designed for the approximate computation of general probability integrals over given domains with locally smooth boundaries but especially for probability integrals occurring in structural reliability. Similar integrals are also found in many other areas, for example, in hydrology, mathematical statistics, control theory, classical hardware reliability and econometrics which are the areas where FORM has already been applied.

The concept of FORM is, essentially, based on

$$P_f = P\left(X_E F\right) = P\left(g(X) \le 0\right) = \int_{G(x)\le 0} dF_X(x) \tag{8}$$

3. Example applications

Some examples are hereby introduced to show particular cases of stochastic modelling as applied to the study of structural analysis of the carrying capacities of piles based on static and dynamic approaches with special consideration to steel and precast concrete as pile types in cohesive, cohesionless, and intermediate soils between sand and clay and layered soils, taken from the author's experience in research in the structural and geotechnical engineering field. Steel and precast concrete piles were analyzed based on static pile capacity equations under various soils such as cohesive, cohesionless, and intermediate soils between sand and clay and layered soils. Likewise, three (3) dynamic formulae (namely: Hiley, Janbu and Gates) were closely examined and analyzed using the first-order reliability concept.

Pile capacity determination is very difficult. A large number of different equations are used, and seldom will any two give the same computed capacity. Organizations which have been using a particular equation tend to stick to it especially when successful data base has been established. It is for this reason that a number of what are believed to be the most widely used (or currently accepted) equations are included in most literature.

Also, the technical literature provides very little information on the structural aspects of pile foundation design, which is a sharp contrast to the mountains of information on the geotechnical aspects. Building codes present design criteria, but they often are inconsistent with criteria for the super structure, and sometimes are incomplete or ambiguous. In many ways this is an orphan topic that neither structural engineers nor geotechnical engineers have claimed as their own.

3.1. Static pile capacity of concrete piling in cohesive soils

The functional relationship between the allowable design load and the allowable pile capacity can be expressed as follows:

G(X) = Allowable Design Load – Allowable Pile Capacity,

So that,

$$\begin{aligned} G(X) &= 0.33 f_{cu} \frac{\pi D_1^2}{4} + 0.4 f_y \frac{\pi D_2^2}{4} - \{\frac{\pi D_1^2}{4}(9S_u) + \pi D_1 L_b \alpha S_u\} / SF \\ &= 0.259 f_{cu} D_1^{\ 2} + 0.314 f_y D_2^{\ 2} - \left(7.07 D_1^{\ 2} S_u + 3.142 D_1 L_b \alpha S_u\right) / SF \end{aligned} \tag{9}$$

In which f_{cu}= characteristic strength of concrete, D_1= pile diameter, D_2= steel diameter, f_y = characteristic strength of steel, S_u= cohesion, L_b= pile length, α= adhesion factor, and SF=factor of safety.

i	xi	pdf	$\bar{x}$	v	$\sigma x = \bar{x}.v$
1	f_{cu}	LN	40 x 10^3	0.15	6.0 x 10^3
2	D_1	N	3 x 10^{-1}m	0.06	1.8 x 10^{-2}
3	f_y	LN	460 x 10^3KN/m^2	0.15	6.9 x 10^4
4	D_2	N	25 x 10^{-3}m	0.06	1.5 x 10^{-3}
5	S_u	LN	12KN/m^2	0.15	1.8
6	α	LN	1.00	0.15	1.5 x 10^{-1}
7	L_b	N	23.0m	0.06	1.38
8	SF	LN	3.0	0.15	4.5 x 10^{-1}

Table 1. Stochastic model for concrete piling in cohesive soils.

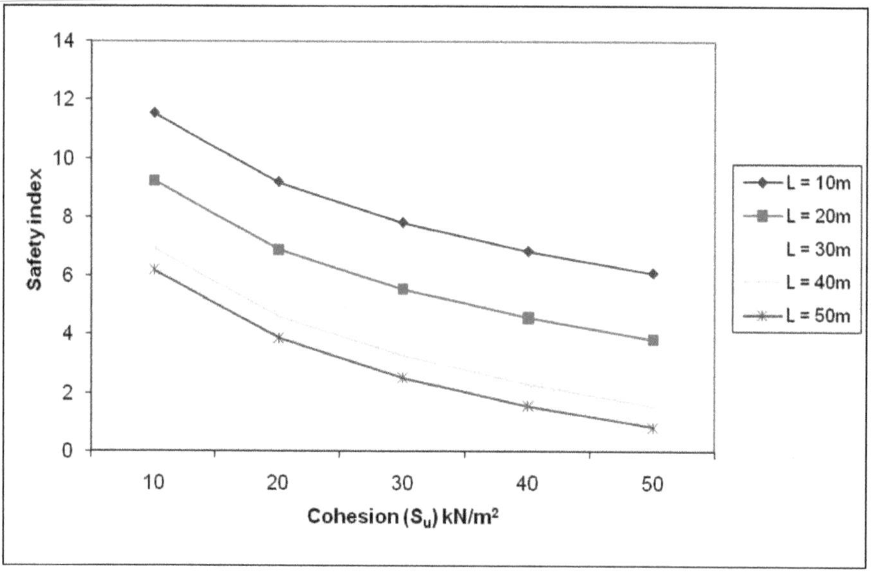

Figure 2. Safety index against Cohesion with varying Diameters in Cohesive Soils.

For concrete in cohesive soils, the safety level associated with piling capacity shows a consistent conservation. At the same time, safety level decreases with increasing length of pile (Fig.3) and the undrained shear strength of the soil (Fig.2). The decrease however depends on the diameter of the pile. Also it can be seen in Fig.1 that the safety level increases with the diameter of the pile, though it decreases with pile length. At any rate, the safety level is generally high for all the range of values for pile length, implying that the estimated pile capacity is highly conservative.

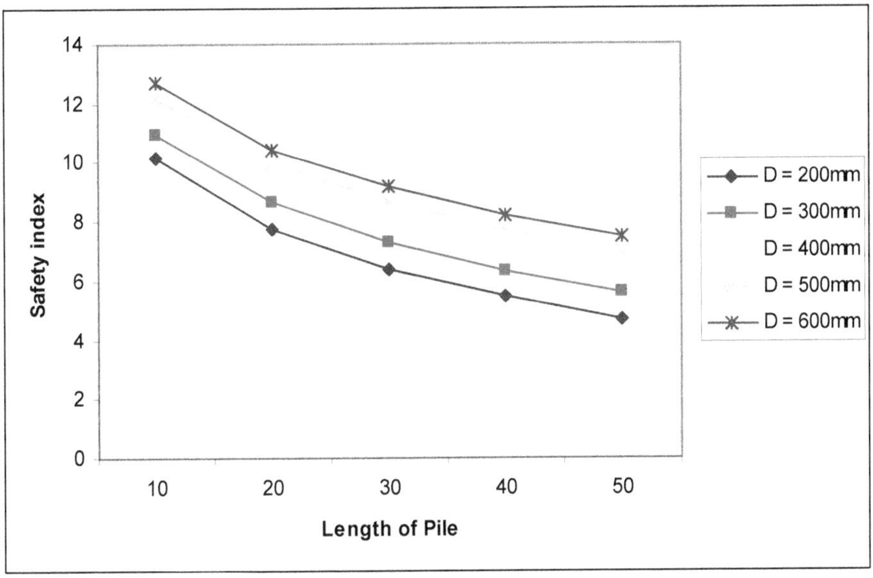

Figure 3. Safety index against Length of Pile with varying Diameters in Cohesive Soils.

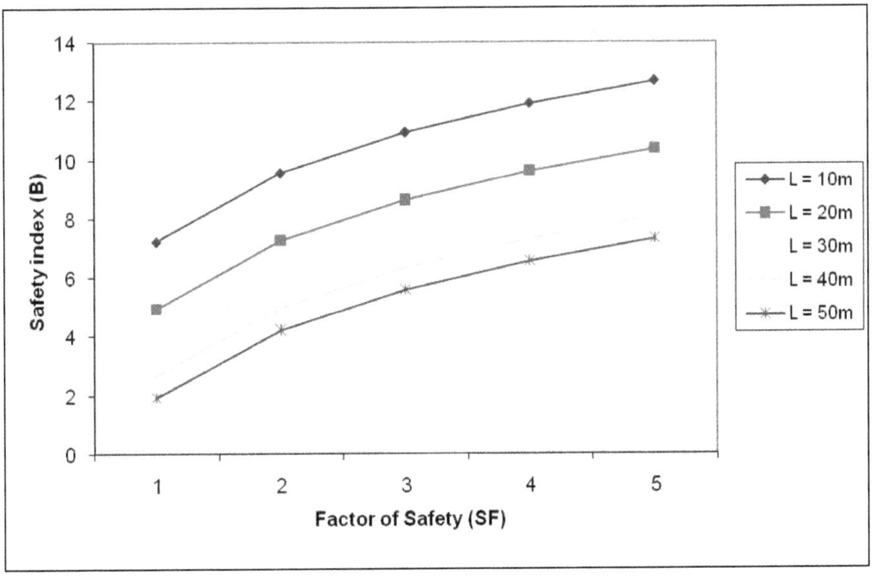

Figure 4. Safety index against Factor of Safety with varying Lengths

3.2. Static pile capacity of concrete piling in cohesionless soils

The Similar to the functional relationship between allowable design load and the allowable pile capacity expressed for cohesive soils, we also have:

G(x) = Allowable Design Load – Allowable Pile Capacity

So that,

$$G(X) = \left[\frac{0.259 f_{cu} D_1^2 + 0.314 f_y D_2^2 - k_f}{SF} \right] \tag{10}$$

in which

$$k_f = 0.785 \gamma L D_1^2 \left(\frac{[e^{(2.36+0.5\phi)\tan\phi}]^2}{2\cos^2(45-0.5\phi)} \right) + 3.142 D_1 L_b + \gamma L(1-\sin\phi')\tan\delta \tag{11}$$

as the expression for assessing the performance of concrete piling in cohesionless soils. In equation (10), the additional variables not in equation (9) are: γ = unit weight of soil, ϕ = drained angle, ϕ' = effective stress angle, and δ = effective friction angle. The assumed statistical values and their corresponding probability distributions are shown in Table2.

i	xi	pdf	$\bar{X}$	V	$\sigma X = \bar{X}.V$
1	f_{cu}	LN	40×10^3KN/m^2	0.15	6.0×10^3
2	D_1	N	4.5×10^{-1}m	0.06	2.7×10^{-2}
3	f_y	LN	460×10^3KN/m^2	0.15	6.9×10^4
4	D_2	N	25×10^{-3}m	0.06	1.5×10^{-3}
5	γ	LN	17.3KN/m^3	0.15	2.60
6	L	N	12m	0.06	7.2×10^{-1}
7	ϕ	LN	34°	0.15	5.1
8	ϕ'	LN	34°	0.15	5.1
9	δ	LN	22°	0.15	3.3
10	SF	LN	3.0	0.15	4.5×10^{-1}

Table 2. Stochastic model for concrete piling in cohesionless soils

The predicted capacity in cohesionless soils leads to a more rapid degeneration of the safety level than in cohesive soils. For instance, it is noted that when piling length is between 10 m and 20 m, there is some margin of safety. At a length of 30 m, the safety level has reduced to zero, implying a total violation of the predicted capacity. This can be seen in Fig. 5. The unit weight of the soil plays a significant role in capacity prediction. The denser the soil, the lower the safety level, no matter the diameter of the pile (see Fig. 6). This similar trend is noted when the angle of internal friction increases (Fig. 7).

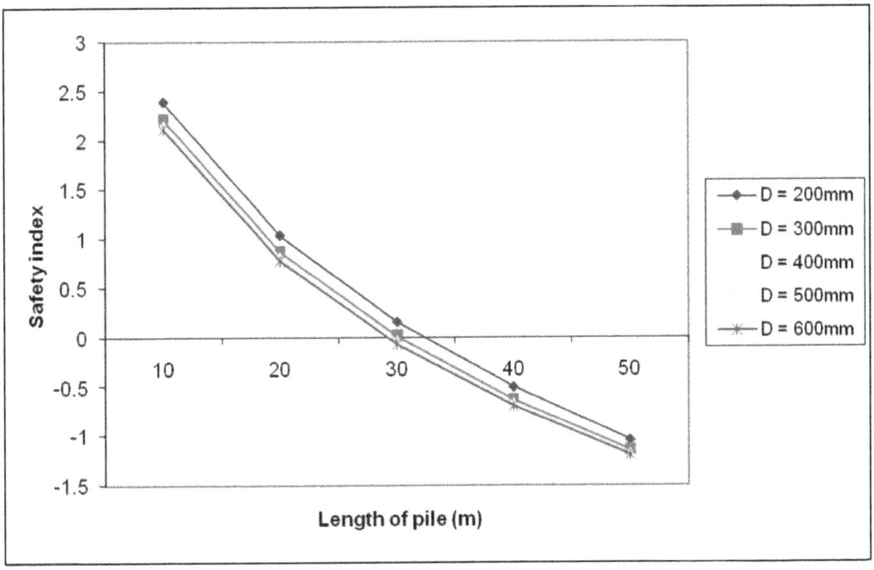

Figure 5. Safety index against Length of Pile with varying Diameters

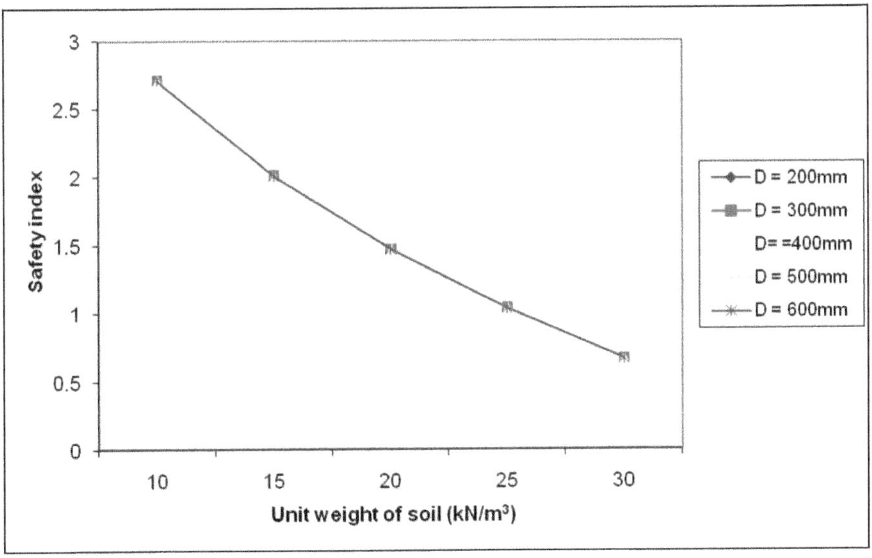

Figure 6. Safety index against Unit weight of soil with varying Diameters

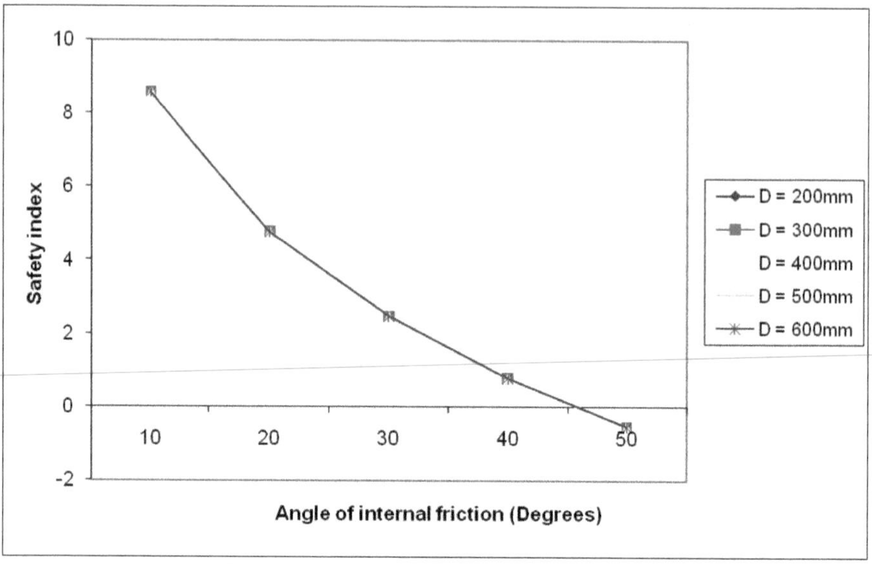

Figure 7. Safety index against Angle of internal friction with varying Diameters

3.3. Static pile static pile capacity of steel piling in cohesive soils

$$G(x) = \text{Allowable Design Load} - \text{Allowable Pile Capacity}$$

$$= 0.35 f_y A_p - \left\{ A_p \left(q S_u \right) + \sum \left(A_s \alpha S_u \right) \right\} / SF$$

$$= 0.35 f_y A_p - \left\{ S_u \left(q A_p + \alpha A_s \right) \right\} / SF \tag{12}$$

i	xi	pd$f$	$\bar{X}$	v	$\sigma x = \bar{X} . v$
1	F_y	LN	$460 \times 10^3 KN/m^2$	0.15	6.9×10^4
2	A_p	N	$1.4 \times 10^{-1}\ m^2$	0.06	8.4×10^{-3}
3	S_u	LN	$28\ KN/m^2$	0.15	4.2
4	α	*LN*	*1.00*	*0.15*	1.5×10^{-1}
5	A_s	N	$48.84\ m^2$	0.06	2.93
6	SF	LN	2.0	0.15	3.0×10^{-1}

Table 3. Stochastic model for steel piling in cohesive soils.

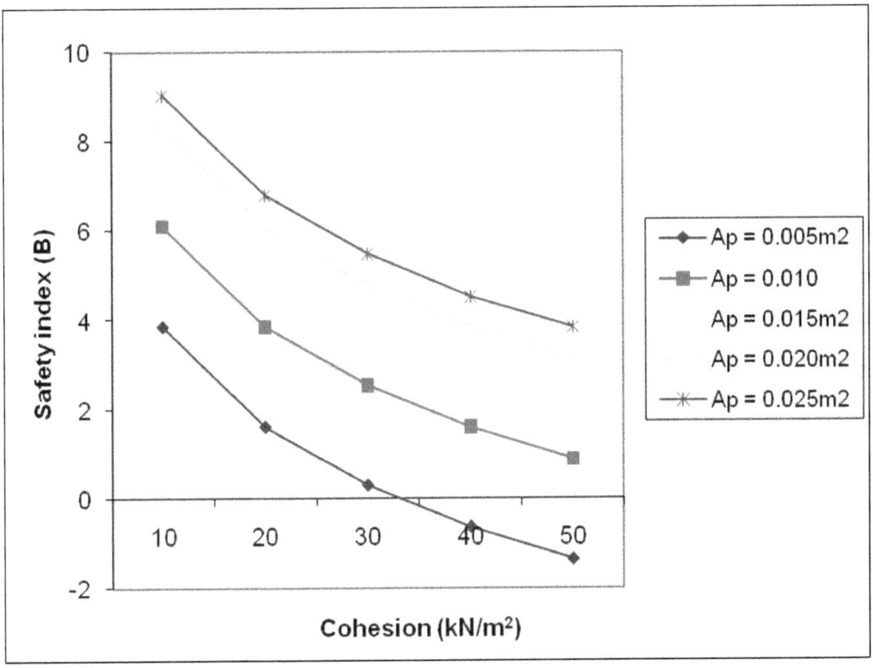

Figure 8. Safety index against Cohesion with varying Pile areas

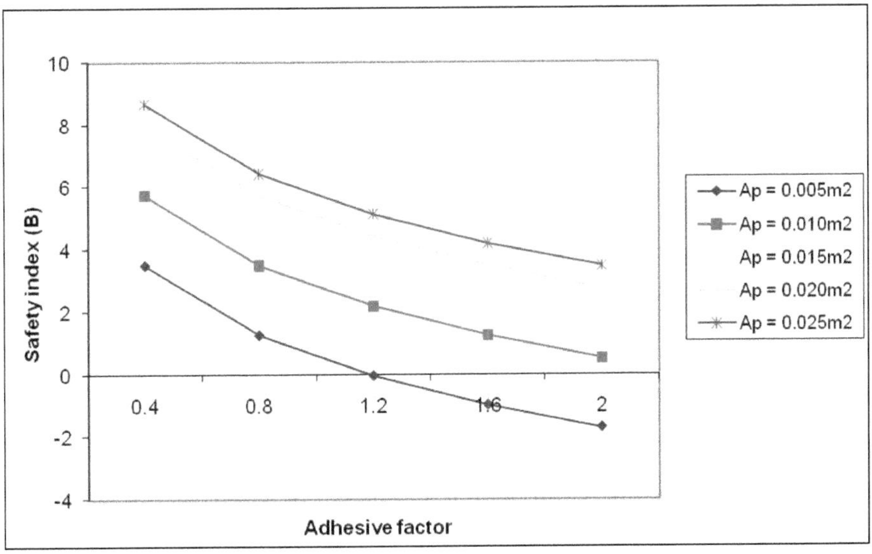

Figure 9. Safety index against Adhesive factor with varying Pile areas

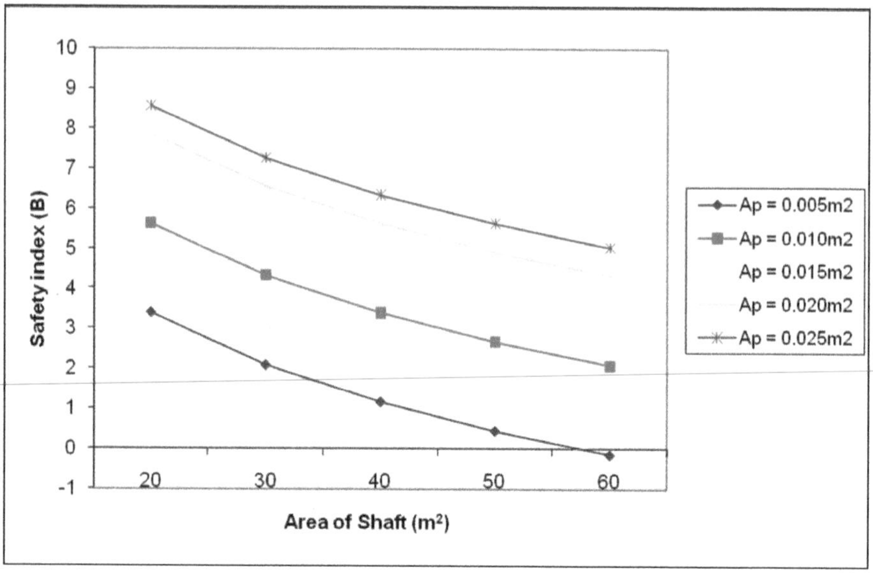

Figure 10. Safety index against Area of shaft with varying Pile areas

The implied safety level associated with piling capacity of steel in cohesive soils shows a consistent conservation. Although, it degenerates with cohesion, adhesion factor and the area of shaft as shown in Figs. 8, 9 and 10 respectively. However, this decrease depends on the area of pile. While the safety level of the piling capacity follows the same trend based on the area of shaft of the pile, but, degenerates less rapidly and higher consistently conservation. It is also noted here that the higher the adhesion factor and the cohesion, the less conservative the static pile capacity.

3.4. Static pile capacity of steel piling in cohesionless soils

$$
\begin{aligned}
G(x) &= \text{Allowable Design Load} - \text{Allowable Pile Capacity} \\
&= 0.35 f_y A_p - \left\{ A_p \bar{q} N'_q + \sum \left(A_s + \bar{q} K \tan\sigma \right) \right\} \Big/ SF \\
&= 0.35 f_y A_p - \left. \left\{ \begin{array}{l} A_p \gamma L \left(\dfrac{\left[e\left(2.356 + 0.5\phi\right) \tan\phi \right]^2}{2\cos^2\left(45 - 0.5\phi\right)} \right) + \\ A_s + \gamma L \left(1 - \sin\phi'\right) \tan\sigma \end{array} \right\} \right/ SF
\end{aligned}
\tag{13}
$$

i	xi	pdf	$\overline{X}$	v	$\sigma x = \overline{x}.\ v$
1	F_y	LN	$460 \times 10^3 KN/m^2$	0.15	6.9×10^4
2	A_p	N	$1.60 \times 10^{-1}\ m^2$	0.06	9.6×10^{-3}
3	γ	LN	$17.3 KN/m^3$	0.15	2.60
4	L	N	12m	0.06	7.2×10^{-1}
5	ϕ	LN	34°	0.15	5.1
6	A_s	N	$16.96\ m^2$	0.06	1.02
7	ϕ'	LN	34°	0.15	5.1
8	δ	LN	22°	0.15	3.3
9	SF	LN	3.0	0.15	4.5×10^{-1}

Table 4. Stochastic model for steel piling in cohesionless soils.

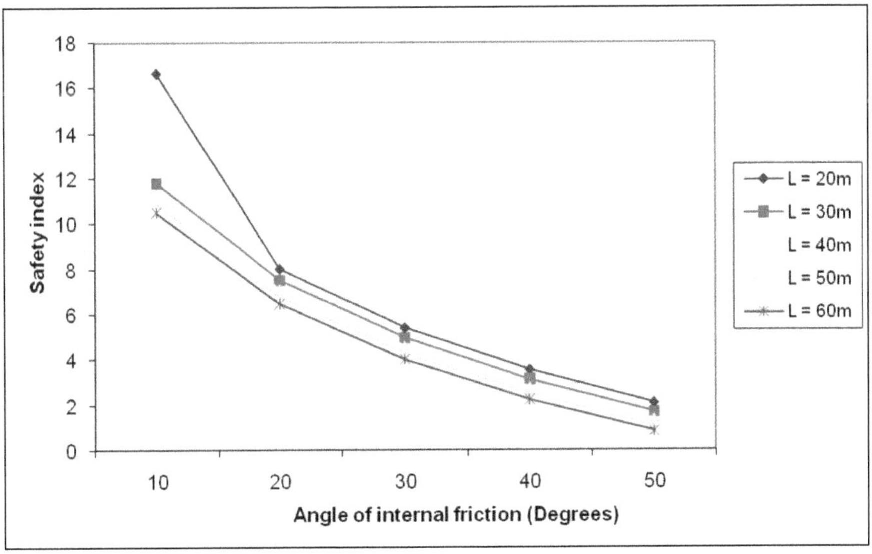

Figure 11. Safety index against Angle of internal friction with varying Lengths

Piling reliability/safety level decreases as pile length, unit weight of soil and the angle of internal friction increases (Figs.11, 12 and 13) but the rate of decrease is more rapid for precast concrete than steel. For concrete piling, pile length greater than 30 m will result in catastrophe while much longer steel piles that economy will permit are admissible.

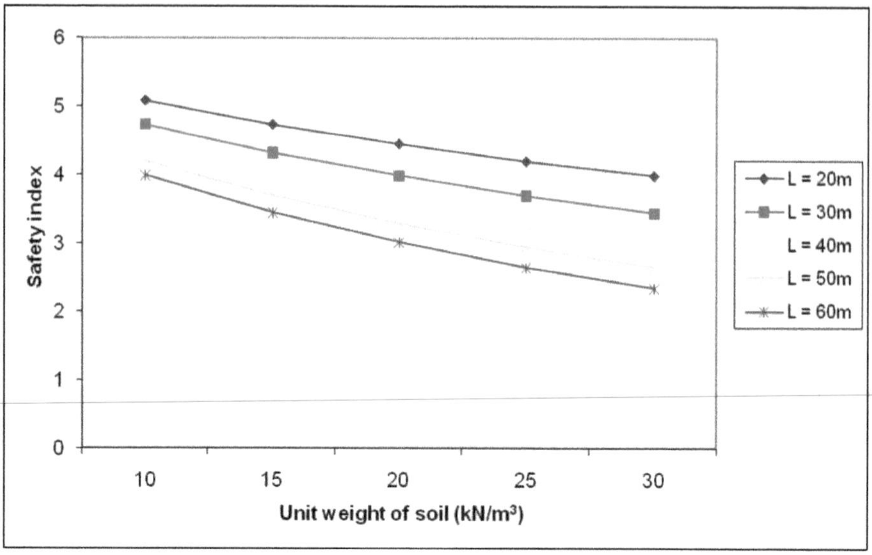

Figure 12. Safety index against Unit weight of soil with varying Lengths

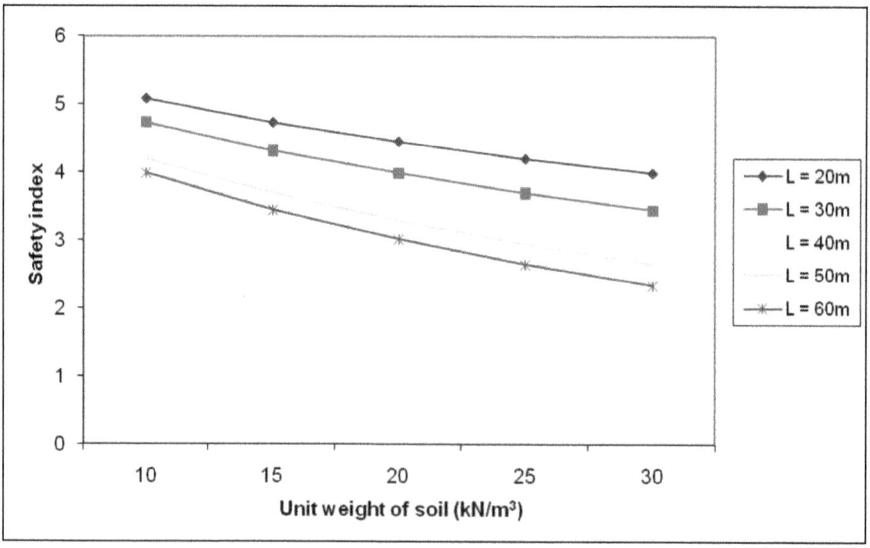

Figure 13. Safety index against Length of Pile with varying Pile areas

3.5. Dynamic pile capacity using Hiley, Janbu and Gates formulae

Estimating the ultimate capacity of a pile while it is being driven in the ground at the site has resulted in numerous equations being presented to the engineering profession. Unfortunately, none of the equations is consistently reliable or reliable over an extended range of pile capacity. Because of this, the best means for predicting pile capacity by dynamic means consists in driving a pile, recording the driving history, and load testing the pile. It would be reasonable to assume that other piles with a similar driving history at that site would develop approximately the same load capacity.

Dynamic formulae have been widely used to predict pile capacity. Some means is needed in the field to determine when a pile has reached a satisfactory bearing value other than by simply driving it to some predetermined depth. Driving the pile to a predetermined depth may or may not obtain the required bearing value because of normal soil variations both laterally and vertically.

3.5.1. Dynamic pile capacity using Hiley formula

G(x) = Allowable Design Load – Allowable Pile Capacity

$$= 0.35 f_y A_p - \left\{ \frac{eh\ Eh}{S + \frac{1}{2}(K_1 + K_2 + K_3)} * \frac{W + n^2 W_p}{W + W_p} \right\} \Big/ SF$$

$$= 0.35 f_y A_p - \left\{ \frac{eh\ Eh}{S + 0.5\left(K_1 + \frac{P_u L}{AE} + K_3\right)} * \frac{W + n^2 W_p}{W + W_p} \right\} \Big/ SF \qquad (14)$$

i	xi	pdf	$\bar{X}$	v	$\sigma x = \bar{X} \cdot v$
1	F_y	LN	460×10^3KN/m^2	0.15	6.9×10^4
2	A_p	N	1.60×10^{-2} m^2	0.06	9.6×10^{-4}
3	e_h	N	0.84	0.06	5.04×10^{-2}
4	E_h	LN	33.12 KN/m	0.15	4.97
5	S	LN	1.79×10^{-2} m	0.15	2.69×10^{-3}
6	K_1	LN	4.06×10^{-3} m	0.15	6.09×10^{-4}
7	P_u	LN	950KN	0.15	1.425×10^2
8	L	N	12.18m	0.06	7.31×10^{-1}
9	E	LN	209×10^6 KN/m^2	0.15	3.14×10^7
10	K_3	LN	2.54×10^{-3} m	0.15	3.81×10^{-4}
11	W	Gumbel	80 KN	0.30	24
12	n	LN	0.5	0.15	7.5×10^{-2}
13	W_p	LN	18.5 KN	0.15	2.775
14	SF	LN	4.0	0.15	6.0×10^{-1}

Table 5. Stochastic model for dynamic pile using Hiley formula.

3.5.2. Dynamic pile capacity using Janbu frmula

G(x) = Allowable Design Load – Allowable Pile Capacity

$$= 0.35 f_y A_p - \left\{\frac{eh\ Eh}{K_u S}\right\} \Big/ SF$$

$$= 0.35 f_y A_p - \left\{ eh\ Eh \Big/ \left(0.75 + 0.15\frac{W_P}{W_r}\right)\left(1 + \sqrt{1 + \frac{eh\ EhL/AES^2}{0.75 + 0.15\frac{W_P}{W_r}}}\right) S \right\} \Big/ SF \tag{15}$$

i	xi	pd*f*	$\overline{X}$	v	$\sigma x = \overline{X} \cdot v$
1	F_y	LN	460 x 10^3KN/m^2	0.15	6.9 x 10^4
2	A_p	N	1.60 x 10^{-2} m^2	0.06	9.6 x 10^{-4}
3	e_h	N	0.84	0.06	5.04 x 10^{-2}
4	E_h	LN	33.12 KN/m	0.15	4.97
5	W_p	LN	18.5 KN	0.15	2.78
6	W_r	Gumbel	35.58 KN	0.30	1.07 x 10^1
7	L	N	12.18m	0.15	7.31 x 10^{-1}
8	E	LN	209 x 10^6 KN/m^2	0.06	3.14 x 10^7
9	S	LN	1.79 x 10^{-2} m	0.15	2.69 x 10^{-3}
10	SF	LN	6.0	0.15	9.0 x 10^{-1}

Table 6. Stochastic model for dynamic pile using Janbu formula.

3.5.3. Dynamic pile capacity using Gates formula

G(x) = Allowable Design Load – Allowable Pile Capacity

$$= 0.35 f_y A_p - \left\{a\sqrt{eh\ Eh}\left(b\text{-}\log S\right)\right\} \Big/ SF$$

$$= 0.35 f_y A_p - \left\{a\left(eh\ Eh\right)^{1/2}\left(b\text{-}\log S\right)\right\} \Big/ SF \tag{16}$$

i	xi	pd*f*	$\overline{X}$	v	$\sigma x = \overline{X} \cdot v$
1	F_y	LN	460 x 10^3KN/m^2	0.15	6.9 x 10^4
2	A_p	N	1.60 x 10^{-2} m^2	0.06	9.6 x 10^{-4}
3	a	N	1.05 x 10^{-1}	0.06	6.3 x 10^{-3}
4	e_h	N	0.85	0.06	5.1 x 10^{-2}
5	E_h	LN	33.12 KNm	0.15	4.97
6	b	N	2.4 x 10^{-3} m	0.06	1.44 x 10^{-4}
7	S	LN	1.79 x 10^{-2} m	0.15	2.69 x 10^{-3}
8	SF	LN	3.0	0.15	4.5 x 10^{-1}

Table 7. Stochastic model for dynamic pile using Gates formula.

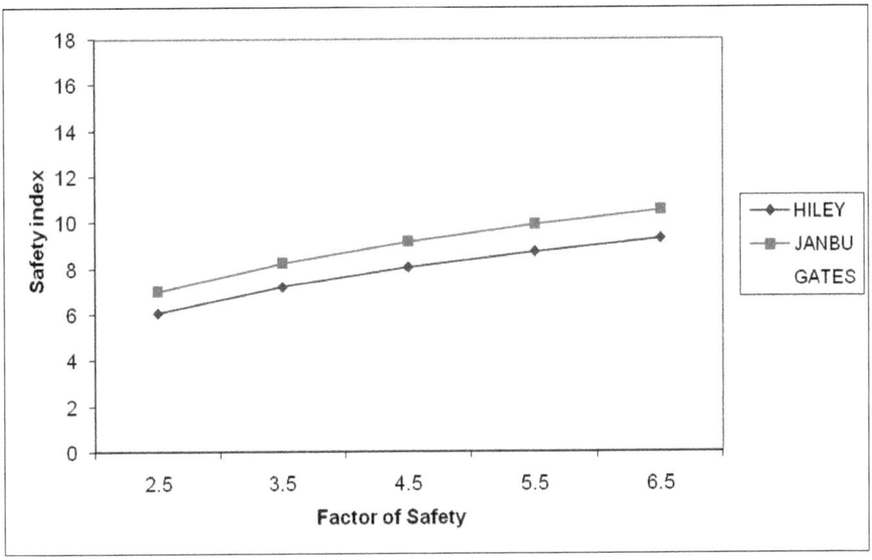

Figure 14. Safety index against Factor of safety using Hiley, Janbu and Gates formulae

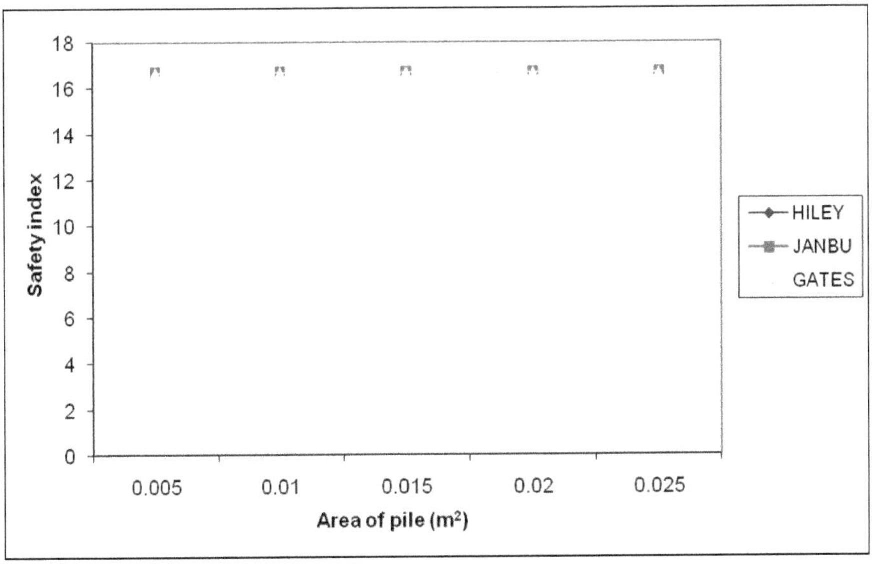

Figure 15. Safety index against Area of pile using Hiley, Janbu and Gates formulae

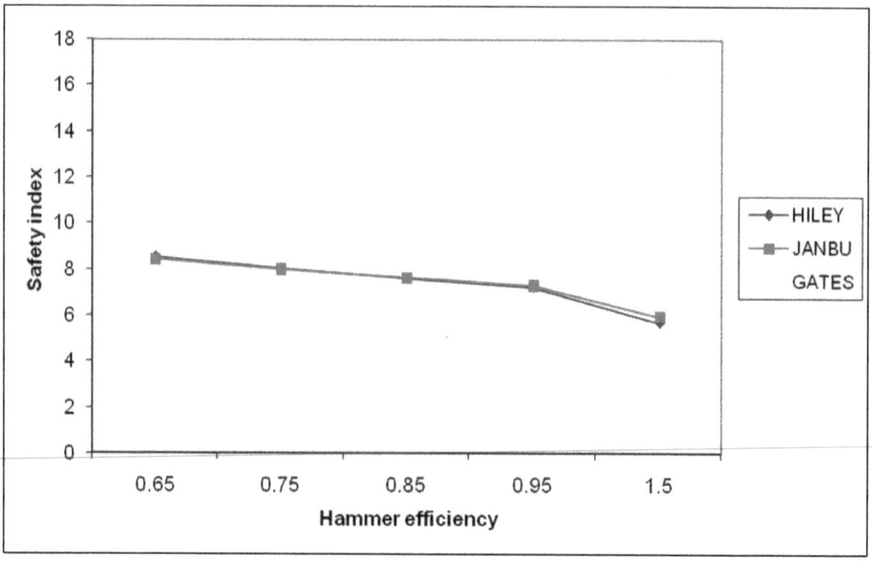

Figure 16. Safety index against Hammer efficiency using Hiley, Janbu and Gates formulae

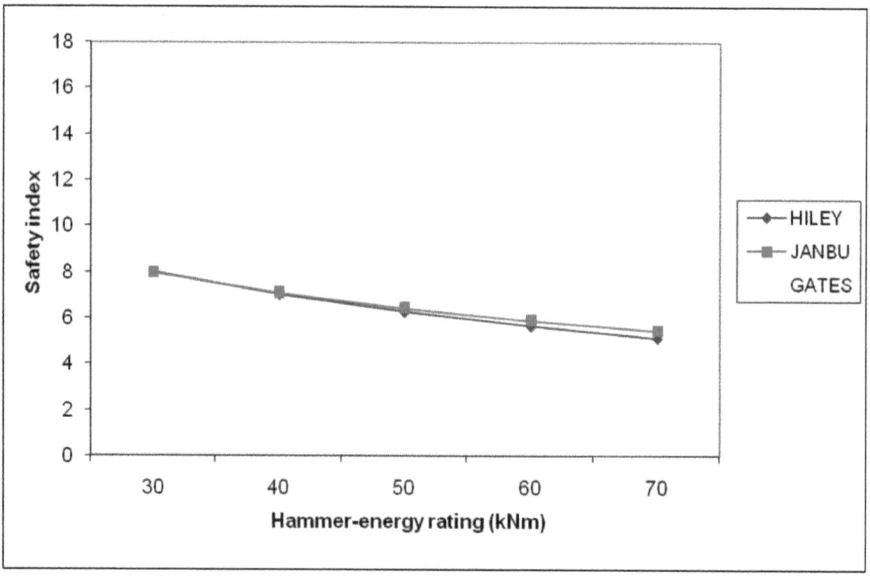

Figure 17. Safety index against Hammer- energy rating using Hiley, Janbu and Gates formulae

The implied safety level associated with piling capacity using Gates' formula is grossly conservative, even much more than Hiley and Janbu formulae. The safety level does not change with the area of pile, hammer efficiency and hammer-energy rating (Figs. 14 to 17). As is common in practice, the areas of piles, hammer efficiency, hammer energy rating and point penetration per blow are subjected to variations and the results of the assessment are as displayed in Figures 14 to 17.

Hiley formula generally and grossly provides a very conservative pile capacity as seen in Figs. 14 to 17. Nevertheless, the safety level does not change with area of pile (Fig. 15). As hammer efficiency and hammer energy rating increase, the safety level reduces significantly as in Fig. 16 and Fig.17 respectively. On the other hand, safety level grows with increasing factor of safety as normally expected (Fig. 14).

Just like the Hiley formula, Janbu formula leads to a grossly conservative pile capacity. However, Janbu's prediction is not as conservative as Hiley's with respect to hammer efficiency and hammer-energy rating. Generally Gates' formula yields the most grossly conservative prediction compared to Hiley and Janbu. It is noted that safety level is not dependent on the area of pile, hammer efficiency and hammer-energy rating (Figs.14 to 17).

4. Conclusion

Reliability-based design methods could be used to address many different aspects of foundation design and construction. However, most of these efforts to date have focused on geotechnical and structural strength requirements, such as the bearing capacity of shallow foundations, the side friction and toe-bearing capacity of deep foundations, and the stresses in deep foundations. All of these are based on the difference between load and capacity, so we can use a more specific definition of reliability as being the probability of the load being less than the capacity for the entire design life of the foundation. Various methods are available to develop reliability-based design of foundations, most notably Stochastic methods, the First-Order Second-Moment Method, and the Load and Resistance Factor Design method.

Deterministic method is not a very cost-effective way to produce reliable structures; therefore, reliability-based design should be adopted to quantify reliability, and thus used to develop balance designs that are both more reliable and less expensive.

From all the preceding sections it is apparent how a stochastic modelling can contribute to the improvement of a structural design, as it can take into account the uncertainties that are present in all human projects.

Author details

David Opeyemi
Rufus Giwa Polytechnic, Owo, Nigeria

Acknowledgement

I like to acknowledge my Supervisor, Professor J.O. Afolayan, for his guidance, teaching, patience, encouragement, suggestions and continued support, particularly for arousing my interest in Risk Analysis (Reliability Analysis of Structures).

5. References

Ayyub, B.M. and Patev, C.R. (1998)."Reliability and Stability Assessment of Concrete Gravity Structures". (RCSLIDE): Theoretical Manual, US Army Corps of Engineers Technical Report ITL – 98 – 6.

Afolayan, J.O. (1999) "Economic Efficiency of Glued Joints in Timber Truss Systems", Building and Environment: The International of Building Science and its Application, Vol. 34, No. 2, pp. 101 – 107.

Afolayan, J. O. (2003) "Improved Design Format for Doubly Symmetric Thin-Walled Structural Steel Beam-Columns", Botswana Journal of Technology, Vol. 12, No.1, pp. 36 – 43

Afolayan, J. O. and Opeyemi, D.A. (2008) "Reliability Analysis of Static Pile Capacity of Concrete in Cohesive and Cohesionless Soils", Research Journal of Applied Sciences, Vol.3, No. 5, pp. 407-411.

Bowles, J.E. (1988) Foundation Analysis and Design. 4th ed., McGraw-Hill Book Company, Singapore.

Coduto, D.P. (2001). Foundation Design: Principles and Practices. 2nd ed., Prentice Hall Inc., New Jersey.

Farid Uddin, A.K.M. (2000) "Risk and Reliability Based Structural and Foundation Design of a Water Reservoir (capacity: 10 million gallon) on the top of Bukit Permatha Hill in Malaysia", 8th ASCE Specialty Conference on Probabilistic Mechanics and Structural Reliability.

Melchers, R.E. (2002). Structural reliability analysis and prediction, John Wiley & Sons, ISBN: 0-471-98771-9, Chichester, West Sussex, UK

Opeyemi, D.A. (2009) "Probabilistic Failure Analysis of Static Pile Capacity for Steel in Cohesive and Cohesionless Soils", Electronic Journal of Geotechnical Engineering, Vol.14, Bund. F, pp. 1-12.

Chapter 3

Coherent Upper Conditional Previsions Defined by Hausdorff Outer Measures to Forecast in Chaotic Dynamical Systems

Serena Doria

Additional information is available at the end of the chapter

1. Introduction

Coherent conditional previsions and probabilities are tools to model and quantify uncertainties; they have been investigated in de Finetti [3], [4], Dubins [10] Regazzini [13], [14] and Williams [20]. Separately coherent upper and lower conditional previsions have been introduced in Walley [18], [19] and models of upper and lower conditional previsions have been analysed in Vicig et al. [17] and Miranda and Zaffalon [12].

In the subjective probabilistic approach coherent probability is defined on an arbitrary class of sets and any coherent probability can be extended to a larger domain. So in this framework no measurability condition is required for random variables. In the sequel, bounded random variables are bounded real-valued functions (these functions are called *gambles* in Walley [19] or *random quantities* in de Finetti [3]). When a measurability condition for a random variable is required, for example to define the Choquet integral, it is explicitly mentioned through the paper.

Separately coherent upper conditional previsions are functionals on a linear space of bounded random variables satisfying the axioms of separate coherence. They cannot always be defined as an extension of conditional expectation of measurable random variables defined by the Radon-Nikodym derivative, according to the axiomatic definition. It occurs because one of the defining properties of the Radon-Nikodym derivative, that is to be measurable with respect to the σ-field of the conditioning events, contradicts a necessary condition for coherence (see Doria [9, Theorem 1], Seidenfeld [16]).

So the necessity to find a new mathematical tool in order to define coherent upper conditional previsions arises.

In Doria [8], [9] a new model of coherent upper conditional prevision is proposed in a metric space. It is defined by the Choquet integral with respect to the s-dimensional Hausdorff outer measure if the conditioning event has positive and finite Hausdorff outer measure in its dimension s. Otherwise if the conditioning event has Hausdorff outer measure in its dimension equal to zero or infinity it is defined by a 0-1 valued finitely, but not countably, additive probability. Coherent upper and lower conditional probabilities are obtained ([6]) when only 0-1 valued random variables are considered.

If the conditioning event B has positive and finite Hausdorff outer measure in its Hausdorff dimension then the given upper conditional prevision defined on a linear lattice of bounded random variables is proven to be a functional, which is monotone, submodular, comonotonically additive and continuous from below. Moreover all these properties are proven to be a sufficient condition under which the upper conditional probability defined by Hausdorff outer measure is the unique monotone set function, which represent a coherent upper conditional prevision as Choquet integral. The given model of coherent upper conditional prevision can be applied to make prevision in chaotic systems.

Many complex systems are strongly dependent on the initial conditions, that is small differences on the initial conditions lead the system to entirely different states. These systems are called *chaotic systems*. Thus uncertainty in the initial conditions produces uncertainty in the final state of the system. Often the final state of the system, called *strange attractor* is represented by a fractal set, i.e., a set with non-integer Hausdorff dimension. The model of coherent upper prevision, introduced in this chapter, can be proposed to forecast in a chaotic system when the conditional prevision of a random variable is conditioned to the attractor of the chaotic system.

The outline of the chapter is the following.

In Section 2 The notion of separately coherent conditional previsions and their properties are recalled.

In Section 3 separately coherent upper conditional previsions are defined in a metric space by the Choquet integral with respect to Hausdorff outer measure if the conditioning event has positive and finite Hausdorff outer measure in its dimension. Otherwise they are defined by a 0-1 valued finitely, but not countably, additive probability.

In Section 4 results are given such that a coherent upper conditional prevision, defined on a linear lattice of bounded random variables containing all constants, is uniquely represented as the Choquet integral with respect to its associated Hausdorff outer measure if and only if it is monotone, submodular and continuous from below.

2. Separately coherent upper conditional previsions

Given a metric space (Ω, d) the Borel σ-field is the σ-field generated by the open sets of Ω. Let $\mathbf{B}$ be a Borel-measurable partition of Ω, i.e. all sets of the partition are Borel sets.

For every $B \in \mathbf{B}$ let us denote by $X|B$ the restriction to B of a random variable defined on Ω and by $\sup(X|B)$ the supremum value that X assumes on B.

Separately coherent upper conditional previsions $\overline{P}(\cdot|B)$ are functionals, defined on a linear space of bounded random variables, i.e. bounded real-valued functions, satisfying the axioms of separate coherence [19].

Definiton 1. Let (Ω, d) be a metric space and let **B** be a Borel-measurable partition of Ω. For every $B \in \mathbf{B}$ let $\mathbf{K}(B)$ be a linear space of bounded random variables on B. Separately coherent upper conditional previsions are functionals $\overline{P}(\cdot|B)$ defined on **K**(B), such that the following conditions hold for every X and Y in $\mathbf{K}(B)$ and every strictly positive constant λ:

1) $\overline{P}\,(X|B) \leq \sup(X|B)$;
2) $\overline{P}(\lambda\, X|B) = \lambda\, \overline{P}(X|B)$ (positive homogeneity);
3) $\overline{P}(X+Y)|B) \leq \overline{P}(X|B) + \overline{P}(Y|B)$ (subadditivity);
4) $\overline{P}(B|B) = 1$.

Coherent upper conditional previsions can always be extended to coherent upper previsions on the class $\mathbf{L}(B)$ of all bounded random variables defined on B. If coherent upper conditional previsions are defined on the class $\mathbf{L}(B)$ no measurability condition is required for the sets B of the partition **B**.

Suppose that $\overline{P}(X|B)$ is a coherent upper conditional prevision on a linear space $\mathbf{K}(B)$ then its conjugate coherent lower conditional prevision is defined by $\underline{P}(X|B) = -\overline{P}(-X|B)$. If for every X belonging to $\mathbf{K}(B)$ we have $P(X|B) = \underline{P}(X|B) = \overline{P}(X|B)$ then $P(X|B)$ is called a coherent *linear* conditional prevision (de Finetti [?]) and it is a linear positive functional on $\mathbf{K}(B)$.

Definition 2. Let (Ω, d) be a metric space and let **B** be a Borel-measurable partition of Ω. For every $B \in \mathbf{B}$ let $\mathbf{K}(B)$ be a linear space of bounded random variables on B. Then linear coherent conditional previsions are functionals $P(\cdot|B)$ defined on $\mathbf{K}(B)$, such that the following conditions hold for every X and Y in $\mathbf{K}(B)$ and every strictly positive constant λ:

1′) if $X \geq 0$ then $P(X|B) \geq 0$ (positivity);
2′) $P(\lambda X|B) = \lambda P(X|B)$ (positive homogeneity);
3′) $P(X+Y)|B) = P(X|B) + P(Y|B)$ (linearity);
4′) $P(B|B) = 1$.

A class of bounded random variables is called a *lattice* if it is closed under point-wise maximum $\vee$ and point-wise minimum $\wedge$.

Two random variables X and Y defined on B are *comonotonic* if, $(X(\omega_1) - X(\omega_2))(Y(\omega_1) - Y(\omega_2)) \geq 0\ \forall \omega_1, \omega_2 \in B$.

Definition 3. Let (Ω, d) be a metric space and let **B** be a Borel-measurable partition of Ω. For every $B \in \mathbf{B}$ let $\mathbf{K}(B)$ be a linear lattice of bounded random variables defined on B and let $\overline{P}(\cdot|B)$ be a coherent upper conditional prevision defined on $\mathbf{K}(B)$ then for every X, Y, X_n in **K**(B) $\overline{P}(\cdot|B)$ is

i) *monotone* iff $X \leq Y$ implies $\overline{P}(X|B) \leq \overline{P}(Y|B)$;
ii) *comonotonically additive* iff $\overline{P}(X+Y|B) = \overline{P}(X|B) + \overline{P}(Y|B)$ if X and Y are comonotonic;
iii) *submodular* iff $\overline{P}(X \vee Y|B) + \overline{P}(X \wedge Y|B) \leq \overline{P}(X|B) + \overline{P}(Y|B)$;
iv) *continuous from below* iff $lim_{n\to\infty}\overline{P}(X_n|B) = \overline{P}(X|B)$ if X_n is an increasing sequence of random variables converging to X.

A bounded random variable is called **B**-*measurable* or measurable with respect to the partition **B** [19, p.291] if it is constant on the atoms B of the partition. Let $G(\mathbf{B})$ be the class of all **B**-measurable random variables.

Denote by $\overline{P}(X|\mathbf{B})$ the random variable equal to $\overline{P}(X|B)$ if $\omega \in B$.

Separately coherent upper conditional previsions $\overline{P}(X|B)$ can be extended to a common domain **H** so that the function $\overline{P}(\cdot|\mathbf{B})$ can be defined from **H** to $G(\mathbf{B})$ to summarize the collection of $\overline{P}(X|B)$ with $B \in \mathbf{B}$.

$\overline{P}(\cdot|\mathbf{B})$ is assumed to be separately coherent if all the $\overline{P}(\cdot|B)$ are separately coherent. In Theorem 1 [9] the function $\overline{P}(X|\mathbf{B})$ is compared with the Radon-Nikodym derivative.

It is proven that, every time that the σ-field of the conditioning events is properly contained in the σ-field of the probability space and it contains all singletons, the Radon-Nikodym derivative cannot be used as a tool to define coherent conditional previsions. This is due to the fact that one of the defining properties of the Radon-Nikodym derivative, that is to be measurable with respect to the σ-field of the conditioning events, contradicts a necessary condition for the coherence.

Analysis done points out the necessity to introduce a different tool to define coherent conditional previsions.

3. Separately coherent upper conditional previsions defined by Hausdorff outer measures

In this section coherent upper conditional previsions are defined by the Choquet integral with respect to Hausdorff outer measures if the conditioning event B has positive and finite Hausdorff outer measure in its dimension. Otherwise if the conditioning event B has Hausdorff outer measure in its dimension equal to zero or infinity they are defined by a 0-1 valued finitely, but not countably, additive probability.

3.1. Hausdorff outer measures

Given a non-empty set Ω, let $\wp(\Omega)$ be the class of all subsets of Ω. An *outer measure* is a function $\mu^* : \wp(\Omega) \rightarrow [0, +\infty]$ such that $\mu^*(\emptyset) = 0$, $\mu^*(A) \leq \mu^*(A')$ if $A \subseteq A'$ and $\mu^*(\bigcup_{i=1}^{\infty} A_i) \leq \sum_{i=1}^{\infty} \mu^*(A_i)$.

Examples of outer set functions or outer measures are the Hausdorff outer measures [11], [15].

Let (Ω, d) be a metric space. A topology, called the *metric topology*, can be introduced into any metric space by defining the open sets of the space as the sets G with the property:

if x is a point of G, then for some $r > 0$ all points y with $d(x, y) < r$ also belong to G.

It is easy to verify that the open sets defined in this way satisfy the standard axioms of the system of open sets belonging to a topology [15, p.26].

The Borel σ-field is the σ-field generated by all open sets. The Borel sets include the closed sets (as complement of the open sets), the F_σ-sets (countable unions of closed sets) and the G_σ-sets (countable intersections of open sets), etc.

The diameter of a non empty set U of Ω is defined as $|U| = \sup\{d(x,y) : x, y \in U\}$ and if a subset A of Ω is such that $A \subset \bigcup_i U_i$ and $0 < |U_i| < \delta$ for each i, the class $\{U_i\}$ is called a δ-cover of A.

Let s be a non-negative number. For $\delta > 0$ we define $h_{s,\delta}(A) = \inf \sum_{ß=1}^{+\infty} |U_i|^s$, where the infimum is over all δ-covers $\{U_i\}$.

The *Hausdorff s-dimensional outer measure* of A, denoted by $h^s(A)$, is defined as

$$h^s(A) = \lim_{\delta \to 0} h_{s,\delta}(A).$$

This limit exists, but may be infinite, since $h_{s,\delta}(A)$ increases as δ decreases because less δ-covers are available. The *Hausdorff dimension* of a set A, $dim_H(A)$, is defined as the unique value, such that

$$h^s(A) = +\infty \quad \text{if} \quad 0 \leq s < \dim_H(A),$$
$$h^s(A) = 0 \quad \text{if} \quad \dim_H(A) < s < +\infty.$$

We can observe that if $0 < h^s(A) < +\infty$ then $dim_H(A) = s$, but the converse is not true.

Hausdorff outer measures are *metric* outer measures:

$$h^s(E \cup F) = h^s(E) + h^s(F) \text{ whenever } d(E,F) = \inf\{d(x,y) : x \in E, y \in F\} > 0.$$

A subset A of Ω is called *measurable* with respect to the outer measure h^s if it decomposes every subset of Ω additively, that is if

$$h^s(E) = h^s(A \cap E) + h^s(E - A) \text{ for all sets } E \subseteq \Omega.$$

All Borel subsets of Ω are measurable with respect to any metric outer measure [11, Theorem 1.5]. So every Borel subset of Ω is measurable with respect to every Hausdorff outer measure h^s since Hausdorff outer measures are metric.

The restriction of h^s to the σ-field of h^s-measurable sets, containing the σ-field of the Borel sets, is called Hausdorff s-dimensional measure. In particular the Hausdorff 0-dimensional measure is the counting measure and the Hausdorff 1-dimensional measure is the Lebesgue measure.

The Hausdorff s-dimensional measures are *modular* on the Borel σ-field, that is

$$h^s(A \cup B) + h^s(A \cap B) = h^s(A) + h^s(B)$$

for every pair of Borelian sets A and B; so that [5, Proposition 2.4] the Hausdorff outer measures are *submodular*

$$h^s(A \cup B) + h^s(A \cap B) \leq h^s(A) + h^s(B).$$

In [15, p.50] and [11, Theorem 1.6 (a)] it has been proven that if A is any subset of Ω there is a G_σ-set G containing A with $h^s(A) = h^s(G)$. In particular h^s is an *outer regular* measure.

Moreover Hausdorff outer measures are *continuous from below* [11, Lemma 1.3], that is for any increasing sequence of sets $\{A_i\}$ we have

$$\lim_{i\to\infty} h^s(A_i) = h^s(\lim_{i\to\infty} A_i).$$

h^s-Measurable sets with finite Hausdorff s-dimensional outer measure can be approximated from below by closed subsets [15, p.50] [11, Theorem 1.6 (b)] or equally the restriction of every Hausdorff outer measure h^s to the class of all h^s-measurable sets with finite Hausdorff outer measure is *inner regular* on the class of all closed subsets of Ω.

In particular any h^s-measurable set with finite Hausdorff s-dimensional outer measure contains an F_σ-set of equal measure, and so contains a closed set differing from it by arbitrary small measure.

Since every metric space is a Hausdorff space then every compact subset of Ω is closed; denote by **O** the class of all open sets of Ω and by **C** the class of all compact sets of Ω, the restriction of each Hausdorff s-dimensional outer measure to the class **H** of all h^s-measurable sets with finite Hausdorff outer measure is *strongly regular* [5, p.43] that is it is regular:

a) $h^s(A) = \inf\{h^s(U)|A \subset U, U \in \mathbf{O}\}$ for all $A \in \mathbf{H}$ (outer regular);

b) $h^s(A) = \sup\{h^s(C)|C \subset A, C \in \mathbf{C}\}$ for all $A \in \mathbf{H}$ (inner regular)

with the additional property:

c) $\inf\{h^s(U - A)|A \subset U, U \in \mathbf{O}\} = 0$ for all $A \in \mathbf{H}$

Any Hausdorff s-dimensional outer measure is translation invariant, that is, $h^s(x + E) = h^s(E)$, where $x + E = \{x + y : y \in E\}$ [11, p.18].

3.2. The Choquet integral

We recall the definition of the Choquet integral [5] with the aim to define upper conditional previsions by Choquet integral with respect to Hausdorff outer measures and to prove their properties. The Choquet integral is an integral with respect to a monotone set function. Given a non-empty set Ω and denoted by S a set system, containing the empty set and properly contained in $\wp(\Omega)$, the family of all subsets of Ω, a monotone set function μ: $S \to \overline{\Re}_+ = \Re_+ \cup \{+\infty\}$ is such that $\mu(\oslash) = 0$ and if A, $B \in S$ with $A \subseteq B$ then $\mu(A) \leq \mu(B)$. Given a monotone set function μ on S, its *outer set function* is the set function μ^* defined on the whole power set $\wp(\Omega)$ by

$$\mu^*(A) = \inf\{\mu(B) : B \supset A; B \in S\}, A \in \wp(\Omega)$$

The inner set function of μ is the set function μ_* defined on the whole power set $\wp(\Omega)$ by

$$\mu_*(A) = \sup\{\mu(B)|B \subset A; B \in S\}, A \in \wp(\Omega)$$

Let μ be a monotone set function defined on S properly contained in $\wp(\Omega)$ and $X : \Omega \to \overline{\Re} = \Re \cup \{-\infty, +\infty\}$ an arbitrary function on Ω. Then the set function

$$G_{\mu,X}(x) = \mu\{\omega \in \Omega : X(\omega) > x\}$$

is decreasing and it is called *decreasing distribution function* of X with respect to μ. If μ is continuous from below then $G_{\mu,X}(x)$ is right continuous. In particular the decreasing distribution function of X with respect to the Hausdorff outer measures is right continuous since these outer measures are continuous from below. A function $X : \Omega \to \overline{\Re}$ is called upper μ-measurable if $G_{\mu^*,X}(x) = G_{\mu_*,X}(x)$. Given an upper μ-measurable function $X : \Omega \to \overline{R}$ with decreasing distribution function $G_{\mu,X}(x)$, if $\mu(\Omega) < +\infty$, the *asymmetric Choquet integral* of X with respect to μ is defined by

$$\int X d\mu = \int_{-\infty}^{0} (G_{\mu,X}(x) - \mu(\Omega))dx + \int_{0}^{\infty} G_{\mu,X}(x)dx$$

The integral is in $\Re$, can assume the values $-\infty$, $+\infty$ or is undefined when the right-hand side is $+\infty - \infty$.

If $X \geq 0$ or $X \leq 0$ the integral always exists. In particular for $X \geq 0$ we obtain

$$\int X d\mu = \int_0^{+\infty} G_{\mu,X}(x)dx$$

If X is bounded and $\mu(\Omega) = 1$ we have that

$$\int X d\mu = \int_{\inf X}^{0} (G_{\mu,X}(x) - 1)dx + \int_0^{\sup X} G_{\mu,X}(x)dx = \int_{\inf X}^{\sup X} G_{\mu,X}(x)dx + \inf X.$$

3.3. The model

A new model of coherent upper conditional prevision is defined in [9, Theorem 2].

Theorem 1. *Let (Ω, d) be a metric space and let $\mathbf{B}$ be a Borel-measurable partition of Ω. For every $B \in \mathbf{B}$ denote by s the Hausdorff dimension of the conditioning event B and by h^s the Hausdorff s-dimensional outer measure. Let $\mathbf{K}(B)$ be a linear space of bounded random variables on B. Moreover, let m be a 0-1 valued finitely additive, but not countably additive, probability on $\wp(B)$ such that a different m is chosen for each B. Then for each $B \in \mathbf{B}$ the functional $\overline{P}(X|B)$ defined on $\mathbf{K}(B)$ by*

$$\overline{P}(X|B) = \frac{1}{h^s(B)} \int_B X dh^s \text{ if } 0 < h^s(B) < +\infty$$

and by

$$\overline{P}(X|B) = m(XB) \text{ if } h^s(B) = 0, +\infty$$

is a coherent upper conditional prevision.

The lower conditional previsions $\underline{P}(A|B)$ can be define as in the previous theorem if h_s denotes the Hausdorff s-dimensional inner measure.

Given an upper conditional prevision $\overline{P}(X|B)$ defined on a linear space the lower conditional prevision $\underline{P}(X|B)$ is obtained as its conjugate, that is $\underline{P}(X|B) = -\overline{P}(-X|B)$. If B has positive and finite Hausdorff outer measure in its Hausdorff dimension s and we denote by h_s the Hausdorff s-dimensional inner measure we have

$$\underline{P}(X|B) = -\overline{P}(-X|B) = -\frac{1}{h^s(B)} \int_B (-X) dh^s =$$

$$\frac{1}{h^s(B)}\int_B Xdh_s = \frac{1}{h_s(B)}\int_B Xdh_s.$$

The last equality holds since each B is h^s-measurable, that is $h^s(B) = h_s(B)$.

The unconditional upper prevision is obtained as a particular case when the conditioning event is Ω, that is $\overline{P}(A) = \overline{P}(A|\Omega)$ and $\underline{P}(A) = \underline{P}(A|\Omega)$.

Coherent upper conditional probabilities are obtained when only 0-1 valued random variables are considered; they have been defined in [6] :

Theorem 2. *Let (Ω, d) be a metric space and let $\mathbf{B}$ be a Borel-measurable partition of Ω. For every $B \in \mathbf{B}$ denote by s the Hausdorff dimension of the conditioning event B and by h^s the Hausdorff s-dimensional outer measure. Let m be a 0-1 valued finitely additive, but not countably additive, probability on $\wp(B)$ such that a different m is chosen for each B. Thus, for each $B \in \mathbf{B}$, the function defined on $\wp(B)$ by*

$$\overline{P}(A|B) = \frac{h^s(AB)}{h^s(B)} \quad \text{if} \quad 0 < \mathrm{h^s(B)} < +\infty$$

and by

$$\overline{P}(A|B) = m(AB) \quad \text{if} \quad \mathrm{h^s(B)} = 0, +\infty$$

is a coherent upper conditional probability.

Let B be a set with positive and finite Hausdorff outer measure in its Hausdorff dimension s. Denote by h^s the s-dimensional Hausdorff outer measure and for every $A \in \wp(B)$ by $\mu_B^*(A) = \overline{P}(A|B) = \frac{h^s(AB)}{h^s(B)}$ the upper conditional probability defined on $\wp(B)$. From Theorem 1 we have that the upper conditional prevision $\overline{P}(\cdot|B)$ is a functional defined on $\mathbf{L}(B)$ with values in $\Re$ and the upper conditional probability μ_B^* integral represents $\overline{P}(X|B)$ since $\overline{P}(X|B) = \int X d\mu_B^* = \frac{1}{h^s(B)} \int X dh^s$. The number $\frac{1}{h^s(B)}$ is a normalizing constant.

4. Examples

Example 1 Let $B = [0,1]$. The Hausdorff dimension of B is 1 and the Hausdorff 1-dimensional measure h^1 is the Lebesgue measure. Moreover $h^1[0,1] = 1$ then the coherent upper conditional prevision is defined for every $X \in \mathbf{K}(B)$ by

$$\overline{P}(X|B) = \frac{1}{h^s(B)} \int_B X dh^s = \int_B X dh^1$$

We recall the definition of the Cantor set. Let E_0 = [0,1], E_1 = [0,1/3] ∪ [2/3,1], E_2 = [0,1/9] ∪ [2/9, 1/3] ∪ [2/3,7/9] ∪ [8/9,1], etc., where E_n is obtained by removing the open middle third of each interval in E_{n-1}, so E_n is the union of 2^n intervals, each of length $\frac{1}{3^n}$. The Cantor's set is the perfect set $E = \bigcap_{n=0}^{\infty} E_n$.

Example 2 Let B be the Cantor set. The Hausdorff dimension of the Cantor set is $s = \frac{ln2}{ln3}$ and $h^s(B) = 1$. Then the coherent upper conditional prevision is defined for every $X \in \mathbf{K}(B)$ by

$$\overline{P}(X|B) = \frac{1}{h^s(B)} \int_B X dh^s = \int_B X dh^{\frac{ln2}{ln3}}$$

Example 3 Let $B = \{\omega_1, \omega_2, ..., \omega_n\}$. The Hausdorff dimension of B is 0 and the Hausdorff 0-dimensional measure h^0 is the counting measure. Moreover $h^0(B) = n$ then the coherent upper conditional prevision is defined for every $X \in \mathbf{K}(B)$ by

$$\overline{P}(X|B) = \frac{1}{h^s(B)} \int_B X dh^s = \frac{1}{n} \sum_1^n X(\omega_i)$$

5. Upper envelope

A necessary and sufficient condition for an upper prevision $\overline{P}$ to be coherent is to be the *upper envelope* of linear previsions, i.e. there is a class M of linear previsions such that $\overline{P}$ =sup$\{P(X) : P \in M\}$ [19, 3.3.3].

Given a coherent upper prevision $\overline{P}$ defined on a domain **K** the maximal coherent extension of $\overline{P}$ to the class of all bounded random variables is called [19, 3.1.1] *natural extension* of $\overline{P}$.

The linear extension theorem [19, 3.4.2] assures that the class of all linear extensions to the class of all bounded random variables of a linear prevision P defined on a linear space **K** is the class $M(P)$ of all linear previsions that are dominated by P on **K**. Moreover the upper and lower envelopes of $M(P)$ are the natural extensions of P [19, Corollary 3.4.3].

Let $P(\cdot|B)$ be the coherent upper conditional prevision on the the class of all bounded Borel-measurable random variables defined on B defined in Theorem 1.

In Doria [9, Theorem 5] it is proven that, for every conditioning event B, the given upper conditional prevision is the upper envelope of all linear extensions of $P(\cdot|B)$ to the class of all bounded random variables on B.

Theorem 3. *Let (Ω, d) be a metric space and let $\mathbf{B}$ be a Borel-measurable partition of Ω. For every conditioning event $B \in \mathbf{B}$ let $\mathbf{L}(B)$ be the class of all bounded random variables defined on B and let $P(\cdot|B)$ be the coherent upper conditional previsionon the class of all bounded Borel-measurable random variables defined in Theorem 1. Then the coherent upper conditional prevision defined on $\mathbf{L}(B)$ as in Theorem 1 is the upper envelope of all linear extensions of $P(\cdot|B)$ to the class $\mathbf{L}(B)$.*

In the same way it can be proven that the conjugate of the coherent upper conditional prevision $\overline{P}(\cdot|B)$ is the *lower envelope* of $M(P)$, the class of all linear extension of $P(\cdot|B)$ dominating $P(\cdot|B)$.

6. Main results

For each B in **B**, denote by s the Hausdorff dimension of B then the Hausdorff s-dimensional outer measure is called the Hausdorff outer measure *associated* with the coherent upper prevision $\overline{P}(\cdot|B)$. Let $B \in \mathbf{B}$ be meausurable with respect to the Hausdorf outer measure associated with $\overline{P}(\cdot|B)$.

The Choquet integral representation of a coherent upper conditional prevision with respect to its associated Hausdorff outer measure has bees investigated in [7]. In [9] necessary and sufficient conditions are given such that a coherent upper conditional prevision is uniquely represented as the Choquet integral with respect to its associated Hausdorff outer measure.

In [9, Theorem 4] it is proven that, if the conditioning event has positive and finite Hausdorff outer measure in its dimension s and $\mathbf{K}(B)$ is a linear lattice of bounded random variables defined on B, necessary conditions for the functional $\overline{P}(X|B)$ to be represented as Choquet integral with respect to the upper conditional probability μ_B^*, i.e. $\overline{P}(X|B) = \frac{1}{h^s(B)} \int X dh^s$, are that $\overline{P}(X|B)$ is monotone, comonotonically additive, submodular and continuous from below.

Theorem 4. *Let (Ω, d) be a metric space and let $\mathbf{B}$ be a Borel-measurable partition of Ω. For every $B \in \mathbf{B}$ denote by s the Hausdorff dimension of the conditioning event B and by h^s the Hausdorff s-dimensional outer measure. Let $\mathbf{K}(B)$ be a linear lattice of bounded random variables defined on B. If the conditioning event B has positive and finite Hausdorff s-dimensional outer measure then the coherent upper conditional prevision $\overline{P}(\cdot|B)$ defined on $\mathbf{K}(B)$ as in Theorem 2 is:*

i) monotone;

ii) comonotonically additive;

iii) submodular;

iv) continuous from below.

Moreover if the conditioning event B has positive and finite Hausdorff s-dimensional outer measure, from the properties of the Choquet integral ([5, Proposition 5.1]) the coherent upper conditional prevision $\overline{P}(\cdot|B)$ is

v) translation invariant;

vi) positively homogeneous;

So the functional $\overline{P}(\cdot|B)$ can be used to defined a *coherent risk measure* [1]. since it is monotone, subadditive, translation invariant and positively homogeneous.

In [9, Theorem 6] sufficient conditions are given for a coherent upper conditional prevision to be uniquely represented as Choquet intergral with respect to its associated Hausdorff outer measure.

Theorem 5. *Let (Ω, d) be a metric space and let $\mathbf{B}$ be a Borel-measurable partition of Ω. For every $B \in \mathbf{B}$ denote by s the Hausdorff dimension of the conditioning event B and by h^s the Hausdorff s-dimensional outer measure. Let $\mathbf{K}(B)$ be a linear lattice of bounded random variables on B containing all constants. If B has positive and finite Hausdorff outer measure in its dimension and the coherent upper conditional prevision $\overline{P}(\cdot|B)$ on $\mathbf{K}(B)$ is monotone, comonotonically additive, submodular and continuous from below then $\overline{P}(\cdot|B)$ is representable as Choquet integral with respect to a monotone, submodular set function which is continuous from below. Furthermore all monotone set functions on $\wp(B)$ with these properties agree on the set system of weak upper level sets $M = \{\{X \geq x\} | X \in \mathbf{K}(B), x \in \Re\}$ with the upper conditional probability $\mu_B^*(A) = \frac{h^s(AB)}{h^s(B)}$ for $A \in \wp(B)$. Let β be a monotone set function on $\wp(B)$, which is submodular, continuous from below and such that represents $\overline{P}(\cdot|B)$ as Choquet integral. Then the following equalities hold*

$$\overline{P}(X|B) = \int_B X d\beta = \int_B X d\mu_B^* = \frac{1}{h^s(B)} \int_B X dh^s.$$

An example is given in the particular case where $\mathbf{K}(B)$ is the linear space of all bounded Borel-measurable random variables on B and the restriction of the Hausdorff s-dimensional outer measure to the Borel σ-field of subsets of B is considered.

Example 2. Let (Ω, d) be a metric space and let **B** be a Borel-measurable partition of Ω. For every $B \in \mathbf{B}$ let $\mathbf{K}(B)$ be the linear space of all bounded Borel-measurable random variables on B and let S be the Borel σ-field of subsets of B. Denote by s the Hausdorff dimension of the conditioning event B and by h^s the Hausdorff s-dimensional outer measure. If $0 < h^s(B) < +\infty$ define $\mu_B(A) = \frac{h^s(AB)}{h^s(B)}$, for every $A \in S$; $\mu_B(A)$ is modular and continuous from below on S since each Hausdorff s-dimensional (outer) measure is σ-additive on the Borel σ-field . Moreover let $P(\cdot|B)$ be a coherent linear conditional prevision, which is continuous from below. Then $P(\cdot|B)$ can be uniquely represented as the Choquet integral with respect to the coherent upper conditional probability μ_B, that is

$$P(X|B) = \int X d\mu_B = \frac{1}{h^s(B)} \int X dh^s.$$

The previous example can be obtained as a consequence of the Daniell-Stone Representation Theorem [5, p. 18].

7. Conclusions

In this chapter a model of coherent upper conditional precision is introduced. It is defined by the Choquet integral with respect to the s-dimensional Hausdorff outer measure if the conditioning event has positive and finite Hausdorff outer measure in its Hausdorff dimension s. Otherwise if the conditioning event has Hausdorff outer measure in its Hausdorff dimension equal to zero or infinity it is defined by a 0-1 valued finitely, but not countably, additive probability. If the conditioning event has positive and finite Hausdorff outer measure in its Hausdorff dimension the given upper conditional prevision, defined on a linear lattice of bounded random variables which contains all constants, is uniquely represented as the Choquet integral with respect Hausdorff outer measure if and only if it is a functional which is monotone, submodular, comonotonically additive and continuous from below.

Coherent upper conditional prevision based on the Hausdorff s-dimensional measure permits to analyze complex systems where information represented by sets with Hausdorff dimension less than s, have no influence on the situation; information represented by sets with the same Hausdorff dimension of the conditioning event can influence the system.

Coherent upper previsions defined by Hausdorff outer measures can also be applied in decision theory, to asses preferences between random variables defined on fractal sets and to defined coherent risk measures.

Author details

Serena Doria
Department of Engineering and Geology, University G.d'Annunzio, Chieti-Pescara, Italy

8. References

[1] P. Artzner, F. Delbaen, J. Elber, D. Heath.(1999) Coherent measures of risk. Mathematical Finance, 9, 203-228.

[2] P. Billingsley. (1986). *Probability and measure*, Wiley, USA.
[3] B. de Finetti. (1970). *Teoria della Probabilita'*, Einaudi Editore, Torino.
[4] B. de Finetti. (1972). Probability, Induction and Statistics, Wiley, New York.
[5] D. Denneberg. (1994). *Non-additive measure and integral*. Kluwer Academic Publishers.
[6] S. Doria. (2007). Probabilistic independence with respect to upper and lower conditional probabilities assigned by Hausdorff outer and inner measures. *International Journal of Approximate Reasoning*, 46, 617-635.
[7] S. Doria. (2010). Coherent upper conditional previsions and their integral representation with respect to Hausdorff outer measures, In Combining Soft Computing and Statistical Methods in Data Analysis (C. Borgelt et al. editors), Advances in Intelligent and Soft Computing 77, 209-216, Springer.
[8] S. Doria. (2011). Coherent upper and lower conditional previsions defined by Hausdorff outer and inner measures, In Modeling, Design and Simulation of Systems with Uncertainties (A. Rauh and E. Auer editors), Mathmatical Engineering, Volume 3, 175-195, Springer.
[9] S. Doria. (2012). Characterization of a coherent upper conditional prevision as the Choquet integral with respect to its associated Hausdorff outer measure. *Annals of Operations Research*, Vol 195, 33-48.
[10] L.E. Dubins. (1975). Finitely additive conditional probabilities, conglomerability and disintegrations. *The Annals of Probability*, Vol. 3, 89-99.
[11] K.J. Falconer. (1986). *The geometry of fractals sets*. Cambridge University Press.
[12] E. Miranda, M.Zaffalon. (2009). Conditional models: coherence and inference trough sequences of joint mass functions, Journal of Statistical Planning and Inference 140(7), 1805-1833.
[13] E. Regazzini. (1985). Finitely additive conditional probabilities, Rend. Sem. Mat. Fis.,Milano, 55,69-89.
[14] E. Regazzini. (1987). De Finetti's coherence and statistical inference, The Annals of Statistics, Vol. 15, No. 2, 845-864.
[15] C.A. Rogers. (1998). *Hausdorff measures*. Cambridge University Press.
[16] T. Seidenfeld, M.J. Schervish, J.B. Kadane. (2001). Improper regular conditional distributions, The Annals of Probability, Vol. 29, No. 4, 1612-1624.
[17] P. Vicig, M. Zaffalon, F.G. Cozman. (2007). Notes on "Notes on conditional previsions", International Journal of Approximate Reasoning, Vol 44, 358-365.
[18] P. Walley. (1981). Coherent lower (and upper) probabilities, Statistics Research Report, University of Warwick.
[19] P. Walley. (1991). *Statistical Reasoning with Imprecise Probabilities*. Chapman and Hall, London.
[20] P.M. Williams. (2007). Notes on conditional previsions, International Journal of Approximate Reasoning, Vol 44, 3, 366-383.

Chapter 4

On Guaranteed Parameter Estimation of Stochastic Delay Differential Equations by Noisy Observations

Uwe Küchler and Vyacheslav A. Vasiliev

Additional information is available at the end of the chapter

1. Introduction

Assume $(\Omega, \mathcal{F}, (\mathcal{F}(t), t \geq 0), P)$ is a given filtered probability space and $W = (W(t), t \geq 0)$, $V = (V(t), t \geq 0)$ are real-valued standard Wiener processes on $(\Omega, \mathcal{F}, (\mathcal{F}(t), t \geq 0), P)$, adapted to $(\mathcal{F}(t))$ and mutually independent. Further assume that $X_0 = (X_0(t), t \in [-1, 0])$ and Y_0 are a real-valued cadlag process and a real-valued random variable on $(\Omega, \mathcal{F}, (\mathcal{F}(t), t \geq 0), P)$ respectively with

$$E \int_{-1}^{0} X_0^2(s)ds < \infty \text{ and } EY_0^2 < \infty.$$

Assume Y_0 and $X_0(s)$ are $\mathcal{F}_0-$measurable, $s \in [-1, 0]$ and the quantities W, V, X_0 and Y_0 are mutually independent.

Consider a two–dimensional random process $(X, Y) = (X(t), Y(t), t \geq 0)$ described by the system of stochastic differential equations

$$dX(t) = aX(t)dt + bX(t-1)dt + dW(t), \tag{1}$$

$$dY(t) = X(t)dt + dV(t), \ t \geq 0 \tag{2}$$

with the initial conditions $X(t) = X_0(t)$, $t \in [-1, 0]$, and $Y(0) = Y_0$. The process X is supposed to be hidden, i.e., unobservable, and the process Y is observed. Such models are used in applied problems connected with control, filtering and prediction of stochastic processes (see, for example, [1, 4, 17–20] among others).

The parameter $\vartheta = (a, b)' \in \Theta$ is assumed to be unknown and shall be estimated based on continuous observation of Y, Θ is a subset of $\mathcal{R}^2$ $((a, b)'$ denotes the transposed $(a, b))$. Equations (1) and (2) together with the initial values $X_0(\cdot)$ and Y_0 respectively have uniquely solutions $X(\cdot)$ and $Y(\cdot)$, for details see [19].

Equation (1) is a very special case of stochastic differential equations with time delay, see [5, 6] and [20] for example.

To estimate the true parameter ϑ with a prescribed least square accuracy ε we shall construct a sequential plan $(T^*(\varepsilon), \vartheta^*(\varepsilon))$ working for all $\vartheta \in \overline{\Theta}$. Here $T^*(\varepsilon)$ is the duration of observations which is a special chosen stopping time and $\vartheta^*(\varepsilon)$ is an estimator of ϑ. The set $\overline{\Theta}$ is defined to be the intersection of the set Θ with an arbitrary but fixed ball $\mathcal{B}_{0,R} \subset \mathcal{R}^2$. Sequential estimation problem has been solved for sets Θ of a different structure in [7]-[9], [11, 13, 14, 16] by observations of the process (1) and in [10, 12, 15] – by noisy observations (2).

In this chapter the set Θ of parameters consists of all $(a,b)'$ from $\mathcal{R}^2$ which do not belong to lines L_1 or L_2 defined in Section 2 below and having Lebesgue measure zero.

This sequential plan is a composition of several different plans which follow the regions to which the unknown true parameter $\vartheta = (a,b)'$ may belong to. Each individual plan is based on a weighted correlation estimator, where the weight matrices are chosen in such a way that this estimator has an appropriate asymptotic behaviour being typical for the corresponding region to which ϑ belongs to. Due to the fact that this behaviour is very connected with the asymptotic properties of the so-called fundamental solution $x_0(\cdot)$ of the deterministic delay differential equation corresponding to (1) (see Section 2 for details), we have to treat different regions of $\Theta = \mathcal{R}^2 \setminus L$, $L = L_1 \cup L_2$, separately. If the true parameter ϑ belongs to L, the weighted correlation estimator under consideration converges weakly only, and thus the assertions of Theorem 3.1 below cannot be derived by means of such estimators. In general, the exception of the set L does not disturb applications of the results below in adaptive filtration, control theory and other applications because of its Lebesgue zero measure.

In the papers [10, 12] the problem described above was solved for the two special sets of parameters Θ_I (a straight line) and Θ_{II} (where $X(\cdot)$ satisfies (1) is stable or periodic (unstable)) respectively. The general sequential estimation problem for all $\vartheta = (a,b)'$ from $\mathcal{R}^2$ except of two lines was solved in [13, 14, 16] for the equation (1) based on the observations of $X(\cdot)$.

In this chapter the sequential estimation method developed in [10, 12] for the system (1), (2) is extended to the case, considered by [13, 14, 16] for the equation (1) (as already mentioned, for all ϑ from $\mathcal{R}^2$ except of two lines for the observations without noises).

A related result in such problem statement was published first for estimators of an another structure and without proofs in [15].

A similar problem for partially observed stochastic dynamic systems without time-delay was solved in [22, 23].

The organization of this chapter is as follows. Section 2 presents some preliminary facts needed for the further studies about we have spoken. In Section 3 we shall present the main result, mentioned above. In Section 4 all proofs are given. Section 5 includes conclusions.

2. Preliminaries

To construct sequential plans for estimation of the parameter ϑ we need some preparation. At first we shall summarize some known facts about the equation (1). For details the reader is referred to [3]. Together with the mentioned initial condition the equation (1) has a uniquely determined solution X which can be represented for $t \geq 0$ as follows:

$$X(t) = x_0(t)X_0(t) + b\int_{-1}^{0} x_0(t-s-1)X_0(s)ds + \int_0^t x_0(t-s)dW(s). \tag{3}$$

Here $x_0 = (x_0(t), t \geq -1)$ denotes the fundamental solution of the deterministic equation

$$x_0(t) = 1 + \int_0^t (ax_0(s) + bx_0(s-1))ds, \quad t \geq 0, \tag{4}$$

corresponding to (1) with $x_0(t) = 0, t \in [-1,0),\ x_0(0) = 1.$

The solution X has the property $E\int_0^T X^2(s)ds < \infty$ for every $T > 0$.

From (3) it is clear, that the limit behaviour for $t \to \infty$ of X very depends on the limit behaviour of $x_0(\cdot)$. The asymptotic properties of $x_0(\cdot)$ can be studied by the Laplace-transform of x_0, which equals $(\lambda - a - be^{-\lambda})^{-1}$, λ any complex number.

Let $s = u(r)$ $(r < 1)$ and $s = w(r)$ $(r \in \mathcal{R}^1)$ be the functions given by the following parametric representation $(r(\xi), s(\xi))$ in $\mathcal{R}^2$:

$$r(\xi) = \xi \cot \xi,\ s(\xi) = -\xi / \sin \xi$$

with $\xi \in (0, \pi)$ and $\xi \in (\pi, 2\pi)$ respectively.

Now we define the parameter set Θ to be the plane $\mathcal{R}^2$ without the lines $L_1 = (a, u(a))_{a\leq 1}$ and $L_2 = (a, w(a))_{a\in\mathcal{R}^1}$ such that $\mathcal{R}^2 = \Theta \cup L_1 \cup L_2$.

It seems not to be possible to construct a general simple sequential procedure which has the desired properties under P_ϑ for all $\vartheta \in \Theta$. Therefore we are going to divide the set Θ into some appropriate smaller regions where it is possible to do. This decomposition is very connected with the structure of the set Λ of all (real or complex) roots of the so-called characteristic equation of (4):

$$\lambda - a - be^{-\lambda} = 0.$$

Put $v_0 = v_0(\vartheta) = \max\{Re\lambda | \lambda \in \Lambda\},\ v_1 = v_1(\vartheta) = \max\{Re\lambda | \lambda \in \Lambda,\ Re\lambda < v_0\}$. Beside of the case $b = 0$ it holds $-\infty < v_1 < v_0 < \infty$. By $m(\lambda)$ we denote the multiplicity of the solution $\lambda \in \Lambda$. Note that $m(\lambda) = 1$ for all $\lambda \in \Lambda$ beside of $(a, b) \in \mathcal{R}^2$ with $b = -e^a$. In this cases we have $\lambda = a - 1 \in \Lambda$ and $m(a-1) = 2$. The values $v_0(\vartheta)$ and $v_1(\vartheta)$ determine the asymptotic behaviour of $x_0(t)$ as $t \to \infty$ (see [3] for details).

Now we are able to divide Θ into some appropriate for our purposes regions. Note, that this decomposition is very related to the classification used in [3]. There the plane $\mathcal{R}^2$ was decomposed into eleven subsets. Here we use another notation.

Definition (Θ). The set Θ of parameters is decomposed as

$$\Theta = \Theta_1 \cup \Theta_2 \cup \Theta_3 \cup \Theta_4,$$

where $\Theta_1 = \Theta_{11} \cup \Theta_{12} \cup \Theta_{13},\ \Theta_2 = \Theta_{21} \cup \Theta_{22},\ \Theta_3 = \Theta_{31},\ \Theta_4 = \Theta_{41} \cup \Theta_{42}$ with

$$\Theta_{11} = \{\vartheta \in \mathcal{R}^2 |\ v_0(\vartheta) < 0\},$$

$$\Theta_{12} = \{\vartheta \in \mathcal{R}^2 |\ v_0(\vartheta) > 0 \text{ and } v_0(\vartheta) \notin \Lambda\},$$

$$\Theta_{13} = \{\vartheta \in \mathcal{R}^2 |\, v_0(\vartheta) > 0;\ v_0(\vartheta) \in \Lambda,\ m(v_0) = 2\},$$
$$\Theta_{21} = \{\vartheta \in \mathcal{R}^2 |\, v_0(\vartheta) > 0, v_0(\vartheta) \in \Lambda,\ m(v_0) = 1,\ \ v_1(\vartheta) > 0 \text{ and } v_1(\vartheta) \in \Lambda\},$$
$$\Theta_{22} = \{\vartheta \in \mathcal{R}^2 |\, v_0(\vartheta) > 0, v_0(\vartheta) \in \Lambda,\ m(v_0) = 1,\ \ v_1(\vartheta) > 0 \text{ and } v_1(\vartheta) \notin \Lambda\},$$
$$\Theta_{31} = \{\vartheta \in \mathcal{R}^2 |\, v_0(\vartheta) > 0,\ \ v_0(\vartheta) \in \Lambda,\ \ m(v_0) = 1 \text{ and } v_1(\vartheta) < 0\},$$
$$\Theta_{41} = \{\vartheta \in \mathcal{R}^2 |\ \ v_0(\vartheta) = 0,\ v_0(\vartheta) \in \Lambda,\ m(v_0) = 1\},$$
$$\Theta_{42} = \{\vartheta \in \mathcal{R}^2 |\ \ v_0(\vartheta) > 0,\ v_0(\vartheta) \in \Lambda,\ m(v_0) = 1,\ \ v_1(\vartheta) = 0 \text{ and } v_1(\vartheta) \in \Lambda\}.$$

It should be noted, that the cases (Q2 ∪ Q3) and (Q5) considered in [3] correspond to our exceptional lines L_1 and L_2 respectively.

Here are some comments concerning the Θ subsets.

The unions $\Theta_1, \ldots, \Theta_4$ are marked out, because the Fisher information matrix and related design matrices which will be considered below, have similar asymptotic properties for all ϑ throughout every Θ_i $(i = 1, \ldots, 4)$.

Obviously, all sets $\Theta_{11}, \ldots, \Theta_{42}$ are pairwise disjoint, the closure of Θ equals to $\mathcal{R}^2$ and the exceptional set $L_1 \cup L_2$ has Lebesgue measure zero.

The set Θ_{11} is the set of parameters ϑ for which there exists a stationary solution of (1).

Note that the one-parametric set Θ_4 is a part of the boundaries of the following regions: Θ_{11}, $\Theta_{12}, \Theta_{21}, \Theta_3$. In this case $b = -a$ holds and (1) can be written as a differential equation with only one parameter and being linear in the parameter.

We shall use a truncation of all the introduced sets. First chose an arbitrary but fixed positive R. Define the set $\overline{\Theta} = \{\vartheta \in \Theta |\ ||\vartheta|| \leq \mathrm{R}\}$ and in a similar way the subsets $\overline{\Theta}_{11}, \ldots, \overline{\Theta}_{42}$.

Sequential estimators of ϑ with a prescribed least square accuracy we have already constructed in [10, 12]. But in these articles the set of possible parameters ϑ were restricted to $\overline{\Theta}_{11} \cup \overline{\Theta}_{12} \cup \{\overline{\Theta}_{41} \setminus \{(0,0)\}\} \cup \overline{\Theta}_{42}$.

To construct a sequential plan for estimating ϑ based on the observation of $Y(\cdot)$ we follow the line of [10, 12]. We shall use a single equation for Y of the form:

$$dY(t) = \vartheta' A(t)dt + \xi(t)dt + dV(t), \tag{5}$$

where $A(t) = (Y(t), Y(t-1))'$,

$$\xi(t) = X(0) - aY(0) - bY(0) + b\int_{-1}^{0} X_0(s)ds - aV(t) - bV(t-1) + W(t).$$

The random variables $A(t)$ and $\xi(t)$ are $\mathcal{F}(t)$-measurable for every fixed $t \geq 1$ and a short calculation shows that all conditions of type (7) in [12], consisting of

$$E\int_1^T (|Y(t)| + |\xi(t)|)dt < \infty \text{ for all } T > 1,$$

$$E[\tilde{\Delta}\xi(t)|\mathcal{F}(t-2)] = 0,\ E[(\tilde{\Delta}\xi(t))^2|\mathcal{F}(t-2)] \leq 1 + R^2$$

hold in our case. Here $\tilde{\Delta}$ denotes the difference operator defined by $\tilde{\Delta}f(t) = f(t) - f(t-1)$.

Using this operator and (5) we obtain the following equation:

$$d\tilde{\Delta}Y(t) = a\tilde{\Delta}Y(t)dt + b\tilde{\Delta}Y(t-1)dt + \tilde{\Delta}\xi(t)dt + d\tilde{\Delta}V(t) \tag{6}$$

with the initial condition $\tilde{\Delta}Y(1) = Y(1) - Y_0$.

Thus we have reduced the system (1), (2) to a single differential equation for the observed process $(\tilde{\Delta}Y(t), t \geq 2)$ depending on the unknown parameters a and b.

3. Construction of sequential estimation plans

In this section we shall construct the sequential estimation procedure for each of the cases $\overline{\Theta}_1 \ldots, \overline{\Theta}_4$ separately. Then we shall define, similar to [11, 13, 14, 16], the final sequential estimation plan, which works in $\overline{\Theta}$ as a sequential plan with the smallest duration of observations.

We shall construct the sequential estimation procedure of the parameter ϑ on the basis of the correlation method in the cases $\overline{\Theta}_1, \overline{\Theta}_4$ (similar to [12, 14, 15]) and on the basis of correlation estimators with weights in the cases $\overline{\Theta}_2 \cup \overline{\Theta}_3$. The last cases and $\overline{\Theta}_{13}$ are new. It should be noted, that the sequential plan, constructed e.g. in [2] does not work for $\overline{\Theta}_3$ here, even in the case if we observe $(X(\cdot))$ instead of $(Y(\cdot))$.

3.1. Sequential estimation procedure for $\vartheta \in \overline{\Theta}_1$

Consider the problem of estimating $\vartheta \in \overline{\Theta}_1$. We will use some modification of the estimation procedure from [12], constructed for the Case II thereon. It can be easily shown, that Proposition 3.1 below can be proved for the cases $\overline{\Theta}_{11} \cup \overline{\Theta}_{12}$ similarly to [12]. Presented below modified procedure is oriented, similar to [16] on all parameter sets $\overline{\Theta}_{11}$, $\overline{\Theta}_{12}$, $\overline{\Theta}_{13}$. Thus we will prove Proposition 3.1 in detail for the case $\overline{\Theta}_{13}$ only. The proofs for cases $\overline{\Theta}_{11} \cup \overline{\Theta}_{12}$ are very similar.

For the construction of the estimation procedure we assume h_{10} is a real number in $(0, 1/5)$ and h_1 is a random variable with values in $[h_{10}, 1/5]$ only, $\mathcal{F}(0)$-measurable and having a known continuous distribution function.

Assume $(c_n)_{n\geq 1}$ is a given unboundedly increasing sequence of positive numbers satisfying the following condition:

$$\sum_{n\geq 1} \frac{1}{c_n} < \infty. \tag{7}$$

This construction follows principally the line of [14, 16] (see [12] as well), for which the reader is referred for details.

We introduce for every $\varepsilon > 0$ and every $s \geq 0$ several quantities:
– the functions

$$\Psi_s(t) = \begin{cases} (\tilde{\Delta}Y(t), \tilde{\Delta}Y(t-s))' & \text{for} \quad t \geq 1+s, \\ (0,0)' \quad \text{for} \quad t < 1+s; \end{cases}$$

– the sequence of stopping times

$$\tau_1(n,\varepsilon) = h_1 \inf\{k \geq 1 : \int_0^{kh_1} ||\Psi_{h_1}(t-2-5h_1)||^2 dt \geq \varepsilon^{-1} c_n\} \quad \text{for} \quad n \geq 1;$$

– the matrices

$$G_1(T,s) = \int_0^T \Psi_s(t-2-5s)\Psi_1'(t)dt, \quad \Phi_1(T,s) = \int_0^T \Psi_s(t-2-5s)d\tilde{\Delta}Y(t),$$

$$G_1(n,k,\varepsilon) = G_1(\tau_1(n,\varepsilon)-kh_1,h_1), \quad \Phi_1(n,k,\varepsilon) = \Phi_1(\tau_1(n,\varepsilon)-kh_1,h_1);$$

– the times

$$k_1(n) = \arg\min_{k=\overline{1,5}} ||G_1^{-1}(n,k,\varepsilon)||, \; n \geq 1;$$

– the estimators

$$\vartheta_1(n,\varepsilon) = G_1^{-1}(n,\varepsilon)\Phi_1(n,\varepsilon), \; n \geq 1, \quad G_1(n,\varepsilon) = G_1(n,k_1(n),\varepsilon), \; \Phi_1(n,\varepsilon) = \Phi_1(n,k_1(n),\varepsilon);$$

– the stopping time

$$\sigma_1(\varepsilon) = \inf\{N \geq 1: \; S_1(N) > (\rho_1\delta_1^{-1})^{1/2}\}, \tag{8}$$

where $S_1(N) = \sum_{n=1}^{N} \beta_1^2(n,\varepsilon)$,

$$\beta_1(n,\varepsilon) = ||\tilde{G}_1^{-1}(n,\varepsilon)||, \quad \tilde{G}_1(n,\varepsilon) = (\varepsilon^{-1}c_n)^{-1}G_1(n,k_1(n),\varepsilon)$$

and $\delta_1 \in (0,1)$ is some fixed chosen number,

$$\rho_1 = 15(3+\mathrm{R}^2)\sum_{n\geq 1}\frac{1}{c_n}.$$

The deviation of the 'first-step estimators' $\vartheta_1(n,\varepsilon)$ has the form:

$$\vartheta_1(n,\varepsilon) - \vartheta = (\varepsilon^{-1}c_n)^{-1/2}\tilde{G}_1^{-1}(n,\varepsilon)\tilde{\zeta}_1(n,\varepsilon), \; n \geq 1, \tag{9}$$

$$\tilde{\zeta}_1(n,\varepsilon) = (\varepsilon^{-1}c_n)^{-1/2} \int_0^{\tau_1(n,\varepsilon)-k_1(n)h_1} \Psi_{h_1}(t-2-5h_1)(\tilde{\Delta}\xi(t)dt + dV(t) - dV(t-1)).$$

By the definition of stopping times $\tau_1(n,\varepsilon) - k_1(n)h_1$ we can control the noise $\tilde{\zeta}_1(n,\varepsilon)$:

$$E_\vartheta||\tilde{\zeta}_1(n,\varepsilon)||^2 \leq 15(3+\mathrm{R}^2), \; n \geq 1, \; \varepsilon > 0$$

and by the definition of the stopping time $\sigma_1(\varepsilon)$ - the first factor $\tilde{G}_1^{-1}(n,\varepsilon)$ in the representation of the deviation (9).

Define the sequential estimation plan of ϑ by

$$T_1(\varepsilon) = \tau_1(\sigma_1(\varepsilon),\varepsilon), \; \vartheta_1(\varepsilon) = \frac{1}{S(\sigma_1(\varepsilon))}\sum_{n=1}^{\sigma_1(\varepsilon)} \beta_1^2(n,\varepsilon)\vartheta_1(n,\varepsilon). \tag{10}$$

We can see that the construction of the sequential estimator $\vartheta_1(\varepsilon)$ is based on the family of estimators $\vartheta(T,s) = G_1^{-1}(T,s)\Phi(T,s)$, $s \geq 0$. We have taken the discretization step h_1 as above, because for $\vartheta \in \overline{\Theta}_{12}$ the functions

$$f(T,s) = e^{2v_0T}\, G_1^{-1}(T,s)$$

for every $s \geq 0$ have some periodic matrix functions as a limit on T almost surely. These limit matrix functions are finite and may be infinite on the norm only for four values of their argument T on every interval of periodicity of the length $\Delta > 1$ (see the proof of Theorem 3.2 in [10, 12]).

In the sequel limits of the type $\lim\limits_{n\to\infty} a(n,\varepsilon)$ or $\lim\limits_{\varepsilon\to 0} a(n,\varepsilon)$ will be used. To avoid repetitions of similar expressions we shall use, similar to [12, 14, 16], the unifying notation $\lim\limits_{n\vee\varepsilon} a(n,\varepsilon)$ for both of those limits if their meaning is obvious.

We state the results concerning the estimation of the parameter $\vartheta \in \overline{\Theta}_1$ in the following proposition.

Proposition 3.1. *Assume that the condition (7) on the sequence (c_n) holds and let the parameter $\vartheta = (a,b)'$ in (1) be such that $\vartheta \in \overline{\Theta}_1$.*

Then:

I. For any $\varepsilon > 0$ and every $\vartheta \in \overline{\Theta}_1$ the sequential plan $(T_1(\varepsilon), \vartheta_1(\varepsilon))$ defined by (10) is closed ($T_1(\varepsilon) < \infty$ P_ϑ – a.s.) and possesses the following properties:

$$1^\circ. \quad \sup_{\vartheta\in\overline{\Theta}_1} E_\vartheta ||\vartheta_1(\varepsilon) - \vartheta||^2 \leq \delta_1 \varepsilon;$$

2°. the inequalities below are valid:

– for $\vartheta \in \overline{\Theta}_{11}$

$$0 < \underline{\lim}_{\varepsilon\to 0} \varepsilon \cdot T_1(\varepsilon) \leq \overline{\lim}_{\varepsilon\to 0} \varepsilon \cdot T_1(\varepsilon) < \infty \quad P_\vartheta - a.s.,$$

– for $\vartheta \in \overline{\Theta}_{12}$

$$0 < \underline{\lim}_{\varepsilon\to 0} [T_1(\varepsilon) - \frac{1}{2v_0} \ln \varepsilon^{-1}] \leq \overline{\lim}_{\varepsilon\to 0} [T_1(\varepsilon) - \frac{1}{2v_0} \ln \varepsilon^{-1}] < \infty \quad P_\vartheta - a.s.,$$

– for $\vartheta \in \overline{\Theta}_{13}$

$$0 < \underline{\lim}_{\varepsilon\to 0} [T_1(\varepsilon) + \frac{1}{v_0} \ln T_1(\varepsilon) - \Psi'_{13}(\varepsilon)], \qquad \overline{\lim}_{\varepsilon\to 0} [T_1(\varepsilon) + \frac{1}{v_0} \ln T_1(\varepsilon) - \Psi''_{13}(\varepsilon)] < \infty \quad P_\vartheta - a.s.,$$

the functions $\Psi'_{13}(\varepsilon)$ and $\Psi''_{13}(\varepsilon)$ are defined in (30).

II. For every $\vartheta \in \overline{\Theta}_1$ the estimator $\vartheta_1(n,\varepsilon)$ is strongly consistent:

$$\lim_{n\vee\varepsilon} \vartheta_1(n,\varepsilon) = \vartheta \ P_\vartheta - a.s.$$

3.2. Sequential estimation procedure for $\vartheta \in \overline{\Theta}_2$

Assume $(c_n)_{n\geq 1}$ is an unboundedly increasing sequence of positive numbers satisfying the condition (7).

We introduce for every $\varepsilon > 0$ several quantities:

– the parameter $\lambda = e^{v_0}$ and its estimator

$$\lambda_t = \frac{\int\limits_2^t \tilde{\Delta}Y(s)\tilde{\Delta}Y(s-1)ds}{\int\limits_2^t (\tilde{\Delta}Y(s-1))^2 ds}, \ t > 2, \ \lambda_t = 0 \text{ otherwise}; \tag{11}$$

– the functions

$$Z(t) = \begin{cases} \tilde{\Delta}Y(t) - \lambda\tilde{\Delta}Y(t-1) & \text{for} \quad t \geq 2, \\ 0 \quad \text{for} \quad t < 2; \end{cases}$$

$$\tilde{Z}(t) = \begin{cases} \tilde{\Delta}Y(t) - \lambda_t\tilde{\Delta}Y(t-1) & \text{for} \quad t \geq 2, \\ 0 \quad \text{for} \quad t < 2, \end{cases}$$

$$\Psi(t) = \begin{cases} (\tilde{\Delta}Y(t), \tilde{\Delta}Y(t-1))' & \text{for} \quad t \geq 2, \\ (0,0)' \quad \text{for} \quad t < 2, \end{cases}$$

$$\tilde{\Psi}(t) = \begin{cases} (\tilde{Z}(t), \tilde{\Delta}Y(t))' & \text{for} \quad t \geq 2, \\ (0,0)' \quad \text{for} \quad t < 2; \end{cases}$$

– the parameter $\alpha = v_0/v_1$ and its estimator

$$\alpha_2(n,\varepsilon) = \frac{\ln \int\limits_4^{\nu_2(n,\varepsilon)} (\tilde{\Delta}Y(t-3))^2 dt}{\delta \ln \varepsilon^{-1} c_n}, \tag{12}$$

where

$$\nu_2(n,\varepsilon) = \inf\{T > 4 : \int_4^T \tilde{Z}^2(t-3)dt = (\varepsilon^{-1}c_n)^\delta\}, \tag{13}$$

$\delta \in (0,1)$ is a given number;
– the sequence of stopping times

$$\tau_2(n,\varepsilon) = h_2 \inf\{k > h_2^{-1}\nu_2(n,\varepsilon) : \int\limits_{\nu_2(n,\varepsilon)}^{kh_2} ||\Psi_2^{-1/2}(n,\varepsilon)\tilde{\Psi}(t-3)||^2 dt \geq 1\},$$

where suppose $h_2 = 1/5$ and

$$\Psi_2(n,\varepsilon) = \text{diag}\{\varepsilon^{-1}c_n, (\varepsilon^{-1}c_n)^{\alpha_2(n,\varepsilon)}\};$$

– the matrices

$$G_2(S,T) = \int_S^T \tilde{\Psi}(t-3)\Psi'(t)dt, \quad \Phi_2(S,T) = \int_S^T \tilde{\Psi}(t-3)d\tilde{\Delta}Y(t),$$

$$G_2(n,k,\varepsilon) = G_2(\nu_2(n,\varepsilon), \tau_2(n,\varepsilon) - kh_2), \quad \Phi_2(n,k,\varepsilon) = \Phi_2(\nu_2(n,\varepsilon), \tau_2(n,\varepsilon) - kh_2);$$

– the times

$$k_2(n) = \arg\min_{k=\overline{1,5}} ||G_2^{-1}(n,k,\varepsilon)||, \ n \geq 1;$$

– the estimators

$$\vartheta_2(n,\varepsilon) = G_2^{-1}(n,\varepsilon)\Phi_2(n,\varepsilon), \ n \geq 1,$$

where

$$G_2(n,\varepsilon) = G_2(n,k_2(n),\varepsilon), \quad \Phi_2(n,\varepsilon) = \Phi_2(n,k_2(n),\varepsilon);$$

– the stopping time

$$\sigma_2(\varepsilon) = \inf\{n \geq 1 : S_2(N) > (\rho_2 \delta_2^{-1})^{1/2}\}, \tag{14}$$

where $S_2(N) = \sum_{n=1}^{N} \beta_2^2(n,\varepsilon), \quad \rho_2 = \rho_1, \quad \delta_2 \in (0,1)$ is some fixed chosen number,

$$\beta_2(n,\varepsilon) = ||\tilde{G}_2^{-1}(n,\varepsilon)||, \quad \tilde{G}_2(n,\varepsilon) = (\varepsilon^{-1}c_n)^{-1/2}\Psi_2^{-1/2}(n,\varepsilon)G_2(n,\varepsilon).$$

In this case we write the deviation of $\vartheta_2(n,\varepsilon)$ in the form

$$\vartheta_2(n,\varepsilon) - \vartheta = (\varepsilon^{-1}c_n)^{-1/2}\tilde{G}_2^{-1}(n,\varepsilon)\tilde{\zeta}_2(n,\varepsilon), \; n \geq 1,$$

where

$$\tilde{\zeta}_2(n,\varepsilon) = \Psi_2^{-1/2}(n,\varepsilon) \int\limits_{\nu_2(n,\varepsilon)}^{\tau_2(n,\varepsilon)-k_2(n)h_2} \tilde{\Psi}(t-3)(\tilde{\Delta}\xi(t)dt + dV(t) - dV(t-1))$$

and we have

$$E_\vartheta||\tilde{\zeta}_2(n,\varepsilon)||^2 \leq 15(3+\mathrm{R}^2), \;\; n \geq 1, \;\; \varepsilon > 0.$$

Define the sequential estimation plan of ϑ by

$$T_2(\varepsilon) = \tau_2(\sigma_2(\varepsilon),\varepsilon), \;\; \vartheta_2(\varepsilon) = \vartheta_2(\sigma_2(\varepsilon),\varepsilon). \tag{15}$$

The construction of the sequential estimator $\vartheta_2(\varepsilon)$ is based on the family of estimators $\vartheta_2(S,T) = G_2^{-1}(S,T)\Phi_2(S,T) = \mathrm{e}^{-v_1 T}\tilde{G}_2(S,T)\tilde{\Phi}_2(S,T), \; T > S \geq 0$, where

$$\tilde{G}_2(S,T) = \mathrm{e}^{-v_1 T}\Psi_2^{-1/2}(T)G_2(S,T), \quad \tilde{\Phi}_2(S,T) = \Psi_2^{-1/2}(T)\Phi_2(S,T)$$

and $\Psi_2(T) = \mathrm{diag}\{\mathrm{e}^{v_1 T}, \mathrm{e}^{v_0 T}\}$. We have taken the discretization step h as above, because for $\vartheta \in \overline{\Theta}_{22}$, similar to the case $\vartheta \in \overline{\Theta}_{12}$, the function

$$f_2(S,T) = \tilde{G}_2^{-1}(S,T)$$

has some periodic (with the period $\Delta > 1$) matrix function as a limit almost surely (see (35)). This limit matrix function may have an infinite norm only for four values of their argument T on every interval of periodicity of the length Δ.

We state the results concerning the estimation of the parameter $\vartheta \in \overline{\Theta}_2$ in the following proposition.

Proposition 3.2. *Assume that the condition (7) on the sequence (c_n) holds as well as the parameter $\vartheta = (a,b)'$ in (1) be such that $\vartheta \in \overline{\Theta}_2$. Then:*

I. For any $\varepsilon > 0$ and every $\vartheta \in \overline{\Theta}_2$ the sequential plan $(T_2(\varepsilon), \vartheta_2(\varepsilon))$ defined by (15) is closed and possesses the following properties:

$$1^\circ. \;\; \sup_{\vartheta \in \overline{\Theta}_2} E_\vartheta||\vartheta_2(\varepsilon) - \vartheta||^2 \leq \delta_2 \varepsilon;$$

2°. *the inequalities below are valid:*

$$0 < \underline{\lim}_{\varepsilon\to 0} [T_2(\varepsilon) - \frac{1}{2v_1}\ln \varepsilon^{-1}] \leq \overline{\lim}_{\varepsilon\to 0} [T_2(\varepsilon) - \frac{1}{2v_1}\ln \varepsilon^{-1}] < \infty \ P_\vartheta - a.s.;$$

II. For every $\vartheta \in \overline{\Theta}_2$ the estimator $\vartheta_2(n,\varepsilon)$ is strongly consistent:

$$\lim_{n\vee\varepsilon} \vartheta_2(n,\varepsilon) = \vartheta \ P_\vartheta - a.s.$$

3.3. Sequential estimation procedure for $\vartheta \in \overline{\Theta}_3$

We shall use the notation, introduced in the previous paragraph for the parameter $\lambda = e^{v_0}$ and its estimator λ_t as well as for the functions $Z(t), \tilde{Z}(t), \Psi(t)$ and $\tilde{\Psi}(t)$.

Chose the non-random functions $\nu_3(n,\varepsilon)$, $n \geq 1$, $\varepsilon > 0$, satisfying the following conditions as $\varepsilon \to 0$ or $n \to \infty$:

$$\nu_3(n,\varepsilon) = o(\varepsilon^{-1}c_n), \quad \frac{\log^{1/2}\nu_3(n,\varepsilon)}{e^{v_0\nu_3(n,\varepsilon)}}\varepsilon^{-1}c_n = o(1). \tag{16}$$

Example: $\nu_3(n,\varepsilon) = \log^2 \varepsilon^{-1}c_n$.

We introduce several quantities:
– the parameter $\alpha_3 = v_0$ and its estimator

$$\alpha_3(n,\varepsilon) = \ln|\lambda_{\nu_3(n,\varepsilon)}|,$$

where λ_t is defined in (11);
– the sequences of stopping times

$$\tau_{31}(n,\varepsilon) = \inf\{T > 0 : \int_{\nu_3(n,\varepsilon)}^{T} \tilde{Z}^2(t-3)dt = \varepsilon^{-1}c_n\}, \tag{17}$$

$$\tau_{32}(n,\varepsilon) = \inf\{T > 0 : \int_{\nu_3(n,\varepsilon)}^{T} (\tilde{\Delta}Y(t-3))^2 dt = e^{2\alpha_3(n,\varepsilon)\varepsilon^{-1}c_n}\}, \tag{18}$$

$$\tau_{min}(n,\varepsilon) = \min\{\tau_{31}(n,\varepsilon), \tau_{32}(n,\varepsilon)\}, \quad \tau_{max}(n,\varepsilon) = \max\{\tau_{31}(n,\varepsilon), \tau_{32}(n,\varepsilon)\},$$

– the matrices

$$\begin{aligned} G_3(S,T) &= \int_S^T \tilde{\Psi}(t)\Psi(t)dt, \\ \Phi_3(S,T) &= \int_S^T \tilde{\Psi}(t)d\tilde{\Delta}Y(t), \\ G_3(n,\varepsilon) &= G_3(\nu_3(n,\varepsilon), \tau_{min}(n,\varepsilon)), \\ \Phi_3(n,\varepsilon) &= \Phi_3(\nu_3(n,\varepsilon), \tau_{min}(n,\varepsilon)); \end{aligned}$$

– the estimators

$$\vartheta_3(n,\varepsilon) = G_3^{-1}(n,\varepsilon)\Phi_3(n,\varepsilon), \ n \geq 1, \ \varepsilon > 0;$$

– the stopping time

$$\sigma_3(\varepsilon) = \inf\{n \geq 1 : S_3(N) > (\rho_3 \delta_3^{-1})^{1/2}\}, \tag{19}$$

where $S_3(N) = \sum_{n=1}^{N} \beta_3^2(n,\varepsilon)$, $\delta_3 \in (0,1)$ is some fixed chosen number,

$$\beta_3(n,\varepsilon) = ||\tilde{G}_3^{-1}(n,\varepsilon)||, \quad \rho_3 = 6(3+R^2)\sum_{n\geq 1}\frac{1}{c_n},$$

$$\tilde{G}_3(n,\varepsilon) = (\varepsilon^{-1}c_n)^{-1/2}\Psi_3^{-1/2}(n,\varepsilon)G_3(n,\varepsilon), \quad \Psi_3(n,\varepsilon) = \mathrm{diag}\{\varepsilon^{-1}c_n, e^{2\alpha_3(n,\varepsilon)\varepsilon^{-1}c_n}\}.$$

In this case we write the deviation of $\vartheta_3(n,\varepsilon)$ in the form

$$\vartheta_3(n,\varepsilon) - \vartheta = (\varepsilon^{-1}c_n)^{-1/2}\tilde{G}_3^{-1}(n,\varepsilon)\tilde{\zeta}_3(n,\varepsilon), \ n \geq 1,$$

where

$$\tilde{\zeta}_3(n,\varepsilon) = \Psi_3^{-1/2}(n,\varepsilon)\int_{\nu_3(n,\varepsilon)}^{\tau_{min}(n,\varepsilon)} \tilde{\Psi}(t-3)(\tilde{\Delta}\xi(t)dt + dV(t) - dV(t-1))$$

and we have

$$E_\vartheta||\tilde{\zeta}_3(n,\varepsilon)||^2 \leq 6(3+R^2), \ n \geq 1, \ \varepsilon > 0.$$

Define the sequential estimation plan of ϑ by

$$T_3(\varepsilon) = \tau_{max}(\sigma_3(\varepsilon),\varepsilon), \ \vartheta_3(\varepsilon) = \vartheta_3(\sigma_3(\varepsilon),\varepsilon). \tag{20}$$

Proposition 3.3. *Assume that the condition (7) on the sequence (c_n) holds and let the parameter $\vartheta = (a,b)'$ in (1) be such that $\vartheta \in \overline{\Theta}_3$. Then:*
I. For every $\vartheta \in \overline{\Theta}_3$ the sequential plan $(T_3(\varepsilon), \vartheta_3(\varepsilon))$ defined in (20) is closed and possesses the following properties:

1°. *for any $\varepsilon > 0$*

$$\sup_{\vartheta\in\overline{\Theta}_3} E_\vartheta||\vartheta_3(\varepsilon) - \vartheta||^2 \leq \delta_3\varepsilon;$$

2°. *the following inequalities are valid:*

$$0 < \underline{\lim}_{\varepsilon\to 0}\, \varepsilon T_3(\varepsilon) \leq \overline{\lim}_{\varepsilon\to 0}\, \varepsilon T_3(\varepsilon) < \infty \ \ P_\vartheta - a.s.;$$

II. For every $\vartheta \in \Theta_3$ the estimator $\vartheta_3(n,\varepsilon)$ is strongly consistent:

$$\lim_{n\vee\varepsilon} \vartheta_3(n,\varepsilon) = \vartheta \ \ P_\vartheta - a.s.$$

3.4. Sequential estimation procedure for $\vartheta \in \overline{\Theta}_4$

In this case $b = -a$ and (6) is the differential equation of the first order:

$$d\tilde{\Delta}Y(t) = aZ^*(t)dt + \tilde{\Delta}\xi(t)dt + dV(t) - dV(t-1), \ t \geq 2,$$

where

$$Z^*(t) = \begin{cases} \tilde{\Delta}Y(t) - \tilde{\Delta}Y(t-1) & \text{for} \quad t \geq 2, \\ 0 \quad \text{for} \quad t < 2. \end{cases}$$

We shall construct sequential plan $(T_4(\varepsilon), \vartheta_4(\varepsilon))$ for estimation of the vector parameter $\vartheta = a(1,-1)'$ with the $(\delta_4\varepsilon)$-accuracy in the sense of the L_2-norm for every $\varepsilon > 0$ and fixed chosen $\delta_4 \in (0,1)$.

First define the sequential estimation plans for the scalar parameter a on the bases of correlation estimators which are generalized least squares estimators:

$$a_4(T) = G_4^{-1}(T)\Phi_4(T),$$
$$G_4(T) = \int_0^T Z^*(t-2)Z^*(t)dt,$$
$$\Phi_4(T) = \int_0^T Z^*(t-2)d\tilde{\Delta}Y(t),\ T > 0.$$

Let $(c_n,\ n \geq 1)$ be an unboundedly increasing sequence of positive numbers, satisfying the condition (7).

We shall define

– the sequence of stopping times $(\tau_4(n,\varepsilon), n \geq 1)$ as

$$\tau_4(n,\varepsilon) = \inf\{T > 2:\ \int_0^T (Z^*(t-2))^2 dt = \varepsilon^{-1}c_n\},\ n \geq 1;$$

– the sequence of estimators

$$a_4(n,\varepsilon) = a_4(\tau_4(n,\varepsilon)) = G_4^{-1}(\tau_4(n,\varepsilon))\Phi_4(\tau_4(n,\varepsilon));$$

– the stopping time

$$\sigma_4(\varepsilon) = \inf\{n \geq 1:\ S_4(N) > (\rho_4\delta_4^{-1})^{1/2}\}, \tag{21}$$

where $S_4(N) = \sum_{n=1}^{N} \tilde{G}_4^{-2}(n,\varepsilon)$, $\rho_4 = \rho_3$, $\tilde{G}_4(n,\varepsilon) = (\varepsilon^{-1}c_n)^{-1}G_4(\tau_4(n,\varepsilon))$. The deviation of $a_4(n,\varepsilon)$ has the form

$$a_4(n,\varepsilon) - a = (\varepsilon^{-1}c_n)^{-1/2}\tilde{G}_4^{-1}(n,\varepsilon)\tilde{\zeta}_4(n,\varepsilon),\ n \geq 1,$$

where

$$\tilde{\zeta}_4(n,\varepsilon) = (\varepsilon^{-1}c_n)^{-1/2} \int_0^{\tau_4(n,\varepsilon)} Z^*(t-2)(\tilde{\Delta}\xi(t)dt + dV(t) - dV(t-1))$$

and we have

$$E_\vartheta||\tilde{\zeta}_4(n,\varepsilon)||^2 \leq 3(3+\mathrm{R}^2),\ n \geq 1,\ \varepsilon > 0.$$

We define the sequential plan $(T_4(\varepsilon), \vartheta_4(\varepsilon))$ for the estimation of ϑ as

$$T_4(\varepsilon) = \tau_4(\sigma_4(\varepsilon),\varepsilon),\ \vartheta_4(\varepsilon) = a_4(\sigma_4(\varepsilon),\varepsilon)(1,-1)'. \tag{22}$$

The following proposition presents the conditions under which $T_4(\varepsilon)$ and $\vartheta_4(\varepsilon)$ are well-defined and have the desired property of preassigned mean square accuracy.

Proposition 3.4. *Assume that the sequence* (c_n) *defined above satisfy the condition* (7). *Then we obtain the following result:*

I. For any $\varepsilon > 0$ *and every* $\vartheta \in \overline{\Theta}_4$ *the sequential plan* $(T_4(\varepsilon), \vartheta_4(\varepsilon))$ *defined by* (22) *is closed and has the following properties:*

$$1^\circ. \quad \sup_{\vartheta \in \overline{\Theta}_4} E_\vartheta ||\vartheta_4(\varepsilon) - \vartheta||^2 \leq \delta_4 \varepsilon;$$

2°. *the following relations hold:*

– *if* $\vartheta \in \overline{\Theta}_{41}$ *then*

$$0 < \underline{\lim}_{\varepsilon \to 0} \varepsilon \cdot T_4(\varepsilon) \leq \overline{\lim}_{\varepsilon \to 0} \varepsilon \cdot T_4(\varepsilon) < \infty \ P_\vartheta - a.s.,$$

– *if* $\vartheta \in \overline{\Theta}_{42}$ *then*

$$0 < \underline{\lim}_{\varepsilon \to 0} [T_4(\varepsilon) - \frac{1}{2v_0} \ln \varepsilon^{-1}] \leq \overline{\lim}_{\varepsilon \to 0} [T_4(\varepsilon) - \frac{1}{2v_0} \ln \varepsilon^{-1}] < \infty \ P_\vartheta - a.s.;$$

II. For every $\vartheta \in \overline{\Theta}_4$ *the estimator* $\vartheta_4(n, \varepsilon)$ *is strongly consistent:*

$$\lim_{n \vee \varepsilon} \vartheta_4(n, \varepsilon) = \vartheta \ P_\vartheta - a.s.$$

3.5. General sequential estimation procedure of the time-delayed process

In this paragraph we construct the sequential estimation procedure for the parameters a and b of the process (1) on the bases of the estimators, presented in subsections 3.1-3.4.

Denote $j^\star = \arg\min_{j=1,4} T_j(\varepsilon)$. We define the sequential plan $(T^\star(\varepsilon),\ \vartheta^\star(\varepsilon))$ of estimation $\vartheta \in \overline{\Theta}$ on the bases of all constructed above estimators by the formulae

$$\text{SEP}^\star(\varepsilon) = (T^\star(\varepsilon), \vartheta^*(\varepsilon)), \quad T^\star(\varepsilon) = T_{j^\star}(\varepsilon), \quad \vartheta^*(\varepsilon) = \vartheta_{j^\star}(\varepsilon).$$

The following theorem is valid.

Theorem 3.1. *Assume that the underlying processes* $(X(t))$ *and* $(Y(t))$ *satisfy the equations* (1), (2), *the parameter* ϑ *to be estimated belongs to the region* $\overline{\Theta}$ *and for the numbers* $\delta_1, \ldots, \delta_4$ *in the definitions* (10), (15), (20) *and* (22) *of sequential plans the condition* $\sum_{j=1}^{4} \delta_j = 1$ *is fulfilled.*

Then the sequential estimation plan $(T^\star(\varepsilon),\ \vartheta^\star(\varepsilon))$ *possess the following properties:*

1°. *for any* $\varepsilon > 0$ *and for every* $\vartheta \in \overline{\Theta}$

$$T^\star(\varepsilon) < \infty \ P_\vartheta - a.s.;$$

2°. *for any* $\varepsilon > 0$

$$\sup_{\vartheta \in \overline{\Theta}} E_\vartheta ||\vartheta^\star(\varepsilon) - \vartheta||^2 \leq \varepsilon;$$

3°. *the following relations hold with* P_ϑ – *probability one:*

– *for* $\vartheta \in \overline{\Theta}_{11} \cup \overline{\Theta}_3 \cup \overline{\Theta}_{41}$

$$\overline{\lim}_{\varepsilon \to 0} \varepsilon \cdot T^*(\varepsilon) < \infty;$$

– for $\vartheta \in \overline{\Theta}_{12} \cup \overline{\Theta}_{42}$

$$\overline{\lim_{\varepsilon \to 0}} \, [T^*(\varepsilon) - \frac{1}{2v_0} \ln \varepsilon^{-1}] < \infty;$$

– for $\vartheta \in \overline{\Theta}_{13}$

$$\overline{\lim_{\varepsilon \to 0}} \, [T^*(\varepsilon) + \frac{1}{v_0} \ln T_1(\varepsilon) - \Psi''_{13}(\varepsilon)] < \infty,$$

the function $\Psi''_{13}(\varepsilon)$ *is defined in (30);*

– for $\vartheta \in \overline{\Theta}_2$

$$\overline{\lim_{\varepsilon \to 0}} \, [T^*(\varepsilon) - \frac{1}{2v_1} \ln \varepsilon^{-1}] < \infty.$$

4. Proofs

Proof of Proposition 3.1. The closeness of the sequential estimation plan, as well as assertions I.2 and II of Proposition 3.1 for the cases $\overline{\Theta}_{11} \cup \overline{\Theta}_{12}$ can be easily verified similar to [10, 12, 14, 16]. Now we verify the finiteness of the stopping time $T_1(\varepsilon)$ in the new case $\overline{\Theta}_{13}$.

By the definition of $\tilde{\Delta}Y(t)$ we have:

$$\tilde{\Delta}Y(t) = \tilde{X}(t) + \tilde{\Delta}V(t), \quad t \geq 1,$$

where

$$\tilde{X}(t) = \int_{t-1}^{t} X(t)dt.$$

It is easy to show that the process $(\tilde{X}(\cdot))$ has the following representation:

$$\tilde{X}(t) = \tilde{x}_0(t)X_0(0) + b\int_{-1}^{0} \tilde{x}_0(t-s-1)X_0(s)ds + \int_0^t \tilde{x}_0(t-s)dW(s)$$

for $t \geq 1$, $\tilde{X}(t) = \int_{t-1}^{0} X_0(s)ds + \int_0^t X(s)ds$ for $t \in [0,1)$ and $\tilde{X}(t) = 0$ for $t \in [-1,0)$. Based on the representation above for the function $x_0(\cdot)$, the subsequent properties of $x_0(t)$ the function $\tilde{x}_0(t) = \int_{t-1}^{t} x_0(s)ds$ can be easily shown to fulfill $\tilde{x}_0(t) = 0,\ t \in [-1,0]$ and as $t \to \infty$

$$\tilde{x}_0(t) = \begin{cases} o(e^{\gamma t}), & \gamma < 0, \quad \vartheta \in \Theta_{11}, \\ \tilde{\phi}_0(t)e^{v_0 t} + o(e^{\gamma_0 t}), & \gamma_0 < v_0,\ \vartheta \in \Theta_{12}, \\ \frac{2}{v_0}[(1-e^{-v_0})t + e^{-v_0} - \frac{1-e^{-v_0}}{v_0}]e^{v_0 t} + o(e^{\gamma_0 t}), & \gamma_0 < v_0,\ \vartheta \in \Theta_{13}, \\ \frac{1-e^{-v_0}}{v_0(v_0-a+1)}e^{v_0 t} + \frac{1-e^{-v_1}}{v_1(a-v_1-1)}e^{v_1 t} + o(e^{\gamma_1 t}), & \gamma_1 < v_1,\ \vartheta \in \Theta_{21}, \\ \frac{1-e^{-v_0}}{v_0(v_0-a+1)}e^{v_0 t} + \tilde{\phi}_1(t)e^{v_1 t} + o(e^{\gamma_1 t}), & \gamma_1 < v_1,\ \vartheta \in \Theta_{22}, \\ \frac{1-e^{-v_0}}{v_0(v_0-a+1)}e^{v_0 t} + o(e^{\gamma t}), & \gamma < 0, \quad \vartheta \in \Theta_3, \\ \frac{1}{1-a} + o(e^{\gamma t}), & \gamma < 0, \quad \vartheta \in \Theta_{41}, \\ \frac{1-e^{-v_0}}{v_0(v_0-a+1)}e^{v_0 t} - \frac{1}{a-1} + o(e^{\gamma t}), & \gamma < 0, \quad \vartheta \in \Theta_{42}, \end{cases}$$

where

$$\tilde{\phi}_i(t) = \tilde{A}_i \cos \xi_i t + \tilde{B}_i \sin \xi_i t$$

and $\tilde{A}_i, \tilde{B}_i, \xi_i$ are some constants (see [10, 12]).

The processes $\tilde{X}(t)$ and $\tilde{\Delta}V(t)$ are mutually independent and the process $\tilde{X}(t)$ has the representation similar to (3). Then, after some algebra similar to those in [10, 12] we get for the processes $\tilde{X}(t), \tilde{Y}(t) = \tilde{X}(t) - \lambda\tilde{X}(t-1), \lambda = e^{v_0}, \tilde{\Delta}Y(t)$ and

$$Z(t) = \begin{cases} \tilde{\Delta}Y(t) - \lambda\tilde{\Delta}Y(t-1) & \text{for} \quad t \geq 2, \\ 0 \quad \text{for} \quad t < 2 \end{cases}$$

in the case $\overline{\Theta}_{13}$ the following limits:

$$\lim_{t\to\infty} t^{-1}e^{-v_0 t}\tilde{\Delta}Y(t) = \lim_{t\to\infty} t^{-1}e^{-v_0 t}\tilde{X}(t) = \tilde{C}_X \quad P_\vartheta - \text{a.s.}, \tag{23}$$

$$\lim_{t\to\infty} e^{-v_0 t}\tilde{Y}(t) = C_Y, \qquad \lim_{t\to\infty} e^{-v_0 t}Z(t) = \tilde{C}_Z \quad P_\vartheta - \text{a.s.},$$

and, as follows, for $u \geq 0$

$$\lim_{T\to\infty} \mid T^{-2}e^{-2v_0 T} \int_1^T \tilde{\Delta}Y(t-u)\tilde{\Delta}Y(t)dt - \frac{\tilde{C}_X^2}{2v_0}\left[1 - \frac{u}{T}\right] e^{-uv_0} \mid = 0 \quad P_\vartheta - \text{a.s.}, \tag{24}$$

$$\lim_{T\to\infty} \mid T^{-1}e^{-2v_0 T} \int_1^T \tilde{\Delta}Y(t-u)Z(t)dt - \frac{\tilde{C}_X\tilde{C}_Z}{2v_0}\left[1 - \frac{u}{T}\right] e^{-uv_0} \mid = 0 \quad P_\vartheta - \text{a.s.},$$

where $\tilde{C}_x, C_Y$ and $\tilde{C}_Z$ are some nonzero constants, which can be found from [10, 12]. From (24) we obtain the limits:

$$\lim_{T\to\infty} \frac{1}{T^2e^{2v_0 T}} G_1(T,s) = G_{13}(s), \quad \lim_{T\to\infty} T^{-1}e^{-4v_0 T}|G_1(T,s)| = G_{13}e^{-(3+11s)v_0} \quad P_\vartheta - a.s.,$$

$$G_{13}(s) = \frac{\tilde{C}_X^2}{2v_0} \begin{pmatrix} e^{-(2+5s)v_0} & e^{-(1+5s)v_0} \\ e^{-2(1+3s)v_0} & e^{-(1+6s)v_0} \end{pmatrix}, \qquad G_{13} = \frac{s\tilde{C}_X^3\tilde{C}_Z}{4v_0^2}$$

and, as follows, we can find

$$\lim_{T\to\infty} T^{-1}e^{2v_0 T}G_1^{-1}(T,s) = \tilde{G}_{13}(s) \;\; P_\vartheta - a.s.,$$

$$\tilde{G}_{13}(s) = \frac{2v_0 e^{(3+11s)v_0}}{s\tilde{C}_X\tilde{C}_Z} \begin{pmatrix} e^{-(1+6s)v_0} & -e^{-(1+5s)v_0} \\ -e^{-2(1+3s)v_0} & e^{-(2+5s)v_0} \end{pmatrix}$$

is a non-random matrix function.

From (23) and by the definition of the stopping times $\tau_1(n,\varepsilon)$ we have

$$\lim_{n\vee\varepsilon} \frac{\tau_1^2(n,\varepsilon)e^{2\tau_1(n,\varepsilon)v_0}}{\varepsilon^{-1}c_n} = g_{13}^* \quad P_\vartheta - \text{a.s.}, \tag{25}$$

where $g_{13}^* = 2v_0\tilde{C}_X^{-2}\left(e^{-2v_0(2+5h_1)} + e^{-4v_0(1+3h_1)}\right)^{-1}$ and, as follows,

$$\lim_{n\vee\varepsilon} [\tau_1(n,\varepsilon) + \frac{1}{v_0}\ln\tau_1(n,\varepsilon) - \frac{1}{2v_0}\ln\varepsilon^{-1}c_n] = \frac{1}{2v_0}\ln g_{13}^* \quad P_\vartheta - \text{a.s.}, \tag{26}$$

$$\lim_{n\vee\varepsilon} \frac{\tau_1(n,\varepsilon)}{\ln\varepsilon^{-1}c_n} = \frac{1}{2v_0} \quad P_\vartheta - \text{a.s.}, \tag{27}$$

$$\lim_{n\vee\varepsilon} [\frac{1}{\ln^3\varepsilon^{-1}c_n}\tilde{G}_1^{-1}(n,\varepsilon) - [(2v_0)^3 g_{13}^*]^{-1} e^{-2v_0k_1(n)h_1}\tilde{G}_{13}(h_1)] = 0 \quad P_\vartheta - \text{a.s.} \tag{28}$$

From (8) and (28) it follows the P_ϑ − a.s. finiteness of the stopping time $\sigma_1(\varepsilon)$ for every $\varepsilon > 0$.

The proof of the assertion I.1 of Proposition 3.1 for the case $\overline{\Theta}_{13}$ is similar e.g. to the proof of corresponding assertion in [14, 16]:

$$E_\vartheta||\vartheta_1(\varepsilon) - \vartheta||^2 = E_\vartheta\frac{1}{S^2(\sigma_1(\varepsilon))}||\sum_{n=1}^{\sigma_1(\varepsilon)} \beta_1^2(n,\varepsilon)(\vartheta_1(n,\varepsilon) - \vartheta)||^2 \le$$

$$\le \varepsilon\frac{\delta_1}{\rho_1}E_\vartheta\sum_{n=1}^{\sigma_1(\varepsilon)}\frac{1}{c_n}\cdot\beta_1^2(n,\varepsilon)\cdot||\tilde{G}_1^{-1}(n,\varepsilon)||^2\cdot||\tilde{\zeta}_1(n,\varepsilon)||^2 \le$$

$$\le \frac{\varepsilon\delta_1}{\rho_1}\sum_{n\ge1}\frac{1}{c_n}E_\vartheta||\tilde{\zeta}_1(n,\varepsilon)||^2 \le \varepsilon\delta_1\frac{15(3+R^2)}{\rho_1}\sum_{n\ge1}\frac{1}{c_n} = \varepsilon\delta_1.$$

Now we prove the assertion I.2 for $\vartheta \in \overline{\Theta}_{13}$. Denote the number

$$\tilde{g}_{13} = [(2v_0)^3 g_{13}^*]^2\rho_1^{-1}\delta_1||\tilde{G}_{13}(h_1)||^{-2}$$

and the times

$$\overline{\sigma}_{13}'(\varepsilon) = \inf\{n \ge 1 : \sum_{n=1}^{N}\ln^6\varepsilon^{-1}c_n > \tilde{g}_{13}e^{4v_0h_1}\},$$

$$\overline{\sigma}_{13}''(\varepsilon) = \inf\{n \ge 1 : \sum_{n=1}^{N}\ln^6\varepsilon^{-1}c_n > \tilde{g}_{13}e^{20v_0h_1}\}.$$

From (8) and (28) it follows, that for ε small enough

$$\overline{\sigma}_{13}'(\varepsilon) \le \sigma_1(\varepsilon) \le \overline{\sigma}_{13}''(\varepsilon) \quad P_\vartheta - \text{a.s.} \tag{29}$$

Denote

$$\Psi_{13}'(\varepsilon) = \frac{1}{2v_0}\ln(\varepsilon^{-1}c_{\overline{\sigma}_{13}'(\varepsilon)}), \quad \Psi_{13}''(\varepsilon) = \frac{1}{2v_0}\ln(\varepsilon^{-1}c_{\overline{\sigma}_{13}''(\varepsilon)}). \tag{30}$$

Then, from (8), (26) and (29) we obtain finally the assertion I.2 of Proposition 3.1:

$$\lim_{\varepsilon\to0} [T_1(\varepsilon) + \frac{1}{v_0}\ln T_1(\varepsilon) - \Psi_{13}'(\varepsilon)] \ge \frac{1}{2v_0}\ln g_{13}^* \quad P_\vartheta - \text{a.s.},$$

$$\overline{\lim_{\varepsilon\to 0}}\,[T_1(\varepsilon)+\frac{1}{v_0}\ln T_1(\varepsilon)-\Psi''_{13}(\varepsilon)]\le\frac{1}{2v_0}\ln g^*_{13}\quad P_\vartheta-\text{a.s.}$$

For the proof of the assertion II of Proposition 3.1 we will use the representation (9) for the deviation

$$\vartheta_1(n,\varepsilon)-\vartheta=\frac{1}{\ln^3\varepsilon^{-1}c_n}\tilde{G}_1^{-1}(n,\varepsilon)\cdot\frac{\tau_1^2(n,\varepsilon)e^{2\tau_1(n,\varepsilon)v_0}}{\varepsilon^{-1}c_n}\left(\cdot\frac{\ln\varepsilon^{-1}c_n}{\tau_1(n,\varepsilon)}\right)^3\cdot\frac{1}{\tau_1^{-1}(n,\varepsilon)e^{2\tau_1(n,\varepsilon)v_0}}\zeta_1(n,\varepsilon),$$

where

$$\zeta_1(n,\varepsilon)=\zeta_1(\tau_1(n,\varepsilon)-k_1(n)h_1,h_1),$$

$$\zeta_1(T,s)=\int_0^T\Psi_s(t-2-5s)(\tilde{\Delta}\xi(t)dt+dV(t)-dV(t-1)).$$

According to (25), (27) and (28) first three factors in the right-hand side of this equality have P_ϑ – a.s. positive finite limits. The last factor vanishes in P_ϑ – a.s. sense by the properties of the square integrable martingales $\zeta_1(T,s)$:

$$\lim_{n\vee\varepsilon}\frac{\zeta_1(n,\varepsilon)}{\tau_1^{-1}(n,\varepsilon)e^{2\tau_1(n,\varepsilon)v_0}}=\lim_{T\to\infty}\frac{\zeta_1(T,h_1)}{T^{-1}e^{2v_0T}}=0\quad P_\vartheta-\text{a.s.}$$

Then the estimators $\vartheta_1(n,\varepsilon)$ are strongly consistent as $\varepsilon\to 0$ or $n\to\infty$ and we obtain the assertion II of Proposition 3.1.

Hence Proposition 3.1 is valid.

Proof of Proposition 3.2.
Similar to the proof of Proposition 3.1 and [7]–[16] we can get the following asymptotic as $t\to\infty$ relations for the processes $\tilde{\Delta}Y(t)$, $Z(t)$ and $\tilde{Z}(t)$:

– for $\vartheta\in\Theta_{21}$

$$\tilde{\Delta}Y(t)=C_Ye^{v_0t}+C_{Y1}e^{v_1t}+o(e^{\gamma t})\quad P_\vartheta-\text{a.s.},$$
$$Z(t)=C_Ze^{v_1t}+o(e^{\gamma t})\quad P_\vartheta-\text{a.s.},$$
$$\lambda_t-\lambda=\frac{2v_0e^{v_0}}{v_0+v_1}C_ZC_Y^{-1}e^{-(v_0-v_1)t}+o(e^{-(v_0-v_1+\gamma)t})\quad P_\vartheta-\text{a.s.},$$
$$\tilde{Z}(t)=\tilde{C}_Ze^{v_1t}+o(e^{\gamma t})\quad P_\vartheta-\text{a.s.};$$

– for $\vartheta\in\Theta_{22}$

$$|\tilde{\Delta}Y(t)-C_Ye^{v_0t}-C_{Y1}(t)e^{v_1t}|=o(e^{\gamma t})\quad P_\vartheta-\text{a.s.}$$
$$|Z(t)-C_Z(t)e^{v_1t}|=o(e^{\gamma t})\quad P_\vartheta-\text{a.s.},$$
$$\lambda_t-\lambda=2v_0e^{v_0}C_Y^{-1}U_Z(t)e^{-(v_0-v_1)t}+o(e^{-(v_0-v_1+\gamma)t})\quad P_\vartheta-\text{a.s.},$$
$$|\tilde{Z}(t)-\tilde{C}_Z(t)e^{v_1t}|=o(e^{\gamma t})\quad P_\vartheta-\text{a.s.},$$

where C_Y and C_{Y1} are some non-zero constants, $0<\gamma<v_1$, $C_Z=C_{Y1}(1-e^{v_0-v_1})$, $\tilde{C}_Z=C_Z\frac{v_1-v_0}{v_1+v_0}$; $C_Z(t)$, $U_Z(t)=\int_0^\infty C_Z(t-u)e^{-(v_0+v_1)u}du$ and $\tilde{C}_Z(t)=C_Z(t)-2v_0U_Z(t)$ are the periodic (with the period $\Delta>1$) functions.

Denote

$$U_{\tilde{Z}}(T) = \int_0^\infty \tilde{C}_Z(T-u)\mathrm{e}^{-(v_0+v_1)u}du,$$

$$U_{\tilde{Z}Z}(S,T) = \int_0^\infty \tilde{C}_Z(T-u)C_Z(S-u)\mathrm{e}^{-2v_1 u}du, \quad \tilde{U}_Z(T) = U_{\tilde{Z}\tilde{Z}}(T,T).$$

It should be noted that the functions $C_Z(t)$, $U_Z(t)$, $\tilde{C}_Z(t)$ and $U_{\tilde{Z}}(T)$ have at most two roots on each interval from $[0,\infty)$ of the length Δ. At the same time the function $U_{\tilde{Z}Z}(S,T)$ - at most four roots.

With P_ϑ-probability one we have:

– for $\vartheta \in \Theta_2$

$$\lim_{T-S\to\infty} \mathrm{e}^{-2v_0 T}\int_S^T (\tilde{\Delta}Y(t-3))^2 dt = \frac{C_Y^2}{2v_0}\mathrm{e}^{-6v_0}, \tag{31}$$

– for $\vartheta \in \Theta_{21}$

$$\lim_{T-S\to\infty} \mathrm{e}^{-2v_1 T}\int_S^T \tilde{Z}^2(t-3)dt = \frac{\tilde{C}_Z^2}{2v_1}\mathrm{e}^{-6v_1}, \tag{32}$$

$$\lim_{T-S\to\infty} \tilde{G}_2^{-1}(S,T) = \tilde{G}_{21}, \tag{33}$$

where

$$\tilde{G}_{21} = \begin{pmatrix} \frac{2v_1(v_1+v_0)^2}{C_Z\tilde{C}_Z(v_1-v_0)^2}\mathrm{e}^{3v_1} & -\frac{4v_0v_1(v_1+v_0)}{C_ZC_Y(v_1-v_0)^2}\mathrm{e}^{3v_0} \\ -\frac{2v_1(v_1+v_0)^2}{C_Z\tilde{C}_Z(v_1-v_0)^2}\mathrm{e}^{v_0+3v_1} & \frac{4v_0v_1(v_1+v_0)}{C_ZC_Y(v_1-v_0)^2}\mathrm{e}^{4v_0} \end{pmatrix},$$

– for $\vartheta \in \Theta_{22}$

$$\lim_{T-S\to\infty} \left| \mathrm{e}^{-2v_1 T}\int_S^T \tilde{Z}^2(t-3)dt - \mathrm{e}^{-6v_1}\tilde{U}_Z(T-3) \right| = 0, \tag{34}$$

$$\lim_{T-S\to\infty} \left| \tilde{G}_2^{-1}(S,T) - \tilde{G}_{22}(T) \right| = 0, \tag{35}$$

where

$$\tilde{G}_{22}(T) = \left[\frac{1}{2v_0}U_{\tilde{Z}Z}(T,T-3) - U_Z(T-3)U_{\tilde{Z}}(T) \right]^{-1} \cdot \begin{pmatrix} \frac{\mathrm{e}^{3v_1}}{2v_0} & -\frac{\mathrm{e}^{3v_0}}{C_Y}U_{\tilde{Z}}(T) \\ -\frac{\mathrm{e}^{v_0+3v_1}}{2v_0} & \frac{\mathrm{e}^{4v_0}}{C_Y}U_{\tilde{Z}}(T) \end{pmatrix}.$$

The matrix $\tilde{G}_{21}$ is constant and non-zero and $\tilde{G}_{22}(T)$ is the periodic matrix function with the period $\Delta > 1$ (see [3], [10, 12, 14]) and may have infinite norm for four points on each interval of periodicity only.

The next step of the proof is the investigation of the asymptotic behaviour of the stopping times $\nu_2(n,\varepsilon)$, $\tau_2(n,\varepsilon)$ and the estimators $\alpha_2(n,\varepsilon)$.

Denote

$$C_{v1} = \mathrm{e}^{-6v_1} \min\left\{ \frac{\tilde{C}_Z^2}{2v_1}, \inf_{T>0} \tilde{U}_Z(T) \right\}, \quad C_{v2} = \mathrm{e}^{-6v_1} \max\left\{ \frac{\tilde{C}_Z^2}{2v_1}, \sup_{T>0} \tilde{U}_Z(T) \right\}.$$

Then for $\vartheta \in \Theta_2$

$$C_{v1} \le \underline{\lim}_{T-S\to\infty} \mathrm{e}^{-2v_1 T} \int_S^T \tilde{Z}^2(t-3)dt \le$$

$$\le \overline{\lim}_{T-S\to\infty} \mathrm{e}^{-2v_1 T} \int_S^T \tilde{Z}^2(t-3)dt \le C_{v2} \quad P_\vartheta - \text{a.s.} \tag{36}$$

and from the definition (13) of $\nu_2(n,\varepsilon)$ and (32), (34) we have

$$C_{v2}^{-1} \le \underline{\lim}_{n\vee\varepsilon} \frac{\mathrm{e}^{2v_1\nu_2(n,\varepsilon)}}{(\varepsilon^{-1}c_n)^\delta} \le \overline{\lim}_{n\vee\varepsilon} \frac{\mathrm{e}^{2v_1\nu_2(n,\varepsilon)}}{(\varepsilon^{-1}c_n)^\delta} \le C_{v1}^{-1} \quad P_\vartheta - \text{a.s.}$$

and thus

$$\frac{1}{2v_1}\ln C_{v2}^{-1} \le \underline{\lim}_{n\vee\varepsilon} [\nu_2(n,\varepsilon) - \frac{\delta}{2v_1}\ln \varepsilon^{-1}c_n] \le$$

$$\le \overline{\lim}_{n\vee\varepsilon} [\nu_2(n,\varepsilon) - \frac{\delta}{2v_1}\ln \varepsilon^{-1}c_n] \le \frac{1}{2v_1}\ln C_{v1}^{-1} \quad P_\vartheta - \text{a.s.} \tag{37}$$

By the definition (12) of $\alpha_2(n,\varepsilon)$ we find the following normalized representation for the deviation $\alpha_2(n,\varepsilon) - \alpha$:

$$\nu_2(n,\varepsilon)(\alpha_2(n,\varepsilon)-\alpha) = \nu_2(n,\varepsilon)\left(\frac{\ln \int_0^{\nu_2(n,\varepsilon)} (\tilde{\Delta}Y(t-3))^2 dt}{\ln \int_0^{\nu_2(n,\varepsilon)} \tilde{Z}^2(t-3)dt} - \frac{v_0}{v_1} \right) =$$

$$= \nu_2(n,\varepsilon)\left(\frac{2v_0\nu_2(n,\varepsilon) + \ln \mathrm{e}^{-2v_0\nu_2(n,\varepsilon)} \int_0^{\nu_2(n,\varepsilon)} (\tilde{\Delta}Y(t-3))^2 dt}{2v_1\nu_2(n,\varepsilon) + \ln \mathrm{e}^{-2v_1\nu_2(n,\varepsilon)} \int_0^{\nu_2(n,\varepsilon)} \tilde{Z}^2(t-3)dt} - \frac{v_0}{v_1} \right) =$$

$$= \nu_2(n,\varepsilon) \frac{v_1 \ln \mathrm{e}^{-2v_0\nu_2(n,\varepsilon)} \int_0^{\nu_2(n,\varepsilon)} (\tilde{\Delta}Y(t-3))^2 dt - v_0 \ln \mathrm{e}^{-2v_1\nu_2(n,\varepsilon)} \int_0^{\nu_2(n,\varepsilon)} \tilde{Z}^2(t-3)dt}{2v_1^2\nu_2(n,\varepsilon) + v_1 \ln \mathrm{e}^{-2v_1\nu_2(n,\varepsilon)} \int_0^{\nu_2(n,\varepsilon)} \tilde{Z}^2(t-3)dt}$$

and using the limit relations (31), (36) and (37) we obtain

$$\alpha_1 \le \underline{\lim}_{n\vee\varepsilon} (\ln \varepsilon^{-1}c_n)\cdot(\alpha - \alpha_2(n,\varepsilon)) \le \overline{\lim}_{n\vee\varepsilon} (\ln \varepsilon^{-1}c_n)\cdot(\alpha - \alpha_2(n,\varepsilon)) \le \alpha_2 \quad P_\vartheta - \text{a.s.},$$

where $\alpha_i = \dfrac{1}{\delta v_1}[v_0 \ln C_{vi} - v_1 \ln \dfrac{C_Y^2}{2v_0}\mathrm{e}^{-6v_0}], \quad i = 1,2.$

Thus for $\vartheta \in \Theta_2$

$$e^{\alpha_1} \leq \underline{\lim}_{n\vee\varepsilon} (\varepsilon^{-1}c_n)^{(\alpha-\alpha_2(n,\varepsilon))} \leq \overline{\lim}_{n\vee\varepsilon} (\varepsilon^{-1}c_n)^{(\alpha-\alpha_2(n,\varepsilon))} \leq e^{\alpha_2} \quad P_\vartheta - \text{a.s.} \tag{38}$$

Let s_1 and s_2 be the positive roots of the following equations

$$C_{v2} \cdot s + \frac{C_Y^2}{2v_0} e^{-6v_0} \cdot e^{\alpha_2} \cdot s^\alpha = 1 \quad \text{and} \quad C_{v1} \cdot s + \frac{C_Y^2}{2v_0} e^{-6v_0} \cdot e^{\alpha_1} \cdot s^\alpha = 1$$

respectively. It is clear that $0 < s_1 \leq s_2 < \infty$.

By the definition of stopping times $\tau_2(n,\varepsilon)$ we have

$$\lim_{n\vee\varepsilon} \left[\frac{1}{\varepsilon^{-1}c_n} \int\limits_{\nu_2(n,\varepsilon)}^{\tau_2(n,\varepsilon)} \tilde{Z}^2(t-3)dt + \frac{1}{(\varepsilon^{-1}c_n)^{\alpha_2(n,\varepsilon)}} \int\limits_{\nu_2(n,\varepsilon)}^{\tau_2(n,\varepsilon)} (\tilde{\Delta}Y(t-3))^2 dt \right] =$$

$$= \lim_{n\vee\varepsilon} \left[\frac{1}{e^{2v_1\tau_2(n,\varepsilon)}} \int\limits_0^{\tau_2(n,\varepsilon)} \tilde{Z}^2(t-3)dt \cdot \frac{e^{2v_1\tau_2(n,\varepsilon)}}{\varepsilon^{-1}c_n} + \right.$$

$$\left. + \frac{1}{e^{2v_0\tau_2(n,\varepsilon)}} \int\limits_0^{\tau_2(n,\varepsilon)} (\tilde{\Delta}Y(t-3))^2 dt \cdot (\varepsilon^{-1}c_n)^{(\alpha-\alpha_2(n,\varepsilon))} \cdot \left(\frac{e^{2v_1\tau_2(n,\varepsilon)}}{\varepsilon^{-1}c_n} \right)^\alpha \right] = 1.$$

Then, using (38), for $\vartheta \in \Theta_2$ we have

$$s_1 \leq \underline{\lim}_{n\vee\varepsilon} \frac{e^{2v_1\tau_2(n,\varepsilon)}}{\varepsilon^{-1}c_n} \leq \overline{\lim}_{n\vee\varepsilon} \frac{e^{2v_1\tau_2(n,\varepsilon)}}{\varepsilon^{-1}c_n} \leq s_2 \quad P_\vartheta - \text{a.s.} \tag{39}$$

and thus

$$\frac{1}{2v_1} \ln s_1 \leq \underline{\lim}_{n\vee\varepsilon} [\tau_2(n,\varepsilon) - \frac{1}{2v_1} \ln \varepsilon^{-1}c_n] \leq$$

$$\leq \overline{\lim}_{n\vee\varepsilon} [\tau_2(n,\varepsilon) - \frac{1}{2v_1} \ln \varepsilon^{-1}c_n] \leq \frac{1}{2v_1} \ln s_2 \quad P_\vartheta - \text{a.s.} \tag{40}$$

From (37) and (40) it follows, in particular, that

$$\lim_{n\vee\varepsilon} [\tau_2(n,\varepsilon) - \nu_2(n,\varepsilon)] = \infty \quad P_\vartheta - \text{a.s.} \tag{41}$$

By the definition of $\tilde{G}_2(n,\varepsilon)$, the following limit relation can be proved

$$\lim_{n\vee\varepsilon} [||\tilde{G}_2^{-1}(n,\varepsilon)||^2 - (1+e^{2v_0})\{ \left(\frac{e^{2v_1\tau_2(n,\varepsilon)}}{\varepsilon^{-1}c_n} \right)^{-2} (< \tilde{G}_2^{-1}(0,\tau_2(n,\varepsilon) - k_2(n)h_2) >_{11})^2 +$$

$$+ \left(\frac{e^{2v_1\tau_2(n,\varepsilon)}}{\varepsilon^{-1}c_n} \right)^{-(1+\alpha)} (\varepsilon^{-1}c_n)^{\alpha_2(n,\varepsilon)-\alpha} (< \tilde{G}_2^{-1}(0,\tau_2(n,\varepsilon) - k_2(n)h_2) >_{12})^2 \}] = 0 \quad P_\vartheta - \text{a.s.},$$

where $< G >_{ij}$ is the ij-th element of the matrix G.

Then, using (33), (35), (38), (39) and (41) we can find, similar to [12, 14], the lower and upper bounds for the limits with P_ϑ-probability one:

$$\tilde{g}_{21} \leq \varliminf_{n \vee \varepsilon} ||\tilde{G}_2^{-1}(n,\varepsilon)|| \leq \varlimsup_{n \vee \varepsilon} ||\tilde{G}_2^{-1}(n,\varepsilon)|| \leq \tilde{g}_{22}, \tag{42}$$

where $\tilde{g}_{21}$ and $\tilde{g}_{22}$ are positive finite numbers.

Thus, by the definition (14) of the stopping time $\sigma_2(\varepsilon)$ and from (42) we have

$$\sigma_{21} \leq \varliminf_{\varepsilon \to 0} \sigma_2(\varepsilon) \leq \varlimsup_{\varepsilon \to 0} \sigma_2(\varepsilon) \leq \sigma_{22} \quad P_\vartheta - \text{a.s.}, \tag{43}$$

where

$$\sigma_{21} = \inf\{n \geq 1: \ N > \tilde{g}_{22}^{-1}(\rho_2 \delta_2^{-1})^{1/2}\}, \qquad \sigma_{22} = \inf\{n \geq 1: \ N > \tilde{g}_{21}^{-1}(\rho_2 \delta_2^{-1})^{1/2}\}$$

and from (40) and (43) we obtain the second property of the assertion I in Proposition 3.2:

$$\frac{1}{2v_1} \ln s_1 \sigma_{21} \leq \varliminf_{\varepsilon \to 0} [T_2(\varepsilon) - \frac{1}{2v_1} \ln \varepsilon^{-1}] \leq \varlimsup_{\varepsilon \to 0} [T_2(\varepsilon) - \frac{1}{2v_1} \ln \varepsilon^{-1}] \leq \frac{1}{2v_1} \ln s_2 \sigma_{22} \quad P_\vartheta - \text{a.s.}$$

The assertions I.1 and II of Proposition 3.2 can be proved similar to the proof of the corresponding statement of Proposition 3.1.

Hence Proposition 3.2 is proven.

Proof of Proposition 3.3.
Similar to the proof of Propositions 3.1, 3.2 and [7]–[16] we get for $\vartheta \in \Theta_3$ the needed asymptotic as $t \to \infty$ relations for the processes $\tilde{\Delta} Y(t)$, $Z(t)$ and $\tilde{Z}(t)$. To this end we introduce the following notation:

$$Z_1(t) = \int_{-\infty}^{t} \tilde{y}_0(t-s) dW(s), \ \ \tilde{y}_0(s) = \tilde{x}_0(s) - \lambda \tilde{x}_0(s-1),$$

$$Z_V(t) = \int_{-\infty}^{t} [\tilde{\Delta} V(s) - \lambda \tilde{\Delta} V(s-1)] e^{-v_0(t-s)} ds, \ \ Z_2(t) = Z_V(t) + Z_3(t),$$

$$Z_3(t) = \int_{-\infty}^{t} Z_1(s) e^{-v_0(t-s)} ds, \ \ \tilde{Z}_1(t) = Z_1(t) - 2v_0 Z_2(t-1),$$

$$\tilde{Z}_2(t) = Z_V(t) + \tilde{Z}_3(t), \ \ \tilde{Z}_3(t) = \int_{-\infty}^{t} \tilde{Z}_1(s) e^{-v_0(t-s)} ds,$$

$$\tilde{C}_Z = 1 + \lambda^2 + 4[\lambda - v_0^{-1}(\lambda - 1)] + E_\vartheta \tilde{Z}_1^2(0),$$

$$C_{\tilde{Z}Z} = 1 + \lambda^2 + 2[\lambda - v_0^{-1}(\lambda - 1)] + E_\vartheta Z_1^2(0) - E_\vartheta Z_1(0) Z_3(-1).$$

It should be noted that in the considered case Θ_3 all the introduced processes $Z_1(\cdot),\ldots,\tilde{Z}_3(\cdot)$ are stationary Gaussian processes, continuous in probability, having a spectral density and, as follows, ergodic, see [21].

According to the definition of the set Θ_3 as $t\to\infty$ we have:

$$\tilde{\Delta}Y(t)=C_Y e^{v_0 t}+o(e^{\gamma t})\ \ P_\vartheta-\text{a.s.},$$

$$|Z(t)-[\tilde{\Delta}V(t)-\lambda\tilde{\Delta}V(t-1)]-Z_1(t)|=o(1)\ \ P_\vartheta-\text{a.s.},$$

where C_Y and $\gamma<v_0$ are some constants.

Using this properties and the representation for the deviation

$$\lambda_t-\lambda=\frac{\int\limits_0^t Z(s)\tilde{\Delta}Y(s-1)ds}{\int\limits_0^t(\tilde{\Delta}Y(s-1))^2ds}$$

of the estimator λ_t defined in (11), it is easy to obtain with P_ϑ-probability one the following limit relations:

$$\lim_{T\to\infty}\frac{1}{e^{2v_0T}}\int\limits_0^T\tilde{\Delta}Y(t-u)\tilde{\Delta}Y(t-s)dt=\frac{C_Y^2}{2v_0}e^{-v_0(u+s)},\ \ u,s\ge 0, \tag{44}$$

$$\lim_{T\to\infty}\left|\frac{1}{e^{v_0T}}\int\limits_0^T Z(t)\tilde{\Delta}Y(t-u)dt-C_Ye^{-v_0u}Z_2(T)\right|=0,\ \ u\ge 0, \tag{45}$$

$$\lim_{t\to\infty}|e^{v_0t}(\lambda_t-\lambda)-2v_0e^{v_0}C_Y^{-1}Z_2(t)|=0, \tag{46}$$

$$\lim_{t\to\infty}|\tilde{Z}(t)-(\tilde{\Delta}V(t)-\lambda\tilde{\Delta}V(t-1))-\tilde{Z}_1(t)|=0,$$

$$\lim_{T\to\infty}\left|\frac{1}{e^{v_0T}}\int\limits_0^T\tilde{Z}(t)\tilde{\Delta}Y(t)dt-C_Y\tilde{Z}_2(T)\right|=0, \tag{47}$$

$$\lim_{T\to\infty}\frac{1}{T}\int\limits_0^T\tilde{Z}(t)Z(t)dt=C_{\tilde{Z}Z}, \tag{48}$$

$$\lim_{T\to\infty}\frac{1}{T}\int\limits_0^T\tilde{Z}^2(t)dt=\tilde{C}_Z. \tag{49}$$

For the investigation of the asymptotic properties of the components of sequential plan we will use Propositions 2 and 3 from [14]. According to these propositions the processes $Z_i(\cdot),\tilde{Z}_i(\cdot),\ i=\overline{1,3}$ and $Z_V(\cdot)$ defined above are $O((\log t)^{\frac{1}{2}})$ as $t\to\infty\ P_\vartheta-\text{a.s.}$

Denote

$$Q = \begin{pmatrix} 1 & 1 \\ -\lambda & 0 \end{pmatrix}, \qquad \varphi(T) = \mathrm{diag}\{T, e^{2v_0 T}\}.$$

From (44), (45), (47), (48) with P_ϑ-probability one holds

$$\lim_{T\to\infty} \varphi^{-1/2}(T) \cdot G_3(0,T) \cdot Q \cdot \varphi^{-1/2}(T) = \mathrm{diag}\{C_{\tilde{Z}Z}, \frac{C_Y^2}{2v_0}\}$$

and, as follows,

$$\lim_{T\to\infty} T \cdot G_3^{-1}(0,T) = C_{\tilde{Z}Z}^{-1} \cdot \begin{pmatrix} 1 & 0 \\ -\lambda & 0 \end{pmatrix} \quad P_\vartheta - \text{a.s.} \tag{50}$$

Further, by the definition (17) of stopping times $\tau_{31}(n,\varepsilon)$, first condition in (16) on the function $\nu_3(n,\varepsilon)$ and from (49) we find

$$\lim_{n\vee\varepsilon} \frac{\tau_{31}(n,\varepsilon)}{\varepsilon^{-1}c_n} = \tilde{C}_Z^{-1} \quad P_\vartheta - \text{a.s.} \tag{51}$$

For the investigation of asymptotic properties of stopping times $\tau_{32}(n,\varepsilon)$ with P_ϑ-probability one we show, using the second condition in (16) on the function $\nu_3(n,\varepsilon)$ and (46), that

$$\lim_{n\vee\varepsilon} \ln \frac{e^{2\alpha_3(n,\varepsilon)\varepsilon^{-1}c_n}}{e^{2\alpha_3\varepsilon^{-1}c_n}} = \lim_{n\vee\varepsilon} 2(\alpha_3(n,\varepsilon)-\alpha)\varepsilon^{-1}c_n = \lim_{n\vee\varepsilon} 2\lambda^{-1}(\lambda_{\nu_3(n,\varepsilon)}-\lambda)\varepsilon^{-1}c_n =$$

$$= 2C_3^{-1} \lim_{n\vee\varepsilon} \frac{Z_2(\nu_3(n,\varepsilon))\varepsilon^{-1}c_n}{e^{v_0\nu_3(n,\varepsilon)}} = 2C_3^{-1} \lim_{n\vee\varepsilon} \frac{Z_2(\nu_3(n,\varepsilon))}{\log^{1/2}\nu_3(n,\varepsilon)} \cdot \frac{\log^{1/2}\nu_3(n,\varepsilon)}{e^{v_0\nu_3(n,\varepsilon)}} \cdot \varepsilon^{-1}c_n = 0$$

and then

$$\lim_{n\vee\varepsilon} \frac{e^{2\alpha_3(n,\varepsilon)\varepsilon^{-1}c_n}}{e^{2v_0\varepsilon^{-1}c_n}} = 1 \quad P_\vartheta - \text{a.s.}$$

Thus, by the definition (18) of stopping times $\tau_{32}(n,\varepsilon)$ and from (44) we find

$$\lim_{n\vee\varepsilon} [\tau_{32}(n,\varepsilon) - \varepsilon^{-1}c_n] = \frac{1}{2v_0} \ln \frac{2v_0 e^{6v_0}}{C_Y^2} \quad P_\vartheta - \text{a.s.} \tag{52}$$

Then, from (50)–(52) with P_ϑ-probability one we obtain

$$\lim_{n\vee\varepsilon} \tilde{G}_3^{-1}(n,\varepsilon) = \{\tilde{C}_Z \vee 1\} C_{\tilde{Z}Z}^{-1} \cdot \begin{pmatrix} 1 & 0 \\ -\lambda & 0 \end{pmatrix},$$

where $a \vee b = \max(a,b)$ and, by the definition (19) of the stopping time $\sigma_3(\varepsilon)$, for ε small enough it follows

$$\sigma_3(\varepsilon) = \bar{\sigma}_3 \quad P_\vartheta - \text{a.s.},$$

where

$$\bar{\sigma}_3 = \inf\{n \geq 1 : N > g_3^{-1}(\rho_3\delta_3^{-1})^{1/2}\}$$

and $g_3 = \{\tilde{C}_Z \vee 1\}^2 C_{\tilde{Z}Z}^{-2}(1+\lambda^2)$.

Thus we obtain the P_ϑ-finiteness of the stopping time $T_3(\varepsilon)$ and the assertion II.2 of Proposition 3.3:

$$\lim_{\varepsilon\to 0} \varepsilon T_3(\varepsilon) = \{\tilde{C}_Z^{-1} \vee 1\} c_{\bar{\sigma}_3} \quad P_\vartheta - \text{a.s.}$$

The assertions I.1 and II of Proposition 3.3 can be proved similar to the proofs of Propositions 3.1 and 3.2.

Hence Proposition 3.3 is proven.

Proof of Proposition 3.4.
This case is a scalar analogue of the case $\Theta_{11} \cup \Theta_{12}$.

By the definition,

$$Z^*(t) = \tilde{X}(t) - \tilde{X}(t-1) + \tilde{\Delta}V(t) - \tilde{\Delta}V(t-1).$$

According to the asymptotic properties of the process $(\tilde{X}(t))$, for $u = 0, 2$ we have:

– for $\vartheta \in \Theta_{41}$:

exist the positive constant limits

$$\lim_{T\to\infty} \frac{1}{T} \int_0^T Z^*(t) Z^*(t-u)dt = C_{41}^*(u) \ P_\vartheta - \text{a.s.}; \tag{53}$$

– for $\vartheta \in \Theta_{42}$:

$$e^{-v_0 t} Z^*(t) = C_{42}^* + o(e^{-(v_0-\gamma)t})) \ \text{ as } t \to \infty \quad P_\vartheta - \text{a.s.},$$

$$\lim_{T\to\infty} \frac{1}{e^{2v_0 T}} \int_0^T Z^*(t) Z^*(t-u)dt = \frac{(C_{42}^*)^2 e^{-v_0 u}}{2v_0} \quad P_\vartheta - \text{a.s.}, \tag{54}$$

where

$$C_{42}^* = \frac{1-e^{v_0}}{v_0(v_0-a+1)}.$$

Assertions I.1 and II of Proposition 3.4 can be proved similar to Proposition 3.1.

Now we prove the closeness of the plan (22) and assertion I.2 of Proposition 3.4. To this end we shall investigate the asymptotic properties of the stopping times $\tau_4(n,\varepsilon)$ and $\sigma_4(\varepsilon)$.

From the definition of $\tau_4(n,\varepsilon)$ and (53), (54) we have

– for $\vartheta \in \Theta_{41}$:

$$\lim_{n\vee\varepsilon} \frac{\tau_4(n,\varepsilon)}{\varepsilon^{-1} c_n} = (C_{41}^*(0))^{-1} \ P_\vartheta - \text{a.s.}; \tag{55}$$

– for $\vartheta \in \Theta_{42}$:

$$\lim_{n\vee\varepsilon} \frac{e^{2v_0\tau_4(n,\varepsilon)}}{\varepsilon^{-1}c_n} = \left(\frac{2v_0}{C_{42}^*}\right)^2 \quad P_\vartheta - \text{a.s.}, \tag{56}$$

$$\lim_{n\vee\varepsilon} [\tau_4(n,\varepsilon) - \frac{1}{2v_0}\ln\varepsilon^{-1}c_n] = \frac{1}{v_0}\ln\frac{2v_0}{C_{42}^*} \quad P_\vartheta - \text{a.s.} \tag{57}$$

As follows, the stopping times $\tau_4(n,\varepsilon)$ are P_ϑ – a.s. finite for all $\vartheta \in \Theta_4$.

Denote

$$\sigma_{41} = \inf\{n \geq 1 : N > (\rho_4\delta_4^{-1})^{1/2}g_{41}^{-1}\}, \qquad \sigma_{42} = \inf\{n \geq 1 : N > (\rho_4\delta_4^{-1})^{1/2}g_{42}^{-1}\},$$

where $g_{41} = [C_{41}^*(2)(C_{41}^*(0))^{-1} \vee 2v_0e^{-2v_0}]$, $g_{42} = [C_{41}^*(2)(C_{41}^*(0))^{-1} \wedge 2v_0e^{-2v_0}]$, where $a \wedge b = \min(a,b)$ and by the definition of the stopping time $\sigma_4(\varepsilon)$ (21) as well as from (53)–(56) follows the P_ϑ – a.s. finiteness of $\sigma_4(\varepsilon)$ and the following inequalities, which hold with P_ϑ-probability one for ε small enough:

$$\sigma_{41} \leq \sigma_4(\varepsilon) \leq \sigma_{42} \quad P_\vartheta - \text{a.s.} \tag{58}$$

Then we obtain the finiteness of the stopping time $T_4(\varepsilon)$ and the assertion I.2 of Proposition 3.4, which follows from (55), (57) and (58):

– for $\vartheta \in \Theta_{41}$:

$$c_{\sigma_{41}}(C_{41}^*(0))^{-1} \leq \underline{\lim}_{n\vee\varepsilon}\, \varepsilon T_4(\varepsilon) \leq \overline{\lim}_{n\vee\varepsilon}\, \varepsilon T_4(\varepsilon) \leq c_{\sigma_{42}}(C_{41}^*(0))^{-1} \quad P_\vartheta - \text{a.s.};$$

– for $\vartheta \in \Theta_{42}$:

$$\frac{1}{2v_0}\ln c_{\sigma_{41}}\left(\frac{2v_0}{C_{42}^*(0)}\right)^2 \leq \underline{\lim}_{n\vee\varepsilon}\, [T_4(\varepsilon) - \frac{1}{2v_0}\ln\varepsilon^{-1}] \leq$$

$$\leq \overline{\lim}_{n\vee\varepsilon}\, [T_4(\varepsilon) - \frac{1}{2v_0}\ln\varepsilon^{-1}] \leq \frac{1}{2v_0}\ln c_{\sigma_{42}}\left(\frac{2v_0}{C_{42}^*(0)}\right)^2 \quad P_\vartheta - \text{a.s.}$$

Hence Proposition 3.4 is valid.

Proof of Theorem 3.1. The closeness of the sequential estimation plan SEP$^\star(\varepsilon)$ (assertion 1) and assertion 3 of Theorem 3.1 follow from Propositions 3.1-3.4 directly.

Now we prove the assertion 2. To this end we show first, that all the stopping times $\tau_i(n,\varepsilon)$, $i = 1,2,4$ and $\tau_{3i}(n,\varepsilon)$, $i = 1,2$ are P_ϑ – a.s.-finite for every $\vartheta \in \overline{\Theta}$. It should be noted, that the integrals

$$\int_0^\infty (\tilde{\Delta}Y(t))^2 dt = \infty \quad \text{and} \quad \int_0^\infty (Z^*(t))^2 dt = \infty \quad P_\vartheta - \text{a.s.} \tag{59}$$

in all the cases $\Theta_1, \ldots, \Theta_4$ and, as follows, all the stopping times $\tau_i(n,\varepsilon)$, $i = 1,2,4$ and $\tau_{32}(n,\varepsilon)$ are P_ϑ-a.s.-finite for every $\varepsilon > 0$ and all $n \geq 1$. The properties (59) can be established by using the asymptotic properties of the process $(X(t), Y(t))$ (see proofs of Propositions 3.1–3.4 and [3], [7]–[16]).

The stopping times $\tau_{31}(n,\varepsilon)$ are finite in the region $\overline{\Theta}_2 \cup \overline{\Theta}_3$ according to Propositions 3.2, 3.3. As follows, it remind only to verify the finiteness of the stopping time $\tau_{31}(n,\varepsilon)$ for $\vartheta \in \overline{\Theta}_1 \cup \overline{\Theta}_4$.

According to the definition (17) of these stopping times it is enough to show the divergence of the following integral

$$\int_0^\infty \tilde{Z}^2(t)dt = \infty \quad P_\vartheta - \text{a.s.},$$

where $\tilde{Z}(t) = \tilde{\Delta}Y(t) - \lambda_t\tilde{\Delta}Y(t-1)$.

This property follows from the following facts:

– for $\vartheta \in \Theta_{11}$

$$\lim_{t\to\infty} \lambda_t = \tilde{\lambda}, \quad P_\vartheta - \text{a.s.},$$

where $\tilde{\lambda}$ is some constant and the process $\tilde{Z}(t)$ can be approximated, similar to the case $\overline{\Theta}_3$ (see the proof of Proposition 3.3) by a Gaussian stationary process;

– for $\vartheta \in \Theta_{12}$

$$\lim_{t\to\infty} |\lambda_t - C_1(t)| = 0 \quad P_\vartheta - \text{a.s.},$$

and then

$$\lim_{t\to\infty} |e^{-v_0 t}\tilde{Z}(t) - C_2(t)| = 0 \quad P_\vartheta - \text{a.s.},$$

where $C_1(t)$ and $C_2(t)$ are some periodic bounded functions;

– for Θ_{13}

$$\lim_{t\to\infty} t(\lambda_t - \lambda) = C_3 \quad P_\vartheta - \text{a.s.}$$

and

$$\lim_{t\to\infty} e^{-v_0 t}\tilde{Z}(t) = C_4 \quad P_\vartheta - \text{a.s.},$$

where C_3 and C_4 are some non-zero constants;

– for Θ_{41}

$$\lim_{t\to\infty} \left| \tilde{Z}(t) - \frac{1-\lambda}{1-a}\left(X_0(0) + b\int_{-1}^{0} X_0(s)ds \right) - \right.$$

$$\left. - \frac{1}{1-a}(W(t) - \lambda W(t-1)) - (\tilde{\Delta}V(t) - \lambda\tilde{\Delta}V(t-1)) \right| = 0 \quad P_\vartheta - \text{a.s.};$$

– for Θ_{42}

$$\lim_{t\to\infty} e^{-v_0 t}\tilde{Z}(t) = C_5 \quad P_\vartheta - \text{a.s.},$$

where C_5 is some non-zero constant.

Denote $\mu_1 = \mu_2 = 1$, $\mu_3 = \mu_4 = 2/5$.

Now we can verify the second property of the estimator $\vartheta(\varepsilon)$. By the definition of stopping times $\sigma_j(\varepsilon)$, $j = \overline{1,4}$, we get

$$\sup_{\vartheta\in\overline{\Theta}} E_\vartheta||\vartheta(\varepsilon) - \vartheta||^2 \le \varepsilon \sup_{\vartheta\in\overline{\Theta}} E_\vartheta \rho_{j^*}^{-1}\delta_{j^*} \sum_{n=1}^{\sigma_{j^*}(\varepsilon)} \frac{1}{c_n}\beta_{j^*}^2(n,\varepsilon)\cdot||\tilde{G}_{j^*}^{-1}(n,\varepsilon)||^2 \cdot ||\tilde{\zeta}_{j^*}(n,\varepsilon)||^2 \le$$

$$\le \varepsilon \sup_{\vartheta\in\overline{\Theta}} E_\vartheta \rho_{j^*}^{-1}\delta_{j^*} \sum_{n\ge 1} \frac{1}{c_n}||\tilde{\zeta}_{j^*}(n,\varepsilon)||^2.$$

Due to the obtained finiteness properties of all the stopping times in these sums all the mathematical expectations are well-defined and we can estimate finally

$$\sup_{\vartheta\in\overline{\Theta}} E_\vartheta||\vartheta(\varepsilon)-\vartheta||^2 \le \varepsilon\sum_{j=1}^{4}\rho_j^{-1}\delta_j\sum_{n\ge1}\frac{1}{c_n}\sup_{\vartheta\in\overline{\Theta}} E_\vartheta||\tilde{\zeta}_j(n,\varepsilon)||^2 \le$$

$$\le 15(3+R^2)\varepsilon\sum_{j=1}^{4}\rho_j^{-1}\delta_j\mu_j\sum_{n\ge1}\frac{1}{c_n} = \varepsilon\sum_{j=1}^{4}\delta_j = \varepsilon.$$

Hence Theorem 3.1 is proven.

5. Conclusion

This chapter presents a sequential approach to the guaranteed parameter estimation problem of a linear stochastic continuous-time system. We consider a concrete stochastic delay differential equation driven by an additive Wiener process with noisy observations.

At the same time for the construction of the sequential estimation plans we used mainly the structure and the asymptotic behaviour of the solution of the system. Analogously, the presented method can be used for the guaranteed accuracy parameter estimation problem of the linear ordinary and delay stochastic differential equations of an arbitrary order with and without noises in observations.

The obtained estimation procedure can be easily generalized, similar to [9, 11, 13, 16], to estimate the unknown parameters with preassigned accuracy in the sense of the L_q-norm ($q \ge 2$). The estimators with such properties may be used in various adaptive procedures (control, prediction, filtration).

Author details

Uwe Küchler
Humboldt University Berlin, Germany

Vyacheslav A. Vasiliev
Tomsk State University, Russia

6. References

[1] Arato, M. [1982]. *Linear Stochastic Systems with Constant Coefficients. A Statistical Approach*, Springer Verlag, Berlin, Heidelberg, New York.
[2] Galtchouk, L. & Konev, V. [2001]. On sequential estimation of parameters in semimartingale regression models with continuous time parameter, *The Annals of Statistics* Vol. (29), 5: 1508–1536.
[3] Gushchin, A. A. & Küchler, U. [1999]. Asymptotic inference for a linear stochastic differential equation with time delay, *Bernoulli* Vol. (5), 6: 1059–1098.
[4] Kallianpur, G. [1987]. *Stochastic Filtering Theory*, Springer Verlag, New York, Heidelberg, Berlin.
[5] Kolmanovskii, V. & A. Myshkis. [1992]. *Applied Theory of Functional Differential Equations*, Kluwer Acad. Pabl.

[6] Küchler, U. & Sørensen, M. [1997]. *Exponential Families of Stochastic Processes*, Springer Verlag, New York, Heidelberg.

[7] Küchler, U. & Vasiliev, V. [2001]. On sequential parameter estimation for some linear stochastic differential equations with time delay, *Sequential Analysis* Vol. (20), 3: 117–146.

[8] Küchler, U. & Vasiliev, V. [2003]. On sequential identification of a diffusion type process with memory, *Proceedings Symp. Int. Fed. Autom. Contr. SYSID-2003*, Rotterdam, Holland, 27-29 August, pp. 1217–1221.

[9] Küchler, U. & Vasiliev, V. [2005a]. Sequential identification of linear dynamic systems with memory, *Statist. Inference for Stochastic Processes* Vol. (8): 1–24.

[10] Küchler, U. & Vasiliev, V. [2005b]. On parameter estimation of stochastic delay differential equations with guaranteed accuracy by noisy observations, *Preprint–15 of the Math. Inst. of Humboldt University*, Berlin, pp. 1–24.

[11] Küchler, U. & Vasiliev, V. [2006]. On Sequential Estimators for an Affine Stochastic Delay Differential Equations, *In Proceedengs of the 5th International Conference, A. Iske, J. Levesley editors, Algorithms for Approximation*, Chester, July 2005, Springer-Verlag, Berlin, Heidelberg, pp. 287–296.

[12] Küchler, U. & Vasiliev, V. [2007]. On parameter estimation of stochastic delay differential equations with guaranteed accuracy by noisy observations, *Journal of Statistical Planning and Inferences*, Elsevier Science Vol. (137): 3007-Ű3023, doi: 10/1016j.jspi.2006.12.001.

[13] Küchler, U. & Vasiliev, V. [2008]. On sequential parameter estimation of a linear regression process, *Proceedings of the 17TH World Congress the Int. Fed. Autom. Contr.*, Seoul, Korea, 5-9 Juli, pp. 10230–10235.

[14] Küchler, U. & Vasiliev, V. [2009a]. On guaranteed parameter estimation of a multiparameter linear regression process, *Preprint–7 of the Math. Inst. of Humboldt University*, Berlin, pp. 1–60.

[15] Küchler, U. & Vasiliev, V. [2009b]. On sequential parameter estimation of stochastic delay differential equations by noisy observations, *Preprints of the 15th IFAC Symposium on System Identification*, Saint-Malo, France, July 6-8, pp. 892–897.

[16] Küchler, U. & Vasiliev, V. [2010a]. On guaranteed parameter estimation of a multiparameter linear regression process, *Automatica, Journal of IFAC*, Elsevier Vol. (46) 4: 637–646.

[17] Küchler, U. & Vasiliev, V. [2010b]. On adaptive control problems of continuous-time stochastic systems, *The 1-st Virtual Control Conference* (VCC-2010, Aalborg University, Denmark), pages 1-5, http://www.vcc-10.org/index_files/papers/p107.pdf.

[18] Liptzer, R. S. & Shiryaev, A. N. [1977]. *Statistics of Random Processes*, Springer-Verlag, New York, Heidelberg.

[19] Mao, X. [1997]. *Stochastic Differential Equations and Application*, Harwood Publishing Chichester, Second Edition.

[20] Mohammed, S. E-A. [1996]. *Stochastic differential systems with memory: Theory, examples and applications*, Probability and Statistics, Birkhäuser Vol. (42): 1–77.

[21] Rozanov, Yu.A. [1967]. *Stationary Gaussian Processes*, Holden Day: San Francisco, CA.

[22] Vasiliev, V.A. & Konev, V.V. [1987]. On sequential identification of linear dynamic systems in continuous time by noisy observations, *Probl. of Contr. and Information Theory* Vol. (16) 2: 101–112.

[23] Vasiliev, V.A. & Konev, V.V. [1991]. Fixed Accuracy Identification of Partly Observable Dynamic Systems, *Proceedings of the 11th World Congress the Int. Fed. Autom. Contr.*, Tallinn, 13-17 Aug., 1990, 2, Oxford etc., pp. 179–183.

Chapter 5

Geometrical Derivation of Equilibrium Distributions in Some Stochastic Systems

Ricardo López-Ruiz and Jaime Sañudo

Additional information is available at the end of the chapter

1. Introduction

Classical statistical physics deals with statistical systems in equilibrium. The ensemble theory offers a useful framework that allows to characterize and to work out the properties of this type of systems [1]. Two fundamental distributions to describe situations in equilibrium are the Boltzmann-Gibbs (exponential) distribution and the Maxwellian (Gaussian) distribution. The first one represents the distribution of the energy states of a system and the second one fits the distribution of velocities in an ideal gas. They can be explained from different perspectives. In the physics of equilibrium, they are usually obtained from the principle of maximum entropy [2]. In the physics out of equilibrium, there have recently been proposed two nonlinear models that explain the decay of any initial distribution to these asymptotic equilibria [3, 4].

In this chapter, these distributions are alternatively obtained from a geometrical interpretation of different multi-agent systems evolving in phase space under the hypothesis of equiprobability. Concretely, an economic context is used to illustrate this derivation. Thus, from a macroscopic point of view, we consider that markets have an intrinsic stochastic ingredient as a consequence of the interaction of an undetermined ensemble of agents that trade and perform an undetermined number of commercial transactions at each moment. A kind of models considering this unknowledge associated to markets are the gas-like models [5, 6]. These random models interpret economic exchanges of money between agents similarly to collisions in a gas where particles share their energy. In order to explain the two before mentioned statistical behaviors, the Boltzmann-Gibbs and Maxwellian distributions, we will not suppose any type of interaction between the agents. The geometrical constraints and the hypothesis of equiprobability will be enough to explain these distributions in a situation of statistical equilibrium.

Thus, the Boltzmann-Gibbs (exponential) distribution is derived in Section 2 from the geometrical properties of the volume of an N-dimensional pyramid or from the properties of the surface of an N-dimensional hyperplane [7, 8]. In both cases, the motivation will be a multi-agent economic system with an open or closed economy, respectively. The Maxwellian (Gaussian) distribution is derived in Section 3 from geometrical arguments over the volume or the surface of an N-sphere [7, 9]. In this case, the motivation will be a multi-particle gas system in contact with a heat reservoir (non-isolated or open system) or with a fixed energy (isolated or closed system), respectively. And finally, in Section 4, the general equilibrium distribution for a set of many identical interacting agents obeying a global additive constraint is also derived [7]. This distribution will be related with the Gamma-like distributions found in several multi-agent economic models. Other two geometrical collateral results, namely the formula for the volume of high-dimensional symmetrical bodies and an alternative image of the canonical ensemble, are proposed in Section 5. And last Section embodies the conclusions.

2. Derivation of the Boltzmann-Gibbs distribution

We proceed to derive here the Boltzmann-Gibbs distribution in two different physical situations with an economic inspiration. The first one considers an ensemble of economic agents that share a variable amount of money (open systems) and the second one deals with the conservative case where the total wealth is fixed (closed systems).

2.1. Multi-agent economic open systems

Here we assume N agents, each one with coordinate x_i, $i = 1, \ldots, N$, with $x_i \geq 0$ representing the wealth or money of the agent i, and a total available amount of money E:

$$x_1 + x_2 + \cdots + x_{N-1} + x_N \leq E. \tag{1}$$

Under random or deterministic evolution rules for the exchanging of money among agents, let us suppose that this system evolves in the interior of the N-dimensional pyramid given by Eq. (1). The role of a heat reservoir, that in this model supplies money instead of energy, could be played by the state or by the bank system in western societies. The formula for the volume $V_N(E)$ of an equilateral N-dimensional pyramid formed by $N+1$ vertices linked by N perpendicular sides of length E is

$$V_N(E) = \frac{E^N}{N!}. \tag{2}$$

We suppose that each point on the N-dimensional pyramid is equiprobable, then the probability $f(x_i)dx_i$ of finding the agent i with money x_i is proportional to the volume formed by all the points into the $(N-1)$-dimensional pyramid having the ith-coordinate equal to x_i. We show now that $f(x_i)$ is the Boltzmann factor (or the Maxwell-Boltzmann distribution), with the normalization condition

$$\int_0^E f(x_i)dx_i = 1. \tag{3}$$

If the ith agent has coordinate x_i, the $N-1$ remaining agents share, at most, the money $E - x_i$ on the $(N-1)$-dimensional pyramid

$$x_1 + x_2 \cdots + x_{i-1} + x_{i+1} \cdots + x_N \leq E - x_i, \tag{4}$$

whose volume is $V_{N-1}(E - x_i)$. It can be easily proved that

$$V_N(E) = \int_0^E V_{N-1}(E - x_i)dx_i. \tag{5}$$

Hence, the volume of the N-dimensional pyramid for which the ith coordinate is between x_i and $x_i + dx_i$ is $V_{N-1}(E - x_i)dx_i$. We normalize it to satisfy Eq. (3), and obtain

$$f(x_i) = \frac{V_{N-1}(E - x_i)}{V_N(E)}, \tag{6}$$

whose final form, after some calculation is

$$f(x_i) = NE^{-1}\left(1 - \frac{x_i}{E}\right)^{N-1}, \tag{7}$$

If we call ϵ the mean wealth per agent, $E = N\epsilon$, then in the limit of large N we have

$$\lim_{N \gg 1} \left(1 - \frac{x_i}{E}\right)^{N-1} \simeq e^{-x_i/\epsilon}. \tag{8}$$

The Boltzmann factor $e^{-x_i/\epsilon}$ is found when $N \gg 1$ but, even for small N, it can be a good approximation for agents with low wealth. After substituting Eq. (8) into Eq. (7), we obtain the Maxwell-Boltzmann distribution in the asymptotic regime $N \to \infty$ (which also implies $E \to \infty$):

$$f(x)dx = \frac{1}{\epsilon} e^{-x/\epsilon} dx, \tag{9}$$

where the index i has been removed because the distribution is the same for each agent, and thus the wealth distribution can be obtained by averaging over all the agents. This distribution has been found to fit the real distribution of incomes in western societies [10, 11].

This geometrical image of the volume-based statistical ensemble [7] allows us to recover the same result than that obtained from the microcanonical ensemble [8] that we show in the next section.

2.2. Multi-agent economic closed systems

Here, we derive the Boltzmann-Gibbs distribution by considering the system in isolation, that is, a closed economy. Without loss of generality, let us assume N interacting economic agents, each one with coordinate x_i, $i = 1, \ldots, N$, with $x_i \geq 0$, and where x_i represents an amount of money. If we suppose now that the total amount of money E is conserved,

$$x_1 + x_2 + \cdots + x_{N-1} + x_N = E, \tag{10}$$

then this isolated system evolves on the positive part of an equilateral N-hyperplane. The surface area $S_N(E)$ of an equilateral N-hyperplane of side E is given by

$$S_N(E) = \frac{\sqrt{N}}{(N-1)!} E^{N-1}. \tag{11}$$

Different rules, deterministic or random, for the exchange of money between agents can be given. Depending on these rules, the system can visit the N-hyperplane in an equiprobable manner or not. If the ergodic hypothesis is assumed, each point on the N-hyperplane is equiprobable. Then the probability $f(x_i)dx_i$ of finding agent i with money x_i is proportional to the surface area formed by all the points on the N-hyperplane having the ith-coordinate equal to x_i. We show that $f(x_i)$ is the Boltzmann-Gibbs distribution (the Boltzmann factor), with the normalization condition (3).

If the ith agent has coordinate x_i, the $N-1$ remaining agents share the money $E - x_i$ on the $(N-1)$-hyperplane

$$x_1 + x_2 \cdots + x_{i-1} + x_{i+1} \cdots + x_N = E - x_i, \tag{12}$$

whose surface area is $S_{N-1}(E - x_i)$. If we define the coordinate θ_N as satisfying

$$\sin\theta_N = \sqrt{\frac{N-1}{N}}, \tag{13}$$

it can be easily shown that

$$S_N(E) = \int_0^E S_{N-1}(E - x_i)\frac{dx_i}{\sin\theta_N}. \tag{14}$$

Hence, the surface area of the N-hyperplane for which the ith coordinate is between x_i and $x_i + dx_i$ is proportional to $S_{N-1}(E - x_i)dx_i/\sin\theta_N$. If we take into account the normalization condition (3), we obtain

$$f(x_i) = \frac{1}{S_N(E)}\frac{S_{N-1}(E - x_i)}{\sin\theta_N}, \tag{15}$$

whose form after some calculation is

$$f(x_i) = (N-1)E^{-1}\left(1 - \frac{x_i}{E}\right)^{N-2}, \tag{16}$$

If we call ϵ the mean wealth per agent, $E = N\epsilon$, then in the limit of large N we have

$$\lim_{N\gg 1}\left(1 - \frac{x_i}{E}\right)^{N-2} \simeq e^{-x_i/\epsilon}. \tag{17}$$

As in the former section, the Boltzmann factor $e^{-x_i/\epsilon}$ is found when $N \gg 1$ but, even for small N, it can be a good approximation for agents with low wealth. After substituting Eq. (1) into Eq. (16), we obtain the Boltzmann distribution (9) in the limit $N \to \infty$ (which also implies $E \to \infty$). This asymptotic result reproduces the distribution of real economic data [10] and also the results obtained in several models of economic agents with deterministic, random or chaotic exchange interactions [6, 12, 13].

Depending on the physical situation, the mean wealth per agent ϵ takes different expressions and interpretations. For instance, we can calculate the dependence of ϵ on the temperature, which in the microcanonical ensemble is defined by the derivative of the entropy with respect to the energy. The entropy can be written as $S = -kN\int_0^\infty f(x)\ln f(x)\,dx$, where $f(x)$ is given by Eq. (9) and k is Boltzmann's constant. If we recall that $\epsilon = E/N$, we obtain

$$S(E) = kN\ln\frac{E}{N} + kN. \tag{18}$$

The calculation of the temperature T gives

$$T^{-1} = \left(\frac{\partial S}{\partial E}\right)_N = \frac{kN}{E} = \frac{k}{\epsilon}. \tag{19}$$

Thus $\epsilon = kT$, and the Boltzmann-Gibbs distribution is obtained in its usual form:

$$f(x)dx = \frac{1}{kT}e^{-x/kT}dx. \tag{20}$$

3. Derivation of the Maxwellian distribution

We proceed to derive here the Maxwellian distribution in two different physical situations with inspiration in the theory of ideal gases. The first one considers an ideal gas with a variable energy (open systems) and the second one deals with the case of a gas with a fixed energy (closed systems).

3.1. Multi-particle open systems

Let us suppose a one-dimensional ideal gas of N non-identical classical particles with masses m_i, with $i = 1, \ldots, N$, and total maximum energy E. If particle i has a momentum $m_i v_i$, we define a kinetic energy:

$$K_i \equiv p_i^2 \equiv \frac{1}{2} m_i v_i^2, \tag{21}$$

where p_i is the square root of the kinetic energy K_i. If the total maximum energy is defined as $E \equiv R^2$, we have

$$p_1^2 + p_2^2 + \cdots + p_{N-1}^2 + p_N^2 \leq R^2. \tag{22}$$

We see that the system has accessible states with different energy, which can be supplied by a heat reservoir. These states are all those enclosed into the volume of the N-sphere given by Eq. (22). The formula for the volume $V_N(R)$ of an N-sphere of radius R is

$$V_N(R) = \frac{\pi^{\frac{N}{2}}}{\Gamma(\frac{N}{2}+1)} R^N, \tag{23}$$

where $\Gamma(\cdot)$ is the gamma function. If we suppose that each point into the N-sphere is equiprobable, then the probability $f(p_i)dp_i$ of finding the particle i with coordinate p_i (energy p_i^2) is proportional to the volume formed by all the points on the N-sphere having the ith-coordinate equal to p_i. We proceed to show that $f(p_i)$ is the Maxwellian distribution, with the normalization condition

$$\int_{-R}^{R} f(p_i) dp_i = 1. \tag{24}$$

If the ith particle has coordinate p_i, the $(N-1)$ remaining particles share an energy less than the maximum energy $R^2 - p_i^2$ on the $(N-1)$-sphere

$$p_1^2 + p_2^2 \cdots + p_{i-1}^2 + p_{i+1}^2 \cdots + p_N^2 \leq R^2 - p_i^2, \tag{25}$$

whose volume is $V_{N-1}(\sqrt{R^2 - p_i^2})$. It can be easily proved that

$$V_N(R) = \int_{-R}^{R} V_{N-1}(\sqrt{R^2 - p_i^2}) dp_i. \tag{26}$$

Hence, the volume of the N-sphere for which the ith coordinate is between p_i and $p_i + dp_i$ is $V_{N-1}(\sqrt{R^2 - p_i^2})dp_i$. We normalize it to satisfy Eq. (24), and obtain

$$f(p_i) = \frac{V_{N-1}(\sqrt{R^2 - p_i^2})}{V_N(R)}, \tag{27}$$

whose final form, after some calculation is

$$f(p_i) = C_N R^{-1}\left(1 - \frac{p_i^2}{R^2}\right)^{\frac{N-1}{2}}, \tag{28}$$

with

$$C_N = \frac{1}{\sqrt{\pi}} \frac{\Gamma(\frac{N+2}{2})}{\Gamma(\frac{N+1}{2})}. \tag{29}$$

For $N \gg 1$, Stirling's approximation can be applied to Eq. (29), leading to

$$\lim_{N \gg 1} C_N \simeq \frac{1}{\sqrt{\pi}} \sqrt{\frac{N}{2}}. \tag{30}$$

If we call ϵ the mean energy per particle, $E = R^2 = N\epsilon$, then in the limit of large N we have

$$\lim_{N \gg 1} \left(1 - \frac{p_i^2}{R^2}\right)^{\frac{N-1}{2}} \simeq e^{-p_i^2/2\epsilon}. \tag{31}$$

The factor $e^{-p_i^2/2\epsilon}$ is found when $N \gg 1$ but, even for small N, it can be a good approximation for particles with low energies. After substituting Eqs. (30)–(31) into Eq. (28), we obtain the Maxwellian distribution in the asymptotic regime $N \to \infty$ (which also implies $E \to \infty$):

$$f(p)dp = \sqrt{\frac{1}{2\pi\epsilon}}\, e^{-p^2/2\epsilon} dp, \tag{32}$$

where the index i has been removed because the distribution is the same for each particle, and thus the velocity distribution can be obtained by averaging over all the particles.

This newly shows that the geometrical image of the volume-based statistical ensemble [7] allows us to recover the same result than that obtained from the microcanonical ensemble [9] that it is presented in the next section.

3.2. Multi-particle closed systems

We start by assuming a one-dimensional ideal gas of N non-identical classical particles with masses m_i, with $i = 1, \ldots, N$, and total energy E. If particle i has a momentum $m_i v_i$, newly we define a kinetic energy K_i given by Eq. (21), where p_i is the square root of K_i. If the total energy is defined as $E \equiv R^2$, we have

$$p_1^2 + p_2^2 + \cdots + p_{N-1}^2 + p_N^2 = R^2. \tag{33}$$

We see that the isolated system evolves on the surface of an N-sphere. The formula for the surface area $S_N(R)$ of an N-sphere of radius R is

$$S_N(R) = \frac{2\pi^{\frac{N}{2}}}{\Gamma(\frac{N}{2})} R^{N-1}, \tag{34}$$

where $\Gamma(\cdot)$ is the gamma function. If the ergodic hypothesis is assumed, that is, each point on the N-sphere is equiprobable, then the probability $f(p_i)dp_i$ of finding the particle i with coordinate p_i (energy p_i^2) is proportional to the surface area formed by all the points on the N-sphere having the ith-coordinate equal to p_i. Our objective is to show that $f(p_i)$ is the Maxwellian distribution, with the normalization condition (24).

If the ith particle has coordinate p_i, the $(N-1)$ remaining particles share the energy $R^2 - p_i^2$ on the $(N-1)$-sphere

$$p_1^2 + p_2^2 \cdots + p_{i-1}^2 + p_{i+1}^2 \cdots + p_N^2 = R^2 - p_i^2, \tag{35}$$

whose surface area is $S_{N-1}(\sqrt{R^2 - p_i^2})$. If we define the coordinate θ as satisfying

$$R^2 \cos^2\theta = R^2 - p_i^2, \tag{36}$$

then

$$R d\theta = \frac{dp_i}{(1 - \frac{p_i^2}{R^2})^{1/2}}. \tag{37}$$

It can be easily proved that

$$S_N(R) = \int_{-\pi/2}^{\pi/2} S_{N-1}(R\cos\theta) R d\theta. \tag{38}$$

Hence, the surface area of the N-sphere for which the ith coordinate is between p_i and $p_i + dp_i$ is $S_{N-1}(R\cos\theta)Rd\theta$. We rewrite the surface area as a function of p_i, normalize it to satisfy Eq. (24), and obtain

$$f(p_i) = \frac{1}{S_N(R)} \frac{S_{N-1}(\sqrt{R^2 - p_i^2})}{(1 - \frac{p_i^2}{R^2})^{1/2}}, \tag{39}$$

whose final form, after some calculation is

$$f(p_i) = C_N R^{-1} \left(1 - \frac{p_i^2}{R^2}\right)^{\frac{N-3}{2}}, \tag{40}$$

with

$$C_N = \frac{1}{\sqrt{\pi}} \frac{\Gamma(\frac{N}{2})}{\Gamma(\frac{N-1}{2})}. \tag{41}$$

For $N \gg 1$, Stirling's approximation can be applied to Eq. (41), leading to

$$\lim_{N \gg 1} C_N \simeq \frac{1}{\sqrt{\pi}} \sqrt{\frac{N}{2}}. \tag{42}$$

If we call ϵ the mean energy per particle, $E = R^2 = N\epsilon$, then in the limit of large N we have

$$\lim_{N \gg 1} \left(1 - \frac{p_i^2}{R^2}\right)^{\frac{N-3}{2}} \simeq e^{-p_i^2/2\epsilon}. \tag{43}$$

As in the former section, the Boltzmann factor $e^{-p_i^2/2\epsilon}$ is found when $N \gg 1$ but, even for small N, it can be a good approximation for particles with low energies. After substituting

Eqs. (42)–(43) into Eq. (40), we obtain the Maxwellian distribution (32) in the asymptotic regime $N \to \infty$ (which also implies $E \to \infty$).

Depending on the physical situation the mean energy per particle ϵ takes different expressions. For an isolated one-dimensional gas we can calculate the dependence of ϵ on the temperature, which in the microcanonical ensemble is defined by differentiating the entropy with respect to the energy. The entropy can be written as $S = -kN\int_{-\infty}^{\infty} f(p)\ln f(p)\,dp$, where $f(p)$ is given by Eq. (32) and k is the Boltzmann constant. If we recall that $\epsilon = E/N$, we obtain

$$S(E) = \frac{1}{2}kN\ln\left(\frac{E}{N}\right) + \frac{1}{2}kN(\ln(2\pi) - 1). \tag{44}$$

The calculation of the temperature T gives

$$T^{-1} = \left(\frac{\partial S}{\partial E}\right)_N = \frac{kN}{2E} = \frac{k}{2\epsilon}. \tag{45}$$

Thus $\epsilon = kT/2$, consistent with the equipartition theorem. If p^2 is replaced by $\frac{1}{2}mv^2$, the Maxwellian distribution is a function of particle velocity, as it is usually given in the literature:

$$g(v)dv = \sqrt{\frac{m}{2\pi kT}}\, e^{-mv^2/2kT} dv. \tag{46}$$

4. General derivation of the equilibrium distribution

In this section, we are interested in the same problem above presented but in a general way. We address this question in the volume-based statistical framework.

Let b be a positive real constant (cases $b = 1, 2$ have been indicated in the former sections). If we have a set of positive variables $(x_1, x_2, \ldots, x_N)$ verifying the constraint

$$x_1^b + x_2^b + \cdots + x_{N-1}^b + x_N^b \leq E \tag{47}$$

with an adequate mechanism assuring the equiprobability of all the possible states $(x_1, x_2, \ldots, x_N)$ into the volume given by expression (47), will we have for the generic variable x the distribution

$$f(x)dx \sim \epsilon^{-1/b} e^{-x^b/b\epsilon} dx, \tag{48}$$

when we average over the ensemble in the limit $N, E \to \infty$, with $E = N\epsilon$, and constant ϵ?. Now it is shown that the answer is affirmative. Similarly, we claim that if the weak inequality (47) is transformed in equality the result will be the same, as it has been proved for the cases $b = 1, 2$ in Refs. [8, 9].

From the cases $b = 1, 2$, (see Eqs. (6) and (27)), we can extrapolate the general formula that will give us the statistical behavior $f(x)$ of the generic variable x, when the system runs equiprobably into the volume defined by a constraint of type (47). The probability $f(x)dx$ of finding an agent with generic coordinate x is proportional to the volume $V_{N-1}((E - x^b)^{1/b})$ formed by all the points into the $(N-1)$-dimensional symmetrical body limited by the constraint $(E - x^b)$. Thus, the N-dimensional volume can be written as

$$V_N(E^{1/b}) = \int_0^{E^{1/b}} V_{N-1}((E - x^b)^{1/b})\, dx. \tag{49}$$

Taking into account the normalization condition $\int_0^{E^{1/b}} f(x)dx = 1$, the expression for $f(x)$ is obtained:

$$f(x) = \frac{V_{N-1}((E - x^b)^{1/b})}{V_N(E^{1/b})}. \quad (50)$$

The N-dimensional volume, $V_N(b, \rho)$, of a b-symmetrical body with side of length ρ is proportional to the term ρ^N and to a coefficient $g_b(N)$ that depends on N:

$$V_N(b, \rho) = g_b(N)\, \rho^N. \quad (51)$$

The parameter b indicates the original equation (47) that defines the boundaries of the volume $V_N(b, \rho)$. Thus, for instance, from Eq. (2), we have $g_{b=1}(N) = 1/N!$.

Coming back to Eq. (50), we can manipulate $V_N((E - x^b)^{1/b})$ to obtain (the index b is omitted in the formule of V_N):

$$V_N((E - x^b)^{1/b}) = g_b(N) \left[(E - x^b)^{1/b}\right]^N = g_b(N)\, E^{\frac{N}{b}} \left(1 - \frac{x^b}{E}\right)^{\frac{N}{b}}. \quad (52)$$

If we suppose $E = N\epsilon$, then ϵ represents the mean value of x^b in the collectivity, that is, $\epsilon =< x^b >$. If N tends toward infinity, it results:

$$\lim_{N \gg 1} \left(1 - \frac{x^b}{E}\right)^{\frac{N}{b}} = e^{-x^b/b\epsilon}. \quad (53)$$

Thus,

$$V_N((E - x^b)^{1/b}) = V_N(E^{1/b})\, e^{-x^b/b\epsilon}. \quad (54)$$

Substituting this last expression in formula (50), the exact form for $f(x)$ is found in the thermodynamic limit $(N, E \to \infty)$:

$$f(x)dx = c_b\, \epsilon^{-1/b}\, e^{-x^b/b\epsilon} dx, \quad (55)$$

with c_b given by

$$c_b = \frac{g_b(N-1)}{g_b(N)\, N^{1/b}}. \quad (56)$$

Hence, the conjecture (48) is proved.

Doing a thermodinamical simile, we can calculate the dependence of ϵ on the temperature by differentiating the entropy with respect to the energy. The entropy can be written as $S = -kN \int_0^{\infty} f(x) \ln f(x)\, dx$, where $f(x)$ is given by Eq. (55) and k is the Boltzmann constant. If we recall that $\epsilon = E/N$, we obtain

$$S(E) = \frac{kN}{b} \ln\left(\frac{E}{N}\right) + \frac{kN}{b}(1 - b \ln c_b), \quad (57)$$

where it has been used that $\epsilon =< x^b >= \int_0^{\infty} x^b f(x) dx$.

The calculation of the temperature T gives

$$T^{-1} = \left(\frac{\partial S}{\partial E}\right)_N = \frac{kN}{bE} = \frac{k}{b\epsilon}. \quad (58)$$

Thus $\epsilon = kT/b$, a result that recovers the theorem of equipartition of energy for the quadratic case $b = 2$. The distribution for all b is finally obtained:

$$f(x)dx = c_b \left(\frac{b}{kT}\right)^{1/b} e^{-x^b/kT}dx. \tag{59}$$

4.1. General relationship between geometry and economic gas models

If we perform the change of variables $y = \epsilon^{-1/b}x$ in the normalization condition of $f(x)$, $\int_0^\infty f(x)dx = 1$, where $f(x)$ is expressed in (55), we find that

$$c_b = \left[\int_0^\infty e^{-y^b/b}\,dy\right]^{-1}. \tag{60}$$

If we introduce the new variable $z = y^b/b$, the distribution $f(x)$ as function of z reads:

$$f(z)dz = \frac{c_b}{b^{1-\frac{1}{b}}}\, z^{\frac{1}{b}-1}\, e^{-z}\, dz. \tag{61}$$

Let us observe that the Gamma function appears in the normalization condition,

$$\int_0^\infty f(z)dz = \frac{c_b}{b^{1-\frac{1}{b}}} \int_0^\infty z^{\frac{1}{b}-1}\, e^{-z}\, dz = \frac{c_b}{b^{1-\frac{1}{b}}}\, \Gamma\left(\frac{1}{b}\right) = 1. \tag{62}$$

This implies that

$$c_b = \frac{b^{1-\frac{1}{b}}}{\Gamma\left(\frac{1}{b}\right)}. \tag{63}$$

By using Mathematica the positive constant c_b is plotted versus b in Fig. 1. We see that $\lim_{b\to 0} c_b = \infty$, and that $\lim_{b\to\infty} c_b = 1$. The minimum of c_b is reached for $b = 3.1605$, taking the value $c_b = 0.7762$. Still further, we can calculate from Eq. (63) the asymptotic dependence of c_b on b:

$$\lim_{b\to 0} c_b = \sqrt{\frac{1}{2\pi}}\,\sqrt{b}\,e^{1/b}\left(1 - \frac{b}{12} + \cdots\right), \tag{64}$$

$$\lim_{b\to\infty} c_b = b^{-1/b}\left(1 + \frac{\gamma}{b} + \cdots\right), \tag{65}$$

where γ is the Euler constant, $\gamma = 0.5772$. The asymptotic function (64) is obtained after substituting in (63) the value of $\Gamma(1/b)$ by $(1/b - 1)!$, and performing the Stirling approximation on this last expression, knowing that $1/b \to \infty$. The function (65) is found after looking for the first Taylor expansion terms of the Gamma function around the origin $x = 0$. They can be derived from the Euler's reflection formula, $\Gamma(x)\Gamma(1-x) = \pi/\sin(\pi x)$. We obtain $\Gamma(x \to 0) = x^{-1} + \Gamma'(1) + \cdots$. From here, recalling that $\Gamma'(1) = -\gamma$, we get $\Gamma(1/b) = b - \gamma + \cdots$, when $b \to \infty$. Although this last term of the Taylor expansion, $-\gamma$, is negligible we maintain it in expression (65). The only minimum of c_b is reached for the solution $b = 3.1605$ of the equation $\psi(1/b) + \log b + b - 1 = 0$, where $\psi(\cdot)$ is the digamma function (see Fig. 1).

Figure 1. Normalization constant c_b versus b, calculated from Eq. (63). The asymptotic behavior is: $\lim_{b\to 0} c_b = \infty$, and $\lim_{b\to\infty} c_b = 1$. This last asymptote is represented by the dotted line. The minimum of c_b is reached for $b = 3.1605$, taking the value $c_b = 0.7762$.

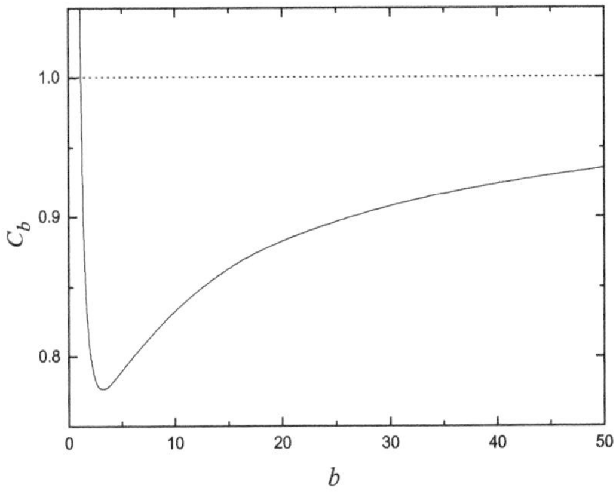

Let us now recall two interesting statistical economic models that display a statistical behavior given by distributions nearly to the form (61), that is, the standard Gamma distributions with shape parameter $1/b$,

$$f(z)dz = \frac{1}{\Gamma(\frac{1}{b})} z^{\frac{1}{b}-1} e^{-z} dz. \tag{66}$$

ECONOMIC MODEL A: The first one is the saving propensity model introduced by Chakraborti and Chakrabarti [11]. In this model a set of N economic agents, having each agent i (with $i = 1, 2, \cdots, N$) an amount of money, u_i, exchanges it under random binary (i, j) interactions, $(u_i, u_j) \to (u_i', u_j')$, by the following the exchange rule:

$$u_i' = \lambda u_i + \epsilon(1-\lambda)(u_i + u_j), \tag{67}$$

$$u_j' = \lambda u_j + \bar{\epsilon}(1-\lambda)(u_i + u_j), \tag{68}$$

with $\bar{\epsilon} = (1-\epsilon)$, and ϵ a random number in the interval $(0, 1)$. The parameter λ, with $0 < \lambda < 1$, is fixed, and represents the fraction of money saved before carrying out the transaction. Let us observe that money is conserved, i.e., $u_i + u_j = u_i' + u_j'$, hence in this model the economy is closed. Defining the parameter $n(\lambda)$ as

$$n(\lambda) = \frac{1+2\lambda}{1-\lambda}, \tag{69}$$

and scaling the wealth of the agents as $\bar{z} = nu/ < u >$, with $< u >$ representing the average money over the ensemble of agents, it is found that the asymptotic wealth distribution in this system is nearly obeying the standard Gamma distribution [14, 15]

$$f(\bar{z})d\bar{z} = \frac{1}{\Gamma(n)}\,\bar{z}^{n-1}\,e^{-\bar{z}}\,d\bar{z}. \tag{70}$$

The case $n = 1$, which means a null saving propensity, $\lambda = 0$, recovers the model of Dragulescu and Yakovenko [10] in which the Gibbs distribution is observed. If we compare Eqs. (70) and (66), a close relationship between this economic model and the geometrical problem solved in the former section can be established. It is enough to make

$$n = 1/b, \tag{71}$$

$$\bar{z} = z, \tag{72}$$

to have two equivalent systems. This means that, from Eq. (71), we can calculate b from the saving parameter λ with the formula

$$b = \frac{1-\lambda}{1+2\lambda}. \tag{73}$$

As λ takes its values in the interval $(0,1)$, then the parameter b also runs in the same interval $(0,1)$. On the other hand, recalling that $z = x^b/b\epsilon$, we can get the equivalent variable x from Eq. (72),

$$x = \left[\frac{\epsilon}{<u>}\,u\right]^{1/b}, \tag{74}$$

where ϵ is a free parameter that determines the mean value of x^b in the equivalent geometrical system. Formula (74) means to perform the change of variables $u_i \to x_i$, with $i = 1, 2, \cdots, N$, for all the particles/agents of the ensemble. Then, we conjecture that the economic system represented by the generic pair (λ, u), when it is transformed in the geometrical system given by the generic pair (b, x), as indicated by the rules (73) and (74), runs in an equiprobable form on the surface defined by the relationship (47), where the inequality has been transformed in equality. This last detail is due to the fact the economic system is closed, and then it conserves the total money, whose equivalent quantity in the geometrical problem is E. If the economic system were open, with an upper limit in the wealth, then the transformed system would evolve in an equiprobable way over the volume defined by the inequality (47), although its statistical behavior would continue to be the same as it has been proved for the cases $b = 1, 2$ in Refs. [8, 9].

ECONOMIC MODEL B: The second one is a model introduced in [16]. In this model a set of N economic agents, having each agent i (with $i = 1, 2, \cdots, N$) an amount of money, u_i, exchanges it under random binary (i, j) interactions, $(u_i, u_j) \to (u_i', u_j')$, by the following the exchange rule:

$$u_i' = u_i - \Delta u, \tag{75}$$

$$u_j' = u_j + \Delta u, \tag{76}$$

where

$$\Delta u = \eta(x_i - x_j)\,\epsilon\omega x_i - [1 - \eta(x_i - x_j)]\,\epsilon\omega x_j, \tag{77}$$

with ϵ a continuous uniform random number in the interval $(0,1)$. When this variable is transformed in a Bernouilli variable, i.e. a discrete uniform random variable taking on the values 0 or 1, we have the model studied by Angle [17], that gives very different asymptotic results. The exchange parameter, ω, represents the maximum fraction of wealth lost by one of the two interacting agents ($0 < \omega < 1$). Whether the agent who is going to loose part of the money is the i-th or the j-th agent, depends nonlinearly on $(x_i - x_j)$, and this is decided by the random dichotomous function $\eta(t)$: $\eta(t > 0) = 1$ (with additional probability 1/2) and $\eta(t < 0) = 0$ (with additional probability 1/2). Hence, when $x_i > x_j$, the value $\eta = 1$ produces a wealth transfer from agent i to agent j with probability 1/2, and when $x_i < x_j$, the value $\eta = 0$ produces a wealth transfer from agent j to agent i with probability 1/2. Defining in this case the parameter $n(\omega)$ as

$$n(\omega) = \frac{3 - 2\omega}{2\omega}, \tag{78}$$

and scaling the wealth of the agents as $\bar{z} = nu/ < u >$, with $< u >$ representing the average money over the ensemble of agents, it is found that the asymptotic wealth distribution in this system is nearly to fit the standard Gamma distribution [15, 16]

$$f(\bar{z})d\bar{z} = \frac{1}{\Gamma(n)}\, \bar{z}^{n-1}\, e^{-\bar{z}}\, d\bar{z}. \tag{79}$$

The case $n = 1$, which means an exchange parameter $\omega = 3/4$, recovers the model of Dragulescu and Yakovenko [10] in which the Gibbs distribution is observed. If we compare Eqs. (79) and (66), a close relationship between this economic model and the geometrical problem solved in the last section can be established. It is enough to make

$$n = 1/b, \tag{80}$$

$$\bar{z} = z, \tag{81}$$

to have two equivalent systems. This means that, from Eq. (80), we can calculate b from the exchange parameter ω with the formula

$$b = \frac{2\omega}{3 - 2\omega}. \tag{82}$$

As ω takes its values in the interval $(0, 1)$, then the parameter b runs in the interval $(0, 2)$. It is curious to observe that in this model the interval $\omega \in (3/4, 1)$ maps on $b \in (1, 2)$, a fact that does not occur in MODEL A. On the other hand, recalling that $z = x^b/b\epsilon$, we can get the equivalent variable x from Eq. (81),

$$x = \left[\frac{\epsilon}{< u >}\, u\right]^{1/b}. \tag{83}$$

where ϵ is a free parameter that determines the mean value of x^b in the equivalent geometrical system. Formula (83) means to perform the change of variables $u_i \to x_i$, with $i = 1, 2, \cdots, N$, for all the particles/agents of the ensemble. Then, also in this case, we conjecture that the economic system represented by the generic pair (λ, u), when it is transformed in the geometrical system given by the generic pair (b, x), as indicated by the rules (82) and (83), runs in an equiprobable form on the surface defined by the relationship (47), where the inequality has been transformed in equality. As explained above, this last detail is due to the fact the economic system is closed, and then it conserves the total money, whose equivalent quantity in the geometrical problem is E. If the economic system were open, with an upper limit in the wealth, then the transformed system would evolve in an equiprobable way over the volume defined by the inequality (47), although its statistical behavior would continue to be the same as it has been proved for the cases $b = 1, 2$ in Refs. [8, 9].

5. Other additional geometrical questions

As two collateral results, we address two additional problems in this section. The first one presents the finding of the general formula for the volume of a high-dimensional symmetrical body and the second one offers an alternative presentation of the canonical ensemble.

5.1. Formula for the volume of a high-dimensional body

We are concerned now with the asymptotic formula ($N \rightarrow \infty$) for the volume of the N-dimensional symmetrical body enclosed by the surface

$$x_1^b + x_2^b + \cdots + x_{N-1}^b + x_N^b = E. \tag{84}$$

The linear dimension ρ of this volume, i.e., the length of one of its sides verifies $\rho \sim E^{1/b}$. As argued in Eq. (51), the N-dimensional volume, $V_N(b,\rho)$, is proportional to the term ρ^N and to a coefficient $g_b(N)$ that depends on N. Thus,

$$V_N(b,\rho) = g_b(N)\,\rho^N, \tag{85}$$

where the characteristic b indicates the particular boundary given by equation (84).

For instance, from Equation (2), we can write in a formal way:

$$g_{b=1}(N) = \frac{1^{\frac{N}{1}}}{\Gamma(\frac{N}{1}+1)}. \tag{86}$$

From Eq. (23), if we take the diameter, $\rho = 2R$, as the linear dimension of the N-sphere, we obtain:

$$g_{b=2}(N) = \frac{\left(\frac{\pi}{4}\right)^{\frac{N}{2}}}{\Gamma\left(\frac{N}{2}+1\right)}. \tag{87}$$

These expressions (86) and (87) suggest a possible general formula for the factor $g_b(N)$, let us say

$$g_b(N) = \frac{a^{\frac{N}{b}}}{\Gamma\left(\frac{N}{b}+1\right)}, \tag{88}$$

where a is a b-dependent constant to be determined. For example, $a = 1$ for $b = 1$ and $a = \pi/4$ for $b = 2$.

In order to find the dependence of a on the parameter b, the regime $N \rightarrow \infty$ is supposed. Applying Stirling approximation for the factorial $(\frac{N}{b})!$ in the denominator of expression (88), and inserting it in expression (56), it is straightforward to find out the relationship:

$$c_b = (ab)^{-1/b}. \tag{89}$$

From here and formula (63), we get:

$$a = \left[\Gamma\left(\frac{1}{b}+1\right)\right]^b, \tag{90}$$

Figure 2. The factor $g_b(N)$ versus b for $N = 10, 40, 100$, calculated from Eq. (91). Observe that $g_b(N) = 0$ for $b = 0$, and $\lim_{b\to\infty} g_b(N) = 1$.

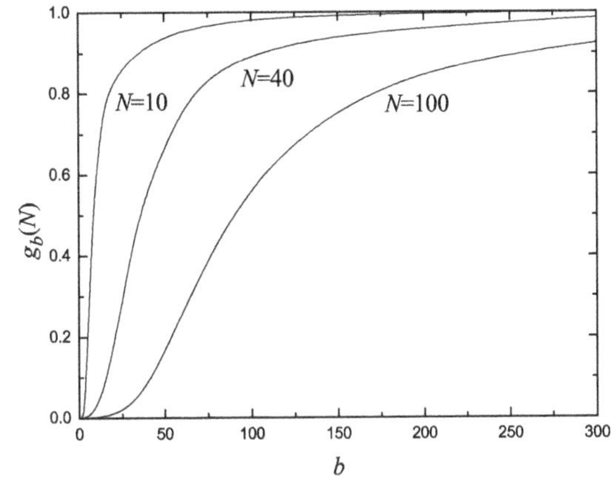

that recovers the exact results for $b = 1, 2$. The behavior of a is monotonous decreasing when b is varied from $b = 0$, where a diverges as $a \sim 1/b + \cdots$, up to the limit $b \to \infty$, where a decays asymptotically toward the value $a_\infty = e^{-\gamma} = 0.5614$.

Hence, the formula for $g_b(N)$ is obtained:

$$g_b(N) = \frac{\Gamma\left(\frac{1}{b}+1\right)^N}{\Gamma\left(\frac{N}{b}+1\right)}, \tag{91}$$

It would be also possible to multiply this last expression (91) by a general polynomial $K(N)$ in the variable N, and all the derivation done from Eq. (88) would continue to be correct. We omit this possibility in our calculations. For a fixed N, we have that $g_b(N)$ increases monotonously from $g_b(N) = 0$, for $b = 0$, up to $g_b(N) = 1$, in the limit $b \to \infty$ (see Fig. 2). For a fixed b, we have that $g_b(N)$ decreases monotonously from $g_b(N) = 1$, for $N = 1$, up to $g_b(N) = 0$, in the limit $N \to \infty$ (see Fig. 3).

The final result, that has been shown to be valid for any N [18], for the volume of an N-dimensional symmetrical body of characteristic b given by the boundary (84) reads:

$$V_N(b,\rho) = \frac{\Gamma\left(\frac{1}{b}+1\right)^N}{\Gamma\left(\frac{N}{b}+1\right)} \rho^N, \tag{92}$$

with $\rho \sim E^{1/b}$.

Figure 3. The factor $g_b(N)$ versus N for $b = 10, 40, 100$, calculated from Eq. (91). Observe that $g_b(N) = 1$ for $N = 1$, and $\lim_{N\to\infty} g_b(N) = 0$.

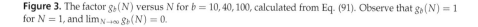

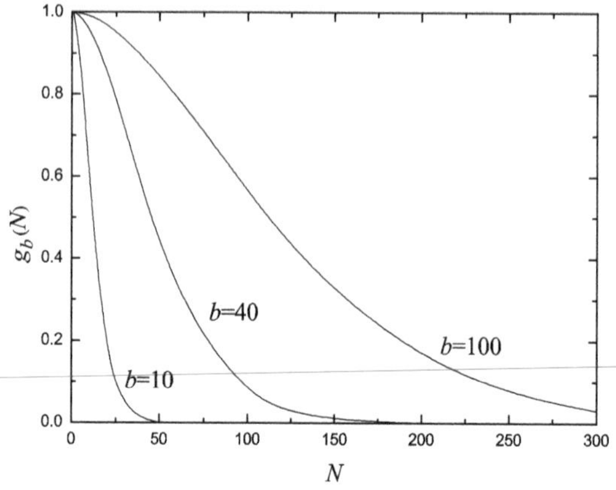

5.2. A microcanonical image of the canonical ensemble

From Section 2, here we offer a different image of the usual presentation that can be found in the literature [1] about the canonical ensemble.

Let us suppose that a system with mean energy $\bar{E}$, and in thermal equilibrium with a heat reservoir, is observed during a very long period τ of time. Let E_i be the energy of the system at time i. Then we have:

$$E_1 + E_2 + \cdots + E_{\tau-1} + E_\tau = \tau \cdot \bar{E}. \tag{93}$$

If we repeat this process of observation a huge number (toward infinity) of times, the different vectors of measurements, $(E_1, E_2, \ldots, E_{\tau-1}, E_\tau)$, with $0 \le E_i \le \tau \cdot \bar{E}$, will finish by covering equiprobably the whole surface of the τ-dimensional hyperplane given by Eq. (93). If it is now taken the limit $\tau \to \infty$, the asymptotic probability $p(E)$ of finding the system with an energy E (where the index i has been removed),

$$p(E) \sim e^{-E/\bar{E}}, \tag{94}$$

is found by means of the geometrical arguments exposed in Section 2 [8]. Doing a thermodynamic simile, the temperature T can also be calculated. It is obtained that

$$\bar{E} = kT. \tag{95}$$

The *stamp* of the canonical ensemble, namely, the Boltzmann factor,

$$p(E) \sim e^{-E/kT}, \tag{96}$$

is finally recovered from this new image of the canonical ensemble.

6. Conclusion

In summary, this work has presented a straightforward geometrical argument that in a certain way recalls us the equivalence between the canonical and the microcanonical ensembles in the thermodynamic limit for the particular context of physical sciences. In the more general context of homogeneous multi-agent systems, we conclude by highlighting the statistical equivalence of the volume-based and surface-based calculations in this type of systems.

Thus, we have shown that the Boltzmann factor or the Maxwellian distribution describe the general statistical behavior of each small part of a multi-component system in equilibrium whose components or parts are given by a set of random linear or quadratic variables, respectively, that satisfy an additive constraint, in the form of a conservation law (closed systems) or in the form of an upper limit (open systems), and that reach the equiprobability when they decay to equilibrium.

Let us remark that these calculations do not need the knowledge of the exact or microscopic randomization mechanisms of the multi-agent system in order to attain the equiprobability. In some cases, it can be reached by random forces [6], in other cases by chaotic [13, 19] or deterministic [12] causes. Evidently, the proof that these mechanisms generate equiprobability is not a trivial task and it remains as a typical challenge in this kind of problems.

The derivation of the equilibrium distribution for open systems in a general context has also been presented by considering a general multi-agent system verifying an additive constraint. Its statistical behavior has been derived from geometrical arguments. Thus, the Maxwellian and the Boltzmann-Gibbs distributions are particular cases of this type of systems. Also, other multi-agent economy models, such as the Dragalescu and Yakovenko's model [10], the Chakraborti and Chakrabarti's model [11] and the modified Angle's model [16], show similar statistical behaviors than our general geometrical system. This fact has fostered our particular geometrical interpretation of all those models.

We hope that this framework can be useful to establish other possible relationships between the statistics of multi-agent systems and the geometry associated to such systems in equilibrium.

Author details

López-Ruiz Ricardo
Department of Computer Science, Faculty of Science, Universidad de Zaragoza, Zaragoza, Spain
Also at BIFI, Institute for Biocomputation and Physics of Complex Systems, Universidad de Zaragoza, Zaragoza, Spain

Sañudo Jaime
Department of Physics, Faculty of Science, Universidad de Extremadura, Badajoz, Spain
Also at BIFI, Institute for Biocomputation and Physics of Complex Systems, Universidad de Zaragoza, Zaragoza, Spain

7. References

[1] Huang K (1987) Statistical Mechanics. John Wiley & Sons, New York; Munster A (1969) Statistical Thermodynamics (volume I). Springer-Verlag, Berlin; Ditlevsen PD (2004) Turbulence and Climate Dynamics. Frydendal, Copenhagen.

[2] Jaynes ET (1957) Information theory and statistical mechanics. Phys Rev E 106:620-630.
[3] Lopez JL, Lopez-Ruiz R, Calbet X (2012) Exponential wealth distribution in a random market. A rigorous explanation. J Math Anal App 386:195-204.
[4] Shivanian E, Lopez-Ruiz R (2012) A new model for ideal gases. Decay to the Maxwellian distribution. Physica A 391:2600-2607.
[5] Mantegna R, Stanley HE (1999) An Introduction to Econophysics: Correlations and Complexity in Finance. Cambridge University Press, Cambridge, U.K.
[6] Yakovenko VM (2009) Econophysics, Statistical Mechanics Approach to. In the Encyclopedia of Complexity and System Science, Meyers, R.A. Ed., Springer, Berlin, pp. 2800-2826.
[7] Lopez-Ruiz R, Sañudo J, Calbet X (2009) Equiprobability, entropy, Gamma distributions and other geometrical questions in multi-agent systems. Entropy 11:959-971.
[8] Lopez-Ruiz R, Sañudo J, Calbet X (2008) Geometrical derivation of the Boltzmann factor. Am J Phys 76:780-781.
[9] Lopez-Ruiz R, Calbet X (2007) Derivation of the Maxwellian distribution from the microcanonical ensemble. Am J Phys 75:752-753.
[10] Dragulescu A, Yakovenko VM (2000,2001) Statistical mechanics of money. Eur Phys J B17:723-729; Evidence for the exponential distribution of income in the USA. Eur Phys J B20:585-589.
[11] Chakraborti A, Chakrabarti BK (2000) Statistical mechanics of money: How saving propensity affects its distribution. Eur Phys J B17:167-170.
[12] Gonzalez-Estevez J, Cosenza MG, Lopez-Ruiz R, Sanchez JR (2008) Pareto and Boltzmann-Gibbs behaviors in a deterministic multi-agent system. Physica A 387:4637-4642.
[13] Pellicer-Lostao C, Lopez-Ruiz R (2010) A chaotic gas-like model for trading markets. J. Computational Science 1:24-32.
[14] Patriarca M, Chakraborti A, Kaski K (2004) Statistical model with a standard Gamma distribution. Phys Rev E 70:016104(5).
[15] Calbet X, Lopez JL, Lopez-Ruiz R (2011) Equilibrium distributions and relaxation times in gas-like economic models: an analytical derivation. Phys Rev E 83:036108(7).
[16] Patriarca M, Heinsalu E, Chakraborti A (2006) The ABCD's of statistical many-agent economy models. Preprint at arXiv:physics/0611245.
[17] Angle J (2006) The inequality process as a wealth maximizing process. Physica A 367:388-414, and references therein.
[18] Toral R (2009) Comment on "Equiprobability, Entropy, Gamma Distributions and Other Geometrical Questions in Multi-Agent Systems. Entropy 11:959-971". Entropy 11:1121-1122.
[19] Bullard JB, Butler A (1992) Nonlinearity and chaos in economic models: Implications for policy decisions. Royal Econ Soc 103:849-867.

Chapter 6

Stochastic Control for Jump Diffusions *

Jingtao Shi

Additional information is available at the end of the chapter

1. Introduction

In this chapter, we will discuss the stochastic optimal control problem for jump diffusions. That is, the controlled stochastic system is driven by both Brownian motion and Poisson random measure and the controller wants to minimize/maximize some cost functional subject to the above stated state equation (stochastic control system) over the admissible control set. This kind of stochastic optimal control problems can be encountered naturally when some sudden and rare breaks take place, such as in the practical stock price market. An admissible control is called optimal if it achieves the infimum/supremum of the cost functional and the corresponding state variable and the cost functional are called the optimal trajectory and the value function, respectively.

It is well-known that Pontryagin's maximum principle (MP for short) and Bellman's dynamic programming principle (DPP for short) are the two principal and most commonly used approaches in solving stochastic optimal control problems. In the statement of maximum principle, the necessary condition of optimality is given. This condition is called the maximum condition which is always given by some Hamiltonian function. The Hamiltonian function is defined with respect to the system state variable and some adjoint variables. The equation that the adjoint variables satisfy is called adjoint equation, which is one or two backward stochastic differential equations (BSDEs for short) of [13]'s type. The system which consists of the adjoint equation, the original state equation, and the maximum condition is referred to as a generalized Hamiltonian system. On the other hand, the basic idea of dynamic programming principle is to consider a family of stochastic optimal control problems with different initial time and states and establish relationships among these problems via the so-called Hailton-Jacobi-Bellman (HJB for short) equation, which is a nonlinear second-order partial differential equation (PDE for short). If the HJB equation is solvable, we can obtain an optimal control by taking the maximizer/miminizer of the generalized Hamiltonian function involved in the HJB equation. To a great extent these two approaches have been developed separately and independently during the research in stochastic optimal control problems.

*The main content of this chapter is from the following published article paper: Shi, J.T., & Wu, Z. (2011). Relationship between MP and DPP for the stochastic optimal control problem of jump diffusions. *Applied mathematics and Optimization*, Vol. 63, 151–189.

Hence, a natural question arises: Are there any relations between these two methods? In fact, the relationship between MP and DPP is essentially the relationship between the adjoint processes and the value function, or the Hamiltonian systems and the HJB equations or even more generally, the relationship between stochastic differential equations (SDEs for short) and PDEs. Such a topic was intuitively discussed by [5], [4] and [9]. However, an important issue in studying the problem is that the derivatives of the value functions are unavoidably involved in these results. In fact, the value functions are usually not necessarily smooth. [19] first obtained the nonsmooth version of the relationship between MP and DPP using the viscosity solution and the second-order adjoint equation. See also the book by [18].

The aim of this chapter is to establish the relationship between MP and DPP within the framework of viscosity solutions in the jump diffusion setting. In this case, the state trajectory is described by a stochastic differential equation with Poisson jumps (SDEP for short). That is to say, the system noise (or the uncertainty of the problem) comes from a Brownian motion and a Poisson random measure. See [15] for theory and applications of this kind of equations. [16] proved the general MP where the control variable is allowed into both diffusion and jump coefficients. HJB equation for optimal control of jump diffusions can be seen in [12], which here is a second-order partial integral-differential equation (PIDE for short). [7] gave a sufficient MP by employing Arrow's generalization of the Mangasarian sufficient condition to the jump diffusion setting. Moreover, on the assumption that the value function is smooth, they showed the adjoint processes' connections to the value function. Let us state some results of [7] in detail with a slight modification to adapt to our setting.

A sufficient MP was proved to say that, for any admissible pair $(\bar{x}^{s,y;\bar{u}}(\cdot),\bar{u}(\cdot))$, if there exists an adapted solution $(p(\cdot),q(\cdot),\gamma(\cdot,\cdot))$ of the following adjoint equtaion

$$\begin{cases} -dp(t) = H_x(t,\bar{x}^{s,y;\bar{u}}(t),\bar{u}(t),p(t),q(t),\gamma(t,\cdot))dt - q(t)dW(t) \\ \qquad\qquad - \int_{\mathbf{E}} \gamma(t,e)\tilde{N}(dedt),\ t\in[0,T], \\ p(T) = -h_x(\bar{x}^{s,y;\bar{u}}(T)), \end{cases} \tag{1}$$

which is a BSDE with Poisson jumps (BSDEP for short) such that

$$H(t,\bar{x}^{s,y;\bar{u}}(t),\bar{u}(t),p(t),q(t),\gamma(t,\cdot)) = \sup_{u\in\mathbf{U}} H(t,\bar{x}^{s,y;\bar{u}}(t),u,p(t),q(t),\gamma(t,\cdot))$$

for all $t\in[0,T]$ and that

$$\hat{H}(x) := \max_{u\in\mathbf{U}} H(t,x,u,p(t),q(t),\gamma(t,\cdot))$$

exists and is a concave function of x for all $t\in[0,T]$, then $(\bar{x}^{s,y;\bar{u}}(\cdot),\bar{u}(\cdot))$ is an optimal pair. In the above, the Hamiltonian function $H:[0,T]\times\mathbf{R}^n\times\mathbf{U}\times\mathbf{R}^n\times\mathbf{R}^{n\times d}\times\mathcal{L}^2(\mathbf{E},\mathcal{B}(\mathbf{E}),\pi;\mathbf{R}^n)\to\mathbf{R}$ is defined as

$$H(t,x,u,p,q,\gamma(\cdot)) := f(t,x,u) + \langle p, b(t,x,u)\rangle + \mathrm{tr}\{\sigma(t,x,u)^\top q\} + \int_{\mathbf{E}} \langle \gamma(e), c(t,x,u,e)\rangle \pi(de).$$

On the other hand, a DPP asserted that if the value function $V(\cdot,\cdot)$ belongs to $C^{1,2}([0,T]\times\mathbf{R}^n)$, then it satisfies the following HJB equation

$$\sup_{u\in\mathbf{U}} F(t,x,u) = F(t,x,\bar{u}(t)) = 0 \tag{2}$$

where

$$F(t,x,u) := f(t,x,u) + V_t(t,x) + \langle V_x(t,x), b(t,x,u)\rangle + \frac{1}{2}\mathrm{tr}\Big\{V_{xx}(t,x)\sigma(t,x,u)\sigma(t,x,u)^\top\Big\}$$
$$+ \int_{\mathbf{E}} \big[V(t,x+c(t,x,u,e)) - V(t,x) - \langle V_x(t,x), c(t,x,u,e)\rangle\big]\pi(de).$$

Moreover, Theorem 2.1 of [7] says that if $V(\cdot,\cdot)$ belongs to $C^{1,3}([0,T]\times\mathbf{R}^n)$, then the processes defined by

$$\begin{cases} p(t) = -V_x(t,\bar{x}^{s,y;\bar{u}}(t)), \\ q(t) = -V_{xx}(t,\bar{x}^{s,y;\bar{u}}(t))\sigma(t,\bar{x}^{s,y;\bar{u}}(t),\bar{u}(t)), \\ \gamma(t,\cdot) = -V_x(t,\bar{x}^{s,y;\bar{u}}(t-) + c(t,\bar{x}^{s,y;\bar{u}}(t-),\bar{u}(t)) + V_x(t,\bar{x}^{s,y;\bar{u}}(t-)), \end{cases} \tag{3}$$

solve the adjoint equation (1).

However, it seems that the above HJB equation (2) and the relationship (3) lack generality, since they require the value function to be smooth, which is not true even in the simplest case; see Example 3.2 of this chapter. This is an important *gap* in the literature [7]. The aim of this chapter is to bridge this gap by employing the notion of semijets evoked in defining the viscosity solutions.

The contribution of this chapter is as follows. Firstly, we give some basic properties of the value function and prove that the DPP still holds in our jump diffusion setting. Then we give the corresponding generalized HJB equation which now is a second-order PIDE. Secondly, we investigate the relationship between MP and DPP without assuming the continuous differentiablity of the value function. We obtain the relationship among the adjoint processes, the generalized Hamiltonian and the value function by employing the notions of the set-valued semijets evoked in defining the viscosity solutions, which is now interpreted as a set inclusion form among subjet, superjet of the value function, set contain adjoint processes and some "$\mathcal{G}$-function" (see the definition in Section 2). It is worth to pointed out that the controlled jump diffusions bring much technique difficulty to obtain the above results. In fact, the solution of the control system is not continuous with jump diffusions. We overcome these difficulty and get the desired results in this chapter which have wide applicable background.

The rest of this chapter is organized as follows. In Section 2, for stochastic optimal control problem of jump diffusions, we give some basic properties of the value function and then set out the corresponding DPP and MP, respectively. In Section 3, the relationship between MP and DPP is proved using the notion of viscosity solutions of PIDEs. Some concluding remarks are given in Section 4.

2. Problem statement and preliminaries

Throughout this chapter, we denote by $\mathbf{R}^n$ the space of n-dimensional Euclidean space, by $\mathbf{R}^{n\times d}$ the space of matrices with order $n\times d$, by $\mathcal{S}^n$ the space of symmetric matrices with order $n\times n$. $\langle\cdot,\cdot\rangle$ and $|\cdot|$ denote the scalar product and norm in the Euclidean space, respectively. $\top$ appearing in the superscripts denotes the transpose of a matrix. $a\vee b$ denotes $\max\{a,b\}$. C always denotes some positive constant.

Let $\mathbf{E} \subset \mathbf{R}^l$ be a nonempty Borel set equipped with its Borel field $\mathcal{B}(\mathbf{E})$. Let $\pi(\cdot)$ be a bounded positive measure on $(\mathbf{E}, \mathcal{B}(\mathbf{E}))$. We denote by $\mathcal{L}^2(\mathbf{E}, \mathcal{B}(\mathbf{E}), \pi; \mathbf{R}^n)$ or $\mathcal{L}^2$ the set of square integrable functions $k(\cdot) : \mathbf{E} \to \mathbf{R}^n$ such that $||k(\cdot)||^2_{\mathcal{L}^2} := \int_{\mathbf{E}} |k(e)|^2 \pi(de) < \infty$.

Let $T > 0$ and let $(\Omega, \mathcal{F}, \mathbf{P})$ be a complete probability space, equipped with a d-dimensional standard Brownian motion $\{W(t)\}_{0 \le t \le T}$ and a Poisson random measure $N(\cdot,\cdot)$ independent of $W(\cdot)$ with the intensity measure $\hat{N}(dedt) = \pi(de)dt$. We write $\tilde{N}(dedt) := N(dedt) - \pi(de)dt$ for the compensated Poisson martingale measure.

For a given $s \in [0, T)$, we suppose the filtration $\{\mathcal{F}^s_t\}_{s \le t \le T}$ is generated as the following

$$\mathcal{F}^s_t := \sigma\{N(\mathbf{A} \times (s, r]); s \le r \le t, \mathbf{A} \in \mathcal{B}(\mathbf{E})\} \bigvee \sigma\{W(r) - W(s); s \le r \le t\} \bigvee \mathcal{N},$$

where $\mathcal{N}$ contains all $\mathbf{P}$-null sets in $\mathcal{F}$ and $\sigma_1 \bigvee \sigma_2$ denotes the σ-field generated by $\sigma_1 \cup \sigma_2$. In particular, if $s = 0$ we write $\mathcal{F}_t \equiv \mathcal{F}^s_t$.

Let $\mathbf{U}$ be a nonempty Borel subset of $\mathbf{R}^k$. For any initial time $s \in [0, T)$ and initial state $y \in \mathbf{R}^n$, we consider the following stochastic control system which is called a *controlled jump diffusion process*

$$\begin{cases} dx^{s,y;u}(t) = b(t, x^{s,y;u}(t), u(t))dt + \sigma(t, x^{s,y;u}(t), u(t))dW(t) \\ \qquad\qquad + \int_{\mathbf{E}} c(t, x^{s,y;u}(t-), u(t), e)\tilde{N}(dedt), \quad t \in (s, T], \\ x^{s,y;u}(s) = y. \end{cases} \tag{4}$$

Here $b : [0, T] \times \mathbf{R}^n \times \mathbf{U} \to \mathbf{R}^n, \sigma : [0, T] \times \mathbf{R}^n \times \mathbf{U} \to \mathbf{R}^{n \times d}, c : [0, T] \times \mathbf{R}^n \times \mathbf{U} \times \mathbf{E} \to \mathbf{R}^n$ are given functions.

For a given $s \in [0, T)$, we denote by $\mathcal{U}[s, T]$ the set of $\mathbf{U}$-valued $\mathcal{F}^s_t$-predictable processes. For given $u(\cdot) \in \mathcal{U}[s, T]$ and $y \in \mathbf{R}^n$, a $\mathbf{R}^n$-valued process $x^{s,y;u}(\cdot)$ is called a solution of (4) if it is an $\mathcal{F}^s_t$-adapted RCLL (i.e., right-continuous with left-hand limits) process such that (4) holds. It is called the *state trajectory* corresponding to the control $u(\cdot) \in \mathcal{U}[s, T]$ with initial state y. We refer to such $u(\cdot) \in \mathcal{U}[s, T]$ as an *admissible control* and $(x^{s,y;u}(\cdot), u(\cdot))$ as an *admissible pair*. For any $s \in [0, T)$, we introduce the following notations.

$$L^2(\Omega, \mathcal{F}^s_T; \mathbf{R}^n) := \Big\{\mathbf{R}^n\text{-valued } \mathcal{F}^s_T\text{-measurable random variables } \xi;\ \mathbb{E}|\xi|^2 < \infty\Big\},$$

$$L^2_{\mathcal{F}}([s, T]; \mathbf{R}^n) := \Big\{\mathbf{R}^n\text{-valued } \mathcal{F}^s_t\text{-adapted processes } \varphi(t);\ \mathbb{E}\int_s^T |\varphi(t)|^2 dt < \infty\Big\},$$

$$L^2_{\mathcal{F},p}([s, T]; \mathbf{R}^n) := \Big\{\mathbf{R}^n\text{-valued } \mathcal{F}^s_t\text{-predictable processes } \phi(t);\ \mathbb{E}\int_s^T |\phi(t)|^2 dt < \infty\Big\},$$

$$F^2_p([s, T]; \mathbf{R}^n) := \Big\{\mathbf{R}^n\text{-valued } \mathcal{F}^s_t\text{-predictable vector processes } \psi(t, e) \text{ defined on}$$
$$\Omega \times [0, T] \times \mathbf{E};\ \mathbb{E}\int_s^T \int_{\mathbf{E}} |\psi(t, e)|^2 \pi(de)dt < \infty\Big\}.$$

We consider the following *cost functional*

$$J(s, y; u(\cdot)) = \mathbb{E}\Big[\int_s^T f(t, x^{s,y;u}(t), u(t))dt + h(x^{s,y;u}(T))\Big], \tag{5}$$

where $f : [0,T] \times \mathbf{R}^n \times \mathbf{U} \to \mathbf{R}, h : \mathbf{R}^n \to \mathbf{R}$ are given functions. For given $(s,y) \in [0,T) \times \mathbf{R}^n$, the *stochastic optimal control problem* is to minimize (5) subject to (4) over $\mathcal{U}[s,T]$. An admissible pair $(\bar{x}^{s,y;\bar{u}}(\cdot), \bar{u}(\cdot))$ is called *optimal* if $\bar{u}(\cdot)$ achieves the infimum of $J(s,y;u(\cdot))$ over $\mathcal{U}[s,T]$. We define the *value function* $V : [0,T] \times \mathbf{R}^n \to \mathbf{R}$ as

$$\begin{cases} V(s,y) = \inf\limits_{u(\cdot)\in\mathcal{U}[s,T])} J(s,y;u(\cdot)), \quad \forall (s,y) \in [0,T) \times \mathbf{R}^n, \\ V(T,y) = h(y), \quad \forall y \in \mathbf{R}^n. \end{cases} \tag{6}$$

In this section, we first give some basic properties of the value function. Then we prove that the DPP still holds and introduce the generalized HJB equation and the generalized Hamiltonian function. The idea of proof is originated from [17], [18] while on some different assumptions. Then we introduce the Hamiltonian function, the $\mathcal{H}$-function and adjoint processes, then give the MP which is a special case of [16].

We first discuss the DPP and make the following assumptions.

(H1) b, σ, c are uniformly continuous in (t,x,u). There exists a constant $C > 0$, such that

$$\begin{cases} |b(t,x,u) - b(t,\hat{x},\hat{u})| + |\sigma(t,x,u) - \sigma(t,\hat{x},\hat{u})| \\ \qquad + ||c(t,x,u,\cdot) - c(t,\hat{x},\hat{u},\cdot)||_{\mathcal{L}^2} \leq C(|x-\hat{x}| + |u-\hat{u}|), \\ |b(t,x,u)| + |\sigma(t,x,u)| + ||c(t,x,u,\cdot)||_{\mathcal{L}^2} \leq C(1+|x|), \quad \forall t \in [0,T], x, \hat{x} \in \mathbf{R}^n, u, \hat{u} \in \mathbf{U}. \end{cases}$$

(H2) f, h are uniformly continuous in (t,x,u). There exists a constant $C > 0$ and an increasing, continuous function $\bar{\omega}_0 : [0,\infty) \times [0,\infty) \to [0,\infty)$ which satisfies $\bar{\omega}_0(r,0) = 0, \forall r \geq 0$, such that

$$\begin{cases} |f(t,x,u) - f(t,\hat{x},u)| + |h(x) - h(\hat{x})| \leq \bar{\omega}_0(|x| \vee |\hat{x}|, |x-\hat{x}|), \\ |f(t,0,u)|, |h(0)| \leq C, \quad \forall t \in [0,T], x, \hat{x} \in \mathbf{R}^n, u \in \mathbf{U}. \end{cases}$$

Obviously, under assumption **(H1)**, for any $u(\cdot) \in \mathcal{U}[s,T]$, SDEP (4) admits a unique solution $x^{s,y;u}(\cdot)$ (see [10]).

Remark 2.1 We point out that assumption **(H2)** on f, h allows them to have polynomial (in particular, quadratic) growth in x, provided that b, σ, c have linear growth. A typical example is the stochastic linear quadratic (LQ for short) problem. Note that **(H2)** is different from assumptions (2.5), pp. 4 of [14] and assumption **(S2)′**, pp. 248 of [18], where global Lipschitz condition is both imposed thus f, h are global linear growth in x.

We need the following lemma.

Lemma 2.1 *Let* **(H1)** *hold. For $k = 2,4$, there exists $C > 0$ such that for any $0 \leq s, \hat{s} \leq T, y, \hat{y} \in \mathbf{R}^n, u(\cdot) \in \mathcal{U}[s,T]$,*

$$\mathbb{E}|x^{s,y,u}(t)|^k \leq C(1+|y|^k), \quad t \in [s,T], \tag{7}$$

$$\mathbb{E}|x^{s,y,u}(t) - y|^k \leq C(1+|y|^k)(T-s)^{\frac{k}{2}}, \quad t \in [s,T], \tag{8}$$

$$\mathbb{E}\Big[\sup_{s\leq t\leq T} |x^{s,y,u}(t) - y|\Big]^k \leq C(1+|y|^k)(T-s)^{\frac{k}{2}}, \tag{9}$$

$$\mathbb{E}|x^{s,y,u}(t) - x^{s,\hat{y},u}(t)|^k \leq C|y-\hat{y}|^k, \quad t \in [s,T], \tag{10}$$

$$\mathbb{E}|x^{s,y,u}(t) - x^{\hat{s},y,u}(t)|^k \leq C(1+|y|)|s-\hat{s}|^{\frac{k}{2}}, \quad t \in [s \vee \hat{s}, T]. \tag{11}$$

Estimates of the moments for SDEPs are proved in Lemma 3.1 of [14] for $k \in [0,2]$. In fact, under assumption **(H1)** we can easily extend his result to the case $k = 2,4$ by virtue of Buckholder-Davis-Gundy's inquality. We leave the detail of the proof to the interested reader.

We give some basic continuity properties of the value function V. The proof is similar to Proposition 2.2, Chapter 2 of [17]. We omit the detail.

Proposition 2.1 *Let* **(H1)**, **(H2)** *hold. Then there exist increasing, continuous functions* $\bar{\omega}_1 : [0,\infty) \to [0,\infty)$, $\bar{\omega}_2 : [0,\infty) \times [0,\infty) \to [0,\infty)$ *which satisfies* $\bar{\omega}_2(r,0) = 0, \forall r \geq 0$, *such that*

$$|V(s,y)| \leq \bar{\omega}_1(|y|), \quad \forall (s,y) \in [0,T) \times \mathbf{R}^n, \tag{12}$$

$$|V(s,y) - V(\hat{s},\hat{y})| \leq \bar{\omega}_2(|y| \vee |\hat{y}|, |y-\hat{y}| + |\hat{s}-s|^{\frac{1}{2}}), \quad \forall s,\hat{s} \in [0,T), y,\hat{y} \in \mathbf{R}^n. \tag{13}$$

The following result is a version of *Bellman's principle of optimality* for jump diffusions.

Theorem 2.1 *Suppose that* **(H1), (H2)** *hold. Then for any* $(s,y) \in [0,T) \times \mathbf{R}^n$,

$$V(s,y) = \inf_{u(\cdot)\in\mathcal{U}[s,T]} \mathbb{E}\Big[\int_s^{\hat{s}} f(t, x^{s,y;u}(t), u(t))dt + V(\hat{s}, x^{s,y;u}(\hat{s}))\Big], \quad \forall 0 \leq s \leq \hat{s} \leq T. \tag{14}$$

The proof is similar to Theorem 3.3, Chapter 4 of [18] or Proposition 3.2 of [14]. We omit it here.

The following result is to get the *generalized HJB equation* and its proof is similar to Proposition 3.4, Chapter 4 of [18].

Theorem 2.2 *Suppose that* **(H1), (H2)** *hold and the value function* $V \in C^{1,2}([0,T] \times \mathbf{R}^n)$. *Then* V *is a solution of the following generalized HJB equation which is a second-order PIDE:*

$$\begin{cases} -V_t(t,x) + \sup_{u\in\mathbf{U}} G(t,x,u,-V(t,x),-V_x(t,x),-V_{xx}(t,x)) = 0, & (t,x) \in [0,T) \times \mathbf{R}^n, \\ V(T,x) = h(x), & x \in \mathbf{R}^n, \end{cases} \tag{15}$$

where, associated with a $\Psi \in C^{1,2}([0,T] \times \mathbf{R}^n)$, *the generalized Hamiltonian function* $G : [0,T] \times \mathbf{R}^n \times \mathbf{U} \times \mathbf{R} \times \mathbf{R}^n \times \mathcal{S}^n \to \mathbf{R}$ is defined by

$$\begin{aligned} &G(t,x,u,\Psi(t,x),\Psi_x(t,x),\Psi_{xx}(t,x)) \\ &:= \langle \Psi_x(t,x), b(t,x,u)\rangle + \frac{1}{2}\mathrm{tr}\Big\{\Psi_{xx}(t,x)\sigma(t,x,u)\sigma(t,x,u)^\top\Big\} \\ &\quad - f(t,x,u) - \int_{\mathbf{E}} \big[\Psi(t,x+c(t,x,u,e)) - \Psi(t,x) + \langle \Psi_x(t,x), c(t,x,u,e)\rangle\big]\pi(de). \end{aligned} \tag{16}$$

In the following, we discuss the stochastic MP and need the following hypothesis.

(H3) b,σ,c,f,h are twice continuously differentiable in x, and $b_x, b_{xx}, \sigma_x, \sigma_{xx}, f_{xx}, h_{xx}$, $||c_x(\cdot)||_{\mathcal{L}^2}, ||c_{xx}(\cdot)||_{\mathcal{L}^2}$ are bounded. There exists a constant $C > 0$ and a modulus of continuity $\bar{\omega} : [0,\infty) \to [0,\infty)$ such that

$$\begin{cases} |b_x(t,x,u)-b_x(t,\hat{x},\hat{u})|+|\sigma_x(t,x,u)-\sigma_x(t,\hat{x},\hat{u})|\leq C|x-\hat{x}|+\bar{\omega}(|u-\hat{u}|),\\ ||c_x(t,x,u,\cdot)-c_x(t,\hat{x},\hat{u},\cdot)||_{\mathcal{L}^2}\leq C|x-\hat{x}|+\bar{\omega}(|u-\hat{u}|),\\ |b_{xx}(t,x,u)-b_{xx}(t,\hat{x},\hat{u})|+|\sigma_{xx}(t,x,u)-\sigma_{xx}(t,\hat{x},\hat{u})|\leq \bar{\omega}(|x-\hat{x}|+|u-\hat{u}|),\\ ||c_{xx}(t,x,u,\cdot)-c_{xx}(t,\hat{x},\hat{u},\cdot)||_{\mathcal{L}^2}\leq \bar{\omega}(|x-\hat{x}|+|u-\hat{u}|),\\ |f_x(t,x,u)|+|h_x(x)|\leq C(1+|x|),\\ |f(t,x,u)|+|h(x)|\leq C(1+|x|^2),\quad \forall t\in[0,T],x,\hat{x}\in\mathbf{R}^n,u,\hat{u}\in\mathbf{U}. \end{cases}$$

For simplicity, we introduce the following notations.

$$\begin{aligned} \bar{b}(t)&:=b(t,\bar{x}^{s,y;\bar{u}}(t),\bar{u}(t)),\quad \bar{\sigma}(t):=\sigma(t,\bar{x}^{s,y;\bar{u}}(t),\bar{u}(t)),\\ \bar{f}(t)&:=f(t,\bar{x}^{s,y;\bar{u}}(t),\bar{u}(t)),\quad \bar{c}(t,\cdot):=c(t,\bar{x}^{s,y;\bar{u}}(t-),\bar{u}(t),\cdot),\\ \Delta\sigma(t;u)&:=\sigma(t,\bar{x}^{s,y;\bar{u}}(t),u)-\bar{\sigma}(t),\quad \Delta c(t,\cdot;u):=c(t,\bar{x}^{s,y;\bar{u}}(t-),u,\cdot)-\bar{c}(t,\cdot), \end{aligned}$$

and similar notations used for all their derivatives, for all $t\in[0,T],u\in\mathbf{U}$.

We define the *Hamiltonian function* $H:[0,T]\times\mathbf{R}^n\times\mathbf{U}\times\mathbf{R}^n\times\mathbf{R}^{n\times d}\times\mathcal{L}^2(\mathbf{E},\mathcal{B}(\mathbf{E}),\pi;\mathbf{R}^n)\to\mathbf{R}$:

$$\begin{aligned} H(t,x,u,p,q,\gamma(\cdot)):=&\langle p,b(t,x,u)\rangle+\mathrm{tr}\{q^\top\sigma(t,x,u)\}-f(t,x,u)\\ &+\int_{\mathbf{E}}\langle\gamma(e),c(t,x,u,e)\rangle\pi(de). \end{aligned}\tag{17}$$

Associated with an optimal pair $(\bar{x}^{s,y;\bar{u}}(\cdot),\bar{u}(\cdot))$, we introduce the following *first-order* and *second-order adjoint equations*, respectively:

$$\begin{cases} -dp(t)=H_x(t,\bar{x}^{s,y;\bar{u}}(t),\bar{u}(t),p(t),q(t),\gamma(t,\cdot))dt-q(t)dW(t)\\ \qquad\qquad -\int_{\mathbf{E}}\gamma(t,e)\tilde{N}(dedt),\ t\in[s,T],\\ p(T)=-h_x(\bar{x}^{s,y;\bar{u}}(T)), \end{cases}\tag{18}$$

$$\begin{cases} -dP(t)=\Big\{\bar{b}_x(t)^\top P(t)+P(t)\bar{b}_x(t)+\bar{\sigma}_x(t)^\top P(t)\bar{\sigma}_x(t)+\bar{\sigma}_x(t)^\top Q(t)+Q(t)\bar{\sigma}_x(t)\\ \qquad +\int_{\mathbf{E}}\big[\bar{c}_x(t,e)^\top P(t)\bar{c}_x(t,e)+\bar{c}_x(t,e)^\top R(t,e)\bar{c}_x(t,e)+\bar{c}_x(t,e)^\top R(t,e)\\ \qquad +R(t,e)\bar{c}_x(t,e)\big]\pi(de)+H_{xx}(t,\bar{x}^{s,y;\bar{u}}(t),\bar{u}(t),p(t),q(t),\gamma(t,\cdot))\Big\}dt\\ \qquad -Q(t)dW(t)-\int_{\mathbf{E}}R(t,e)\tilde{N}(dedt),\qquad t\in[s,T],\\ P(T)=-h_{xx}(\bar{x}^{s,y;\bar{u}}(T)), \end{cases}\tag{19}$$

Under **(H1)**$\sim$**(H3)**, by Lemma 2.4 of [16], we know that BSDEPs (18) and (19) admit unique adapted solutions $(p(\cdot),q(\cdot),\gamma(\cdot,\cdot))$ and $(P(\cdot),Q(\cdot),R(\cdot,\cdot))$ satisfying

$$\begin{cases} (p(\cdot),q(\cdot),\gamma(\cdot,\cdot))\in L^2_{\mathcal{F}}([s,T];\mathbf{R}^n)\times L^2_{\mathcal{F},p}([s,T];\mathbf{R}^{n\times d})\times F^2_p([s,T];\mathbf{R}^n),\\ (P(\cdot),Q(\cdot),R(\cdot,\cdot))\in L^2_{\mathcal{F}}([s,T];\mathcal{S}^n)\times\big(L^2_{\mathcal{F},p}([s,T];\mathcal{S}^n)\big)^d\times F^2_p([s,T];\mathcal{S}^n). \end{cases}\tag{20}$$

Note that $p(\cdot)$ and $P(\cdot)$ are RCLL processes. Associated with an optimal pair $(\bar{x}^{s,y;\bar{u}}(\cdot),\bar{u}(\cdot))$ and its corresponding adjoint processes $(p(\cdot),q(\cdot),\gamma(\cdot,\cdot))$ and $(P(\cdot),Q(\cdot),R(\cdot,\cdot))$ satisfying (20), we define an $\mathcal{H}$-function $\mathcal{H}:[0,T]\times\mathbf{R}^n\times\mathbf{U}\to\mathbf{R}$ as

$$\begin{aligned}
\mathcal{H}(t,x,u) &:= H(t,x,u,p(t),q(t),\gamma(t,\cdot)) - \frac{1}{2}\mathrm{tr}\Big\{P(t)\Big[\bar{\sigma}(t)\bar{\sigma}(t)^\top + \int_{\mathbf{E}}\bar{c}(t,e)\bar{c}(t,e)^\top\pi(de)\Big]\Big\}\\
&\quad + \frac{1}{2}\mathrm{tr}\Big\{P(t)\Big[\Delta\sigma(t;u)\Delta\sigma(t;u)^\top + \int_{\mathbf{E}}\Delta c(t,e;u)\Delta c(t,e;u)^\top\pi(de)\Big]\Big\}\\
&\quad - \frac{1}{2}\mathrm{tr}\Big\{\int_{\mathbf{E}}R(t,e)\Big[\bar{c}(t,e)\bar{c}(t,e)^\top - \Delta c(t,e;u)\Delta c(t,e;u)^\top\Big]\pi(de)\Big\}\\
&\equiv \langle p(t),b(t,x,u)\rangle + \mathrm{tr}\{q(t),\sigma(t,x,u)\} + \int_{\mathbf{E}}\langle\gamma(t,e),c(t,x,u,e)\rangle\pi(de)\\
&\quad - f(t,x,u) - \frac{1}{2}\mathrm{tr}\Big\{P(t)\Big[\bar{\sigma}(t)\bar{\sigma}(t)^\top + \int_{\mathbf{E}}\bar{c}(t,e)\bar{c}(t,e)^\top\pi(de)\Big]\Big\}\\
&\quad + \frac{1}{2}\mathrm{tr}\Big\{P(t)\Big[\Delta\sigma(t;u)\Delta\sigma(t;u)^\top + \int_{\mathbf{E}}\Delta c(t,e;u)\Delta c(t,e;u)^\top\pi(de)\Big]\Big\}\\
&\quad - \frac{1}{2}\mathrm{tr}\Big\{\int_{\mathbf{E}}R(t,e)\Big[\bar{c}(t,e)\bar{c}(t,e)^\top - \Delta c(t,e;u)\Delta c(t,e;u)^\top\Big]\pi(de)\Big\}.
\end{aligned} \tag{21}$$

The following result is the general stochastic MP for jump diffusions.

Theorem 2.3 *Suppose that* **(H1)**$\sim$**(H3)** *hold. Let $(s,y)\in[0,T)\times\mathbf{R}^n$ be fixed and $(\bar{x}^{s,y;\bar{u}}(\cdot),\bar{u}(\cdot))$ be an optimal pair of our stochastic optimal control problem. Then there exist triples of processes $(p(\cdot),q(\cdot),\gamma(\cdot,\cdot))$ and $(P(\cdot),Q(\cdot),R(\cdot,\cdot))$ satisfying (20) and they are solutions of first-order and second-order adjoint equations (18) and (19), respectively, such that the following maximum condition holds:*

$$\begin{aligned}
&H(t,\bar{x}^{s,y;\bar{u}}(t),\bar{u}(t),p(t),q(t),\gamma(t,\cdot)) - H(t,\bar{x}^{s,y;\bar{u}}(t),u,p(t),q(t),\gamma(t,\cdot))\\
&- \frac{1}{2}\mathrm{tr}\Big\{P(t)\Big[\Delta\sigma(t;u)\Delta\sigma(t;u)^\top + \int_{\mathbf{E}}\Delta c(t,e;u)\Delta c(t,e;u)^\top\pi(de)\Big]\Big\}\\
&- \frac{1}{2}\mathrm{tr}\Big\{\int_{\mathbf{E}}R(t,e)\Big[\Delta c(t,e;u)\Delta c(t,e;u)^\top\Big]\pi(de)\Big\} \geq 0, \quad \forall u\in\mathbf{U}, a.e. t\in[s,T], \mathbf{P}\text{-}a.s.,
\end{aligned} \tag{22}$$

or equivalently,

$$\mathcal{H}(t,\bar{x}(t),\bar{u}(t)) = \max_{u\in\mathbf{U}}\mathcal{H}(t,\bar{x}(t),u), \quad a.e. t\in[s,T], \quad \mathbf{P}\text{-}a.s. \tag{23}$$

Proof It is an immediate consequence of Theorem 2.1 of [16]. The equivalence of (22) and (23) is obvious. □

Remark 2.2 Note that the integrand with respect to the compensated martingale measure $\tilde{N}$ in the second-order adjoint equation enters into the above maximum condition, while the counterpart in the diffusion case does not! This marks one essential difference of the maximum principle between an optimally controlled diffusion (*continuous*) process and an optimally controlled jump (*discontinuous*) process.

3. Relationship between Stochastic MP and DPP

In this section, we will establish the relationship between stochastic MP and DPP in the language of viscosity solutions. That is to say, we will consider the viscosity solutions of the generalized HJB equation (15). In our jump diffusion setting, we need use the viscosity solution theory for second-order PIDEs. For convenience, we refer to [1], [2], [14], [3], [11], [6] for a deep investigation of PIDEs in the framework of viscosity solutions. In Subsection 3.1, we first present some preliminary results concerning viscosity solutions and semijets. Then we give the relationship between stochastic MP and DPP in Subsection 3.2. Special cases on the assumption that the value function is smooth are given as corollaries. Some examples are also given to illustrate our results.

3.1. Preliminary results: viscosity solutions and semijets

To make the chapter self-contained, we present the definition of viscosity solutions and semijets, which is frequently seen in the literature and we state it adapting to the generalized HJB equation (15) in our jump diffusion setting.

Definition 3.1 *(i) A function $v \in C([0,T] \times \mathbf{R}^n)$ is called a viscosity subsolution of (15) if $v(T,x) \leq h(x), \forall x \in \mathbf{R}^n$, and for any test function $\psi \in C^{1,2}([0,T] \times \mathbf{R}^n)$, whenever $v - \psi$ attains a global maximum at $(t,x) \in [0,T) \times \mathbf{R}^n$, then*

$$-\psi_t(t,x) + \sup_{u \in \mathbf{U}} G(t,x,u,-\psi(t,x),-\psi_x(t,x),-\psi_{xx}(t,x)) \leq 0. \tag{24}$$

(ii) A function $v \in C([0,T] \times \mathbf{R}^n)$ is called a viscosity supersolution of (15) if $v(T,x) \geq h(x), \forall x \in \mathbf{R}^n$, and for any test function $\psi \in C^{1,2}([0,T] \times \mathbf{R}^n)$, whenever $v - \psi$ attains a global minimum at $(t,x) \in [0,T] \times \mathbf{R}^n$, then

$$-\psi_t(t,x) + \sup_{u \in \mathbf{U}} G(t,x,u,-\psi(t,x),-\psi_x(t,x),-\psi_{xx}(t,x)) \geq 0. \tag{25}$$

(iii) If $v \in C([0,T] \times \mathbf{R}^n)$ is both a viscosity subsolution and viscosity supersolution of (15), then it is called a viscosity solution of (15).

In order to give the existence and uniqueness result for viscosity solution of the generalized HJB equation (15), it is convenient to give an intrinsic characterization of viscosity solutions. Let us recall the right parabolic super-subjets of a continuous function on $[0,T] \times \mathbf{R}^n$ (see [18] or [3]). For $v \in C([0,T] \times \mathbf{R}^n)$ and $(\hat{t},\hat{x}) \in [0,T) \times \mathbf{R}^n$, the *right parabolic superjet* of v at $(\hat{t},\hat{x})$ is the set triple

$$\begin{aligned} \mathcal{P}^{1,2,+}_{t+,x} v(\hat{t},\hat{x}) := \Big\{ (q,p,P) \in \mathbf{R} \times \mathbf{R}^n \times \mathcal{S}^n \big| v(t,x) \leq v(\hat{t},\hat{x}) + q(t-\hat{t}) + \langle p, x-\hat{x} \rangle \\ + \frac{1}{2}(x-\hat{x})^\top P(x-\hat{x}) + o(|t-\hat{t}| + |x-\hat{x}|^2), \text{ as } t \downarrow \hat{t}, x \to \hat{x} \Big\}, \end{aligned} \tag{26}$$

and the *right parabolic subjet* of v at $(\hat{t},\hat{x})$ is the set

$$\begin{aligned} \mathcal{P}^{1,2,-}_{t+,x} v(\hat{t},\hat{x}) := \Big\{ (q,p,P) \in \mathbf{R} \times \mathbf{R}^n \times \mathcal{S}^n \big| v(t,x) \geq v(\hat{t},\hat{x}) + q(t-\hat{t}) + \langle p, x-\hat{x} \rangle \\ + \frac{1}{2}(x-\hat{x})^\top P(x-\hat{x}) + o(|t-\hat{t}| + |x-\hat{x}|^2), \text{ as } t \downarrow \hat{t}, x \to \hat{x} \Big\}. \end{aligned} \tag{27}$$

From the above definitions, we see immediately that

$$\begin{cases} \mathcal{P}^{1,2,+}_{t+,x} v(\hat{t},\hat{x}) + [0,\infty) \times \{0\} \times \mathcal{S}^n_+ = \mathcal{P}^{1,2,+}_{t+,x} v(\hat{t},\hat{x}), \\ \mathcal{P}^{1,2,-}_{t+,x} v(\hat{t},\hat{x}) - [0,\infty) \times \{0\} \times \mathcal{S}^n_+ = \mathcal{P}^{1,2,-}_{t+,x} v(\hat{t},\hat{x}), \end{cases}$$

where $\mathcal{S}^n_+ := \{S \in \mathcal{S}^n | S \geq 0\}$, and $A \pm B := \{a \pm b | a \in A, b \in B\}$ for any subsets A and B in a same Euclidean space.

Remark 3.1 Suppose that $\psi \in C^{1,2}([0,T] \times \mathbf{R}^n)$. If $v - \psi$ attains a global maximum at $(\hat{t},\hat{x}) \in [0,T) \times \mathbf{R}^n$, then

$$\big(\psi_t(\hat{t},\hat{x}), \psi_x(\hat{t},\hat{x}), \psi_{xx}(\hat{t},\hat{x})\big) \in \mathcal{P}^{1,2,+}_{t+,x} v(\hat{t},\hat{x}).$$

If $v - \psi$ attains a global minimum at $(\hat{t},\hat{x}) \in [0,T) \times \mathbf{R}^n$, then

$$\big(\psi_t(\hat{t},\hat{x}), \psi_x(\hat{t},\hat{x}), \psi_{xx}(\hat{t},\hat{x})\big) \in \mathcal{P}^{1,2,-}_{t+,x} v(\hat{t},\hat{x}).$$

The following result is useful and whose proof for diffusion case can be found, for instance, in Lemma 5.4, Chapter 4 of [18].

Proposition 3.1 *Let $v \in C([0,T] \times \mathbf{R}^n)$ and $(t_0,x_0) \in [0,T) \times \mathbf{R}^n$ be given. Then*

(i) $(q,p,P) \in \mathcal{P}^{1,2,+}_{t+,x} v(t_0,x_0)$ if and only if there exist a function $\psi \in C^{1,2}([0,T] \times \mathbf{R}^n)$, such that

$$\begin{cases} (\psi(t_0,x_0), \psi_t(t_0,x_0), \psi_x(t_0,x_0), \psi_{xx}(t_0,x_0)) = (v(t_0,x_0), q, p, P), \\ \psi(t,x) > v(t,x), \quad \forall (t_0,x_0) \neq (t,x) \in [t_0,T] \times \mathbf{R}^n. \end{cases} \tag{28}$$

(ii) $(q,p,P) \in \mathcal{P}^{1,2,-}_{t+,x} v(t_0,x_0)$ if and only if there exist a function $\psi \in C^{1,2}([0,T] \times \mathbf{R}^n)$, such that

$$\begin{cases} (\psi(t_0,x_0), \psi_t(t_0,x_0), \psi_x(t_0,x_0), \psi_{xx}(t_0,x_0)) = (v(t_0,x_0), q, p, P), \\ \psi(t,x) < v(t,x), \quad \forall (t_0,x_0) \neq (t,x) \in [t_0,T] \times \mathbf{R}^n. \end{cases} \tag{29}$$

We will also make use of the *partial* super-subjets with respect to one of the variables t and x. Therefore, we need the following definitions.

$$\begin{cases} \mathcal{P}^{2,+}_{x} v(\hat{t},\hat{x}) := \Big\{(p,P) \in \times\mathbf{R}^n \times \mathcal{S}^n \big| v(\hat{t},x) \leq v(\hat{t},\hat{x}) + \langle p, x-\hat{x}\rangle \\ \qquad + \frac{1}{2}(x-\hat{x})^\top P(x-\hat{x}) + o(|x-\hat{x}|^2), \text{ as } x \to \hat{x}\Big\}, \\ \mathcal{P}^{2,-}_{x} v(\hat{t},\hat{x}) := \Big\{(p,P) \in \times\mathbf{R}^n \times \mathcal{S}^n \big| v(\hat{t},x) \geq v(\hat{t},\hat{x}) + \langle p, x-\hat{x}\rangle \\ \qquad + \frac{1}{2}(x-\hat{x})^\top P(x-\hat{x}) + o(|x-\hat{x}|^2), \text{ as } x \to \hat{x}\Big\}, \end{cases} \tag{30}$$

and

$$\begin{cases} \mathcal{P}^{1,+}_{t+} v(\hat{t},\hat{x}) := \Big\{q \in \mathbf{R} \big| v(t,\hat{x}) \leq v(\hat{t},\hat{x}) + q(t-\hat{t}) + o(|t-\hat{t}|), \text{ as } t \downarrow \hat{t}\Big\}, \\ \mathcal{P}^{1,-}_{t+} v(\hat{t},\hat{x}) := \Big\{q \in \mathbf{R} \big| v(t,\hat{x}) \geq v(\hat{t},\hat{x}) + q(t-\hat{t}) + o(|t-\hat{t}|), \text{ as } t \downarrow \hat{t}\Big\}. \end{cases} \tag{31}$$

Using the above definitions, we can give the following intrinsic formulation of viscosity solution of the generalized HJB equation (15).

Definition 3.2 *(i) A function $v \in C([0,T] \times \mathbf{R}^n)$ satisfying $v(T,x) \leq h(x), \forall x \in \mathbf{R}^n$ is a viscosity subsolution of (15) if, for any test function $\psi \in C^{1,2}([0,T] \times \mathbf{R}^n)$, if $v - \psi$ attains a global maximum at $(t,x) \in [0,T) \times \mathbf{R}^n$ and if $(q,p,P) \in \mathcal{P}^{1,2,+}_{t+,x} v(t,x)$ with $q = \psi_t(t,x), p = \psi_x(t,x), P \leq \psi_{xx}(t,x)$, then*

$$-q + \sup_{u \in \mathbf{U}} \{G(t,x,u,-\psi(t,x),-p,-P)\} \leq 0. \tag{32}$$

(ii) A function $v \in C([0,T] \times \mathbf{R}^n)$ satisfying $v(T,x) \geq h(x), \forall x \in \mathbf{R}^n$ is a viscosity supersolution of (15) if, for any test function $\psi \in C^{1,2}([0,T] \times \mathbf{R}^n)$, if $v - \psi$ attains a global minimum at $(t,x) \in [0,T) \times \mathbf{R}^n$ and if $(q,p,P) \in \mathcal{P}^{1,2,-}_{t+,x} v(t,x)$ with $q = \psi_t(t,x), p = \psi_x(t,x), P \geq \psi_{xx}(t,x)$, then

$$-q + \sup_{u \in \mathbf{U}} \{G(t,x,u,-\psi(t,x),-p,-P)\} \geq 0. \tag{33}$$

(iii) If $v \in C([0,T] \times \mathbf{R}^n)$ is both a viscosity subsolution and viscosity supersolution of (15), then it is called a viscosity solution of (15).

Proposition 3.2 *Definitions 3.1 and 3.2 are equivalent.*

Proof The result is immediate in view of Proposition 3.1. In fact it is a special case of Proposition 1 of [3]. □

The following result is the existence and uniqueness of viscosity solution of the generalized HJB equation (15).

Theorem 3.1 *Suppose* **(H1)**~**(H2)** *hold. Then we have the following equivalent results.*

(i) The value function $V \in C([0,T] \times \mathbf{R}^n)$ defined by (6) is a unique viscosity solution of (15) in the class of functions satisfying (12), (13).

(ii) The value function $V \in C([0,T] \times \mathbf{R}^n)$ is the only function that satisfies (12), (13) and the following: For all $(t,x) \in [0,T) \times \mathbf{R}^n$,

$$\begin{cases} -q + \sup_{u \in \mathbf{U}} \{G(t,x,u,-\psi_1(t,x),-p,-P)\} \leq 0, \quad \forall (q,p,P) \in \mathcal{P}^{1,2,+}_{t+,x} V(t,x), \\ \psi_1 \in C^{1,2}([0,T] \times \mathbf{R}^n) \text{ such that } \psi_1(t',x') > v(t',x'), \ \forall (t',x') \neq (t,x) \in [t,T] \times \mathbf{R}^n, \\ -q + \sup_{u \in \mathbf{U}} \{G(t,x,u,-\psi_2(t,x),-p,-P)\} \geq 0, \quad \forall (q,p,P) \in \mathcal{P}^{1,2,-}_{t+,x} V(t,x), \\ \psi_2 \in C^{1,2}([0,T] \times \mathbf{R}^n) \text{ such that } \psi_2(t',x') < v(t',x'), \ \forall (t',x') \neq (t,x) \in [t,T] \times \mathbf{R}^n, \\ V(T,x) = h(x). \end{cases}$$

Proof Result (i) is a special case of Theorems 3.1 and 4.1 of [14]. Result (ii) is obvious by virtue of Propositions 3.1 and 3.2. The equivalence between (i) and (ii) is obvious. □

3.2. Main results: Relationship between Stochastic MP and DPP

Proposition 2.1 tell us that the value function has nice continuity properties. But in general we cannot obtain the differentiablity of it. Therefore we should not suppose the generalized HJB equation (15) always admits an enough smooth (classic) solution. In fact it is not true even in the simplest case; see Example 3.2 in this subsection. This is an important *gap* in the literature (see Section 2, [7], for example). Fortunately, this gap can be bridged by the theory of viscosity solutions. This is one of the main contributions of this paper.

The following result shows that the adjoint process p, P and the value function V relate to each other within the framework of the superjet and the subjet in the state variable x along an optimal trajectory.

Theorem 3.2 *Suppose* **(H1)**$\sim$**(H3)** *hold and let* $(s,y) \in [0,T) \times \mathbf{R}^n$ *be fixed. Let* $(\bar{x}^{s,y;\bar{u}}(\cdot), \bar{u}(\cdot))$ *be an optimal pair of our stochastic optimal control problem. Let* $(p(\cdot), q(\cdot), \gamma(\cdot,\cdot))$ *and* $(P(\cdot), Q(\cdot), R(\cdot,\cdot))$ *are first-order and second-order adjoint processes, respectively. Then*

$$\{-p(t)\} \times [-P(t), \infty) \subseteq \mathcal{P}_x^{2,+} V(t, \bar{x}^{s,y;\bar{u}}(t)), \quad \forall t \in [s,T], \quad \mathbf{P}\text{-}a.s., \tag{34}$$

$$\mathcal{P}_x^{2,-} V(t, \bar{x}^{s,y;\bar{u}}(t)) \subseteq \{-p(t)\} \times (-\infty, P(t)], \quad \forall t \in [s,T], \quad \mathbf{P}\text{-}a.s. \tag{35}$$

We also have

$$\mathcal{P}_x^{1,-} V(t, \bar{x}^{s,y;\bar{u}}(t)) \subseteq \{-p(t)\} \subseteq \mathcal{P}_x^{1,+} V(t, \bar{x}^{s,y;\bar{u}}(t)), \quad \forall t \in [s,T], \quad \mathbf{P}\text{-}a.s. \tag{36}$$

Proof Fix a $t \in [s,T]$. For any $z \in \mathbf{R}^n$, denote by $x^{t,z;\bar{u}}(\cdot)$ the solution of the following SDEP:

$$\begin{aligned} x^{t,z;\bar{u}}(r) = z &+ \int_t^r b(r, x^{t,z;\bar{u}}(r), \bar{u}(r))dr + \int_t^r \sigma(r, x^{t,z;\bar{u}}(r), \bar{u}(r))dW(r) \\ &+ \int_{\mathbf{E}} \int_t^r c(r, x^{t,z;\bar{u}}(r-), \bar{u}(r), e)\tilde{N}(dedr), \quad r \in [t,T]. \end{aligned} \tag{37}$$

It is clear that (37) can be regarded as an SDEP on $(\Omega, \mathcal{F}, \{\mathcal{F}_r^s\}_{r \geq s}, \mathbf{P}(\cdot|\mathcal{F}_t^s)(\omega))$ for $\mathbf{P}$-$a.s.\omega$, where $\mathbf{P}(\cdot|\mathcal{F}_t^s)(\omega)$ is the regular conditional probability given $\mathcal{F}_t^s$ defined on $(\Omega, \mathcal{F})$ (see pp. 12-16 of [10]). In probability space $(\Omega, \mathcal{F}, \mathbf{P}(\cdot|\mathcal{F}_t^s)(\omega))$, random variable $\bar{x}^{s,y;\bar{u}}(t,\omega)$ is almost surely a constant vector in $\mathbf{R}^n$ (we still denote it by $\bar{x}^{s,y;\bar{u}}(t,\omega)$).

Set $\xi^{t,z;\bar{u}}(r) := x^{t,z;\bar{u}}(r) - \bar{x}^{t,\bar{x}^{s,y;\bar{u}}(t);\bar{u}}(r), t \leq r \leq T$. Thus by Lemma 2.1 we have

$$\mathbb{E}\big[\sup_{t \leq r \leq T} |\xi^{t,z;\bar{u}}(r)|^k | \mathcal{F}_t^s\big](\omega) \leq C|z - \bar{x}^{s,y;\bar{u}}(t,\omega)|^k, \quad \mathbf{P}\text{-}a.s.\omega., \quad \text{for } k = 2,4. \tag{38}$$

Now we rewrite the equation for $\xi^{t,z;\bar{u}}(\cdot)$ in two different ways based on different orders of expansion, which called the *first-order and second-order variational equations*, respectively:

$$\begin{cases} d\xi^{t,z;\bar{u}}(r) = \bar{b}_x(r)\xi^{t,z;\bar{u}}(r)dr + \bar{\sigma}_x(r)\xi^{t,z;\bar{u}}(r)dW(r) + \int_{\mathbf{E}} \bar{c}_x(r,e)\xi^{t,z;\bar{u}}(r-)\tilde{N}(dedr) \\ \qquad\qquad + \varepsilon_{z1}(r)dr + \varepsilon_{z2}(r)dW(r) + \int_{\mathbf{E}} \varepsilon_{z3}(r,e)\tilde{N}(dedr), \quad r \in [t,T], \\ \xi^{t,z;\bar{u}}(t) = z - \bar{x}^{s,y;\bar{u}}(t), \end{cases} \tag{39}$$

where

$$\begin{cases} \varepsilon_{z1}(r) := \int_0^1 \left[b_x(r, \bar{x}^{s,y;\bar{u}}(r) + \theta\xi^{t,z;\bar{u}}(r), \bar{u}(r)) - \bar{b}_x(r)\right]\xi^{t,z;\bar{u}}(r)d\theta, \\ \varepsilon_{z2}(r) := \int_0^1 \left[\sigma_x(r, \bar{x}^{s,y;\bar{u}}(r) + \theta\xi^{t,z;\bar{u}}(r), \bar{u}(r)) - \bar{\sigma}_x(r)\right]\xi^{t,z;\bar{u}}(r)d\theta, \\ \varepsilon_{z3}(r,\cdot) := \int_0^1 \left[c_x(r, \bar{x}^{s,y;\bar{u}}(r-) + \theta\xi^{t,z;\bar{u}}(r-), \bar{u}(r), \cdot) - \bar{c}_x(r,\cdot)\right]\xi^{t,z;\bar{u}}(r-)d\theta, \end{cases}$$

and

$$\begin{cases} d\xi^{t,z;\bar{u}}(r) = \left\{\bar{b}_x(r)\xi^{t,z;\bar{u}}(r) + \frac{1}{2}\xi^{t,z;\bar{u}}(r)^\top \bar{b}_{xx}(r)\xi^{t,z;\bar{u}}(r)\right\}dr \\ \qquad + \left\{\bar{\sigma}_x(r)\xi^{t,z;\bar{u}}(r) + \frac{1}{2}\xi^{t,z;\bar{u}}(r)^\top \bar{\sigma}_{xx}(r)\xi^{t,z;\bar{u}}(r)\right\}dW(r) \\ \qquad + \int_{\mathbf{E}}\left\{\bar{c}_x(r,e)\xi^{t,z;\bar{u}}(r-) + \frac{1}{2}\xi^{t,z;\bar{u}}(r-)^\top \bar{c}_{xx}(r,e)\xi^{t,z;\bar{u}}(r-)\right\}\tilde{N}(dedr) \\ \qquad + \varepsilon_{z4}(r)dr + \varepsilon_{z5}(r)dW(r) + \int_{\mathbf{E}} \varepsilon_{z6}(r,e)\tilde{N}(dedr), \quad r \in [t,T], \\ \xi^{t,z;\bar{u}}(t) = z - x^{s,y;\bar{u}}(t), \end{cases} \tag{40}$$

where

$$\begin{cases} \varepsilon_{z4}(r) := \int_0^1 (1-\theta)\xi^{t,z;\bar{u}}(r)^\top \left[b_{xx}(r, \bar{x}^{s,y;\bar{u}}(r) + \theta\xi^{t,z;\bar{u}}(r), \bar{u}(r)) - \bar{b}_{xx}(r)\right]\xi^{t,z;\bar{u}}(r)d\theta, \\ \varepsilon_{z5}(r) := \int_0^1 (1-\theta)\xi^{t,z;\bar{u}}(r)^\top \left[\sigma_{xx}(r, \bar{x}^{s,y;\bar{u}}(r) + \theta\xi^{t,z;\bar{u}}(r), \bar{u}(r)) - \bar{\sigma}_{xx}(r)\right]\xi^{t,z;\bar{u}}(r)d\theta, \\ \varepsilon_{z6}(r,\cdot) := \int_0^1 (1-\theta)\xi^{t,z;\bar{u}}(r-)^\top \left[c_{xx}(r, \bar{x}^{s,y;\bar{u}}(r-) + \theta\xi^{t,z;\bar{u}}(r-), \bar{u}(r), \cdot) - \bar{c}_{xx}(r,\cdot)\right]\xi^{t,z;\bar{u}}(r-)d\theta. \end{cases}$$

We are going to show that, there exists a deterministic continuous and increasing function $\delta : [0,\infty) \to [0,\infty)$, independent of $z \in \mathbf{R}^n$, with $\frac{\delta(r)}{r} \to 0$ as $r \to 0$, such that

$$\begin{cases} \mathbb{E}\left[\int_t^T |\varepsilon_{z1}(r)|^2 dr \middle| \mathcal{F}_t^s\right](\omega) \le \delta(|z - \bar{x}^{s,y;\bar{u}}(t,\omega)|^2), \quad \mathbf{P}\text{-}a.s.\omega, \\ \mathbb{E}\left[\int_t^T |\varepsilon_{z2}(r)|^2 dr \middle| \mathcal{F}_t^s\right](\omega) \le \delta(|z - \bar{x}^{s,y;\bar{u}}(t,\omega)|^2), \quad \mathbf{P}\text{-}a.s.\omega, \\ \mathbb{E}\left[\int_t^T ||\varepsilon_{z3}(r,\cdot)||^2_{\mathcal{L}^2} dr \middle| \mathcal{F}_t^s\right](\omega) \le \delta(|z - \bar{x}^{s,y;\bar{u}}(t,\omega)|^2), \quad \mathbf{P}\text{-}a.s.\omega, \end{cases} \tag{41}$$

and

$$\begin{cases} \mathbb{E}\left[\int_t^T |\varepsilon_{z4}(r)|^2 dr \middle| \mathcal{F}_t^s\right](\omega) \le \delta(|z - \bar{x}^{s,y;\bar{u}}(t,\omega)|^4), \quad \mathbf{P}\text{-}a.s.\omega, \\ \mathbb{E}\left[\int_t^T |\varepsilon_{z5}(r)|^2 dr \middle| \mathcal{F}_t^s\right](\omega) \le \delta(|z - \bar{x}^{s,y;\bar{u}}(t,\omega)|^4), \quad \mathbf{P}\text{-}a.s.\omega, \\ \mathbb{E}\left[\int_t^T ||\varepsilon_{z6}(r,\cdot)||^2_{\mathcal{L}^2} dr \middle| \mathcal{F}_t^s\right](\omega) \le \delta(|z - \bar{x}^{s,y;\bar{u}}(t,\omega)|^4), \quad \mathbf{P}\text{-}a.s.\omega. \end{cases} \tag{42}$$

We start to prove (41). To this end, let us fixed an $\omega \in \Omega$ such that (38) holds. Then, by setting $b_x(r,\theta) := b_x(r, \bar{x}^{s,y;\bar{u}}(r) + \theta\xi^{t,z;\bar{u}}(r), \bar{u}(r))$ and in virtue of **(H3)**, we have

$$\begin{aligned}
&\mathbb{E}\Big[\int_t^T |\varepsilon_{z1}(r)|^2 dr \big| \mathcal{F}_t^s\Big](\omega)\\
&\leq \int_t^T \mathbb{E}\Big\{\int_0^1 |b_x(r,\theta) - \bar{b}_x(r)|^2 d\theta \cdot |\xi^{t,z;\bar{u}}(r)|^2 \big| \mathcal{F}_t^s\Big\}(\omega) dr\\
&\leq C \int_t^T \mathbb{E}\Big[|\xi^{t,z;\bar{u}}(r)|^4 \big| \mathcal{F}_t^s\Big](\omega) dr \leq C|z - \bar{x}^{s,y;\bar{u}}(t,\omega)|^4.
\end{aligned}$$

Thus, the first inequality in (41) follows if we choose $\delta(r) \equiv Cr^2, r \geq 0$. The second inequality in (41) can be proved similarly. Setting $c_x(r,\theta,\cdot) := c_x(r, \bar{x}^{s,y;\bar{u}}(r-) + \theta\xi^{t,z;\bar{u}}(r-), \bar{u}(r), \cdot)$ and using **(H3)**, we have

$$\begin{aligned}
&\mathbb{E}\Big[\int_t^T ||\varepsilon_{z3}(r,\cdot)||^2_{\mathcal{L}^2} dr \big| \mathcal{F}_t^s\Big](\omega)\\
&\leq \int_t^T \mathbb{E}\Big\{\int_0^1 ||c_x(r,\theta,\cdot) - \bar{c}_x(r,\cdot)||^2_{\mathcal{L}^2} d\theta \cdot |\xi^{t,z;\bar{u}}(r-)|^2 \big| \mathcal{F}_t^s\Big\}(\omega) dr\\
&\leq C \int_t^T \mathbb{E}\Big[|\xi^{t,z;\bar{u}}(r-)|^4 \big| \mathcal{F}_t^s\Big](\omega) dr\\
&= C \int_t^T \mathbb{E}\Big[|\xi^{t,z;\bar{u}}(r)|^4 \big| \mathcal{F}_t^s\Big](\omega) dr \leq C|z - \bar{x}^{s,y;\bar{u}}(t,\omega)|^4.
\end{aligned} \tag{43}$$

The equality in (43) holds because the discontinuous points of $\xi^{t,z;\bar{u}}(\cdot)$ are at most countable. Thus, the third inequality in (41) follows for an obvious $\delta(\cdot)$ as above.

We continue to prove (42). Let $b_{xx}(r,\theta) := b_{xx}(r, \bar{x}^{s,y;\bar{u}}(r) + \theta\xi^{t,z;\bar{u}}(r), \bar{u}(r))$. Using **(H3)**, we can show that

$$\begin{aligned}
&\mathbb{E}\Big[\int_t^T |\varepsilon_{z4}(r)|^2 dr \big| \mathcal{F}_t^s\Big](\omega)\\
&\leq \int_t^T \mathbb{E}\Big[\int_0^1 |b_{xx}(r,\theta) - \bar{b}_{xx}(r)|^2 d\theta \cdot |\xi^{t,z;\bar{u}}(r)|^4 \big| \mathcal{F}_t^s\Big](\omega) dr\\
&\leq \int_t^T \Big\{\mathbb{E}\big[\bar{\omega}(|\xi^{t,z;\bar{u}}(r)|^4) \big| \mathcal{F}_t^s\big](\omega)\Big\}^{\frac{1}{2}} \Big\{\mathbb{E}\big[|\xi^{t,z;\bar{u}}(r)|^8 \big| \mathcal{F}_t^s\big](\omega)\Big\}^{\frac{1}{2}} dr\\
&\leq C \int_t^T \Big\{\mathbb{E}\big[\bar{\omega}(|\xi^{t,z;\bar{u}}(r)|^4) \big| \mathcal{F}_t^s\big](\omega)\Big\}^{\frac{1}{2}} dr \cdot |z - \bar{x}^{s,y;\bar{u}}(t,\omega)|^4.
\end{aligned} \tag{44}$$

This yields the first inequality in (42) if we choose $\delta(r) \equiv Cr\sqrt{\bar{\omega}(r)}, r \geq 0$. Noting that the modulus of continuity $\bar{\omega}(\cdot)$ is defined in **(H3)**. The second inequality in (42) can be proved similarly. Setting $c_{xx}(r,\theta,\cdot) := c_{xx}(r, \bar{x}^{s,y;\bar{u}}(r-) + \theta\xi^{t,z;\bar{u}}(r-), \bar{u}(r), \cdot)$ and by virtue of **(H3)**, noting the remark following (43), we show that

$$\mathbb{E}\Big[\int_t^T ||\varepsilon_{z6}(r,\cdot)||^2_{\mathcal{L}^2}dr\big|\mathcal{F}_t^s\Big](\omega)$$
$$\leq \int_t^T \mathbb{E}\Big[\int_0^1 ||c_{xx}(r,\theta,\cdot)-\bar{c}_{xx}(r,\cdot)||^2_{\mathcal{L}^2}d\theta\cdot|\xi^{t,z;\bar{u}}(r-)|^4\big|\mathcal{F}_t^s\Big](\omega)dr$$
$$\leq \int_t^T \Big\{\mathbb{E}\big[\bar{\omega}(|\xi^{t,z;\bar{u}}(r-)|^4)\big|\mathcal{F}_t^s\big](\omega)\Big\}^{\frac{1}{2}}\Big\{\mathbb{E}\big[|\xi^{t,z;\bar{u}}(r-)|^8\big|\mathcal{F}_t^s\big](\omega)\Big\}^{\frac{1}{2}}dr$$
$$\leq C\int_t^T \Big\{\mathbb{E}\big[\bar{\omega}(|\xi^{t,z;\bar{u}}(r)|^4)\big|\mathcal{F}_t^s\big](\omega)\Big\}^{\frac{1}{2}}dr\cdot|z-\bar{x}^{s,y;\bar{u}}(t,\omega)|^4.$$

This yields the third inequality in (42) for an obvious $\delta(\cdot)$ as above. Finally, we can select the largest $\delta(\cdot)$ obtained in the above six calculations. For example, we can choose an enough large constant $C>0$ and define $\delta(r)\equiv Cr(r\vee\sqrt{\bar{\omega}(r)}), r\geq 0$. Then (41), (42) follows with a $\delta(\cdot)$ independent of $z\in\mathbf{R}^n$.

Applying Itô's formula to $\langle\xi^{t,z;\bar{u}}(\cdot),p(\cdot)\rangle$, noting (18) and (40), we have

$$\begin{aligned}
&\mathbb{E}\Big\{\int_t^T\langle\bar{f}_x(r),\xi^{t,z;\bar{u}}(r)\rangle dr+\langle h_x(\bar{x}^{s,y;\bar{u}}(T)),\xi^{t,z;\bar{u}}(T)\rangle\big|\mathcal{F}_t^s\Big\}\\
&=\langle -p(t),\xi^{t,z;\bar{u}}(t)\rangle-\mathbb{E}\Big\{\frac{1}{2}\int_t^T\Big[\langle p(r),\xi^{t,z;\bar{u}}(r)^\top\bar{b}_{xx}(r)\xi^{t,z;\bar{u}}(r)\rangle\\
&\quad+\langle q(r),\xi^{t,z;\bar{u}}(r)^\top\bar{\sigma}_{xx}(r)\xi^{t,z;\bar{u}}(r)\rangle+\int_{\mathbf{E}}\langle\gamma(r,e),\xi^{t,z;\bar{u}}(r)^\top\bar{c}_{xx}(r,e)\xi^{t,z;\bar{u}}(r)\rangle\pi(de)\Big]dr\\
&\quad-\int_t^T\Big[\langle p(r),\varepsilon_{z4}(r)\rangle+\langle q(r),\varepsilon_{z5}(r)\rangle+\int_{\mathbf{E}}\langle\gamma(r,e),\varepsilon_{z6}(r,e)\rangle\pi(de)\Big]dr\Big|\mathcal{F}_t^s\Big\},\quad \mathbf{P}\text{-}a.s.
\end{aligned}\tag{45}$$

On the other hand, apply Itô's formula to $\Phi^{t,z;\bar{u}}(r):=\xi^{t,z;\bar{u}}(r)\xi^{t,z;\bar{u}}(r)^\top$, noting (39), we get

$$\begin{cases}
d\Phi^{t,z;\bar{u}}(r)=\Big\{\bar{b}_x(r)\Phi^{t,z;\bar{u}}(r)+\Phi^{t,z;\bar{u}}(r)\bar{b}_x(r)^\top+\bar{\sigma}_x(r)\Phi^{t,z;\bar{u}}(r)\bar{\sigma}_x(r)^\top\\
\qquad+\int_{\mathbf{E}}\bar{c}_x(r,e)\Phi^{t,z;\bar{u}}(r)\bar{c}_x(r,e)^\top\pi(de)+\varepsilon_{z7}(r)\Big\}dr\\
\qquad+\Big\{\bar{\sigma}_x(r)\Phi^{t,z;\bar{u}}(r)+\Phi^z(r)\bar{\sigma}_x(r)^\top+\varepsilon_{z8}(r)\Big\}dW(r)\\
\qquad+\int_{\mathbf{E}}\Big\{\bar{c}_x(r,e)\Phi^{t,z;\bar{u}}(r-)\bar{c}_x(r,e)^\top+\bar{c}_x(r,e)\Phi^{t,z;\bar{u}}(r-)\\
\qquad+\Phi^{t,z;\bar{u}}(r-)\bar{c}_x(r,e)^\top+\varepsilon_{z9}(r,e)\Big\}\tilde{N}(dedr),\quad r\in[t,T],\\
\Phi^{t,z;\bar{u}}(t)=\xi^{t,z;\bar{u}}(t)\xi^{t,z;\bar{u}}(t)^\top,
\end{cases}\tag{46}$$

where

$$
\begin{cases}
\varepsilon_{z7}(r) := \varepsilon_{z1}(r)\xi^{t,z;\bar{u}}(r)^\top + \xi^{t,z;\bar{u}}(r)\varepsilon_{z1}(r)^\top + \bar{\sigma}_x(r)\xi^{t,z;\bar{u}}(r)\varepsilon_{z2}(r)^\top \\
\qquad + \varepsilon_{z2}(r)\xi^{t,z;\bar{u}}(r)^\top \bar{\sigma}_x(r)^\top + \varepsilon_{z2}(r)\varepsilon_{z2}(r)^\top + \int_{\mathbf{E}} \Big\{ \bar{c}_x(r,e)\xi^{t,z;\bar{u}}(r)\varepsilon_{z3}(r,e)^\top \\
\qquad + \varepsilon_{z3}(r,e)\xi^{t,z;\bar{u}}(r)^\top \bar{c}_x(r,e)^\top + \varepsilon_{z3}(r,e)\varepsilon_{z3}(r,e)^\top \Big\} \pi(de), \\
\varepsilon_{z8}(r) := \varepsilon_{z2}(r)\xi^{t,z;\bar{u}}(r)^\top + \xi^{t,z;\bar{u}}(r)\varepsilon_{z2}(r)^\top, \\
\varepsilon_{z9}(r,\cdot) := \bar{c}_x(r,\cdot)\xi^{t,z;\bar{u}}(r-)\varepsilon_{z3}(r,\cdot)^\top + \varepsilon_{z3}(r,\cdot)\xi^{t,z;\bar{u}}(r-)^\top \bar{c}_x(r,\cdot)^\top \\
\qquad + \xi^{t,z;\bar{u}}(r-)\varepsilon_{z3}(r,\cdot)^\top + \varepsilon_{z3}(r,\cdot)\xi^{t,z;\bar{u}}(r-)^\top + \varepsilon_{z3}(r,\cdot)\varepsilon_{z3}(r,\cdot)^\top.
\end{cases}
$$

Once more applying Itô's formula to $\Phi^{t,z;\bar{u}}(\cdot)^\top P(\cdot)$, noting (19) and (46), we get

$$
\begin{aligned}
&-\mathbb{E}\Big\{ \int_t^T \xi^{t,z;\bar{u}}(r)^\top \bar{H}_{xx}(r)\xi^{t,z;\bar{u}}(r)dr + \xi^{t,z;\bar{u}}(T)^\top h_{xx}(\bar{x}^{s,y;\bar{u}}(T))\xi^{t,z;\bar{u}}(T) \big| \mathcal{F}_t^s \Big\} \\
&= -\xi^{t,z;\bar{u}}(t)^\top P(t)\xi^{t,z;\bar{u}}(t) \qquad (47) \\
&\quad -\mathbb{E}\Big\{ \int_t^T \mathrm{tr}\big\{P(r)\varepsilon_{z7}(r) + Q(r)\varepsilon_{z8}(r) + \int_{\mathbf{E}} R(r,e)\varepsilon_{z9}(r,e)\pi(de)\big\} dr \big| \mathcal{F}_t^s \Big\}, \quad \mathbf{P}\text{-}a.s.
\end{aligned}
$$

Let us call a $z \in \mathbf{R}^n$ *rational* if all its coordinate are rational. Since the set of all rational $z \in \mathbf{R}^n$ is countable, we may find a *common* subset $\Omega_0 \subseteq \Omega$ with $P(\Omega_0) = 1$ such that for any $\omega_0 \in \Omega_0$,

$$
\begin{cases}
V(t, \bar{x}^{s,y;\bar{u}}(t,\omega_0)) = \mathbb{E}\Big[\int_s^T \bar{f}(r)dr + h(\bar{x}^{s,y;\bar{u}}(T)) \big| \mathcal{F}_t^s \Big](\omega_0), \\
(38), (41), (42), (45), (47) \text{ are satisfied for any rational } z, \text{ and } \bar{u}(\cdot)|_{[t,T]} \in \mathcal{U}[t,T].
\end{cases}
$$

Let $\omega_0 \in \Omega_0$ be fixed, then for any rational $z \in \mathbf{R}^n$, noting (41) and (42), we have

$$
\begin{aligned}
&V(t,z) - V(t, \bar{x}^{s,y;\bar{u}}(t,\omega_0)) \\
&= \mathbb{E}\Big\{ \int_t^T [f(r, x^{t,z;\bar{u}}(r), \bar{u}(r)) - \bar{f}(r)]dr + h(x^{t,z;\bar{u}}(T)) - h(\bar{x}^{s,y;\bar{u}}(T)) \big| \mathcal{F}_t^s \Big\}(\omega_0) \\
&= \mathbb{E}\Big\{ \int_t^T \langle \bar{f}_x(r), \xi^{t,z;\bar{u}}(r)\rangle + \langle h_x(\bar{x}^{s,y;\bar{u}}(T)), \xi^{t,z;\bar{u}}(T)\rangle \big| \mathcal{F}_t^s \Big\}(\omega_0) \qquad (48) \\
&\quad + \frac{1}{2}\mathbb{E}\Big\{ \int_t^T \xi^{t,z;\bar{u}}(r)^\top \bar{f}_{xx}(r)\xi^{t,z;\bar{u}}(r)dr + \xi^{t,z;\bar{u}}(T)^\top h_{xx}(\bar{x}^{s,y;\bar{u}}(T))\xi^{t,z;\bar{u}}(T) \big| \mathcal{F}_t^s \Big\}(\omega_0) \\
&\quad + o(|z - \bar{x}^{s,y;\bar{u}}(t,\omega_0)|^2).
\end{aligned}
$$

By virtue of (45) and (47), we have

$$
\begin{aligned}
& V(t,z) - V(t,\bar{x}^{s,y;\bar{u}}(t,\omega_0)) \\
\leq & - \langle p(t,\omega_0), \xi^{t,z;\bar{u}}(t,\omega_0)\rangle - \frac{1}{2}\mathbb{E}\Big\{ \int_t^T \xi^{t,z;\bar{u}}(r)^\top \bar{H}_{xx}(r)\xi^{t,z;\bar{u}}(r)dr \\
& + \xi^{t,z;\bar{u}}(T)^\top h_{xx}(\bar{x}^{s,y;\bar{u}}(T))\xi^{t,z;\bar{u}}(T)\big|\mathcal{F}_t^s\Big\}(\omega_0) + o(|z - \bar{x}^{s,y;\bar{u}}(t,\omega_0)|^2) \\
= & - \langle p(t,\omega_0), \xi^{t,z;\bar{u}}(t,\omega_0)\rangle - \frac{1}{2}\xi^{t,z;\bar{u}}(t,\omega_0)^\top P(t,\omega_0)\xi^{t,z;\bar{u}}(t,\omega_0) + o(|z - \bar{x}^{s,y;\bar{u}}(t,\omega_0)|^2) \\
= & - \langle p(t,\omega_0), z - \bar{x}^{s,y;\bar{u}}(t,\omega_0)\rangle - \frac{1}{2}(z - \bar{x}^{s,y;\bar{u}}(t,\omega_0))^\top P(t,\omega_0)(z - \bar{x}^{s,y;\bar{u}}(t,\omega_0)) \\
& + o(|z - \bar{x}^{s,y;\bar{u}}(t,\omega_0)|^2).
\end{aligned}
\tag{49}
$$

Note that the term $o(|z - \bar{x}^{s,y;\bar{u}}(t,\omega_0)|^2)$ depends only on the size of $|z - \bar{x}^{s,y;\bar{u}}(t,\omega_0)|^2$, and it is independent of z. Therefore, by the continuity of $V(t,\cdot)$, we see that (49) holds for all $z \in \mathbf{R}^n$, which by definition (30) proves

$$
(-p(t), -P(t)) \in \mathcal{P}_x^{2,+}V(t,\bar{x}^{s,y;\bar{u}}(t)), \quad \forall t \in [s,T], \quad \mathbf{P}\text{-}a.s.
$$

Then (34) holds. Let us now show (35). Fix an $\omega \in \Omega$ such that (49) holds for any $z \in \mathbf{R}^n$. For any $(p,P) \in \mathcal{P}_x^{2,-}V(t,\bar{x}^{s,y;\bar{u}}(t))$, by definition (30) and (49) we have

$$
\begin{aligned}
0 \leq & V(t,z) - V(t,\bar{x}(t,\omega)) - \langle p, z - \bar{x}^{s,y;\bar{u}}(t,\omega)\rangle - \frac{1}{2}(z - \bar{x}^{s,y;\bar{u}}(t,\omega))^\top P(z - \bar{x}^{s,y;\bar{u}}(t,\omega)) \\
= & -\langle p(t,\omega) + p, z - \bar{x}^{s,y;\bar{u}}(t,\omega)\rangle - \frac{1}{2}(z - \bar{x}^{s,y;\bar{u}}(t,\omega))^\top (P(t,\omega) + P)(z - \bar{x}^{s,y;\bar{u}}(t,\omega)) \\
& + o(|z - \bar{x}^{s,y;\bar{u}}(t,\omega)|^2).
\end{aligned}
$$

Then, it is necessary that

$$
p = -p(t), \quad P \leq -P(t), \quad \mathbf{P}\text{-}a.s.
$$

Thus, (35) holds. (36) is immediate. The proof is complete. □

Remark 3.2 It is interesting to note that if $V \in C^{1,2}([0,T] \times \mathbf{R}^n)$, then (34)~(35) reduce to

$$
\begin{cases}
V_x(t,\bar{x}^{s,y;\bar{u}}(t)) = -p(t), \\
V_{xx}(t,\bar{x}^{s,y;\bar{u}}(t)) \leq -P(t), \quad \forall t \in [s,T], \quad \mathbf{P}\text{-}a.s.
\end{cases}
\tag{50}
$$

We point out that in our jump diffusion setting the strict inequality $V_{xx}(t,\bar{x}^{s,y;\bar{u}}(t)) < -P(t)$ may happen in some cases, as shown in the following example.

Example 3.1 Consider the following linear stochastic control system with Poisson jumps $(n = 1)$:

$$
\begin{cases}
dx^{s,y;u}(t) = u(t)dt + \delta u(t)\tilde{N}(dt), \quad t \in [s,T], \\
x^{s,y;u}(s) = y,
\end{cases}
$$

for some $\delta \neq 0$. Here N is a Poisson process with the intensity λdt and $\tilde{N}(dt) := N(dt) - \lambda dt(\lambda > 0)$ is the compensated martingale measure. The quadratic cost functional is taken as

$$
J(s,y;u(\cdot)) = \frac{1}{2}\mathbb{E}\int_s^T \Big(\frac{1}{\lambda} - \frac{1-\lambda}{\lambda}e^{\frac{t-T}{\delta^2}}\Big)x^{s,y;u}(t)^2 dt.
$$

Define

$$\phi(t) = \delta^2\big(1 - e^{\frac{t-T}{\delta^2}}\big), \quad \forall t \in [0, T].$$

For any $u(\cdot) \in \mathcal{U}[s, T]$, applying Itô's formula to $\phi(t)x^{s,y;u}(t)^2$, we have

$$d\phi(t)x^{s,y;u}(t)^2 = \phi(t)\big(2x^{s,y;u}(t)u(t) + \lambda\delta^2 u(t)^2\big)dt + x^{s,y;u}(t)^2\Big(\frac{\phi(t)}{\delta^2} - 1\Big)dt$$
$$+ \phi(t)\big(2\delta x^{s,y;u}(t-)u(t) + \delta^2 u(t)^2\big)\tilde{N}(dt).$$

Integrating from s to T, taking expectation on both sides, we have

$$\begin{aligned} 0 &= \mathbb{E}[\phi(T)x^{s,y;u}(T)^2] \\ &= \phi(s)y^2 + \mathbb{E}\int_s^T \big[2\phi(t)x^{s,y;u}(t)u(t) + \lambda\delta^2\phi(t)u(t)^2 + x^{s,y;u}(t)^2\big(\frac{1}{\delta^2}\phi(t) - 1\big)\big]dt. \end{aligned}$$

Thus

$$J(s, y; u(\cdot)) = \frac{1}{2}\phi(s)y^2 + \mathbb{E}\int_s^T \lambda\delta^2\phi(t)\Big[u(t) + \frac{x^{s,y;u}(t)}{\lambda\delta^2}\Big]^2 dt.$$

This implies that

$$\bar{u}(t) = -\frac{x^{s,y;u}(t)}{\lambda\delta^2}, \quad \forall t \in [s, T],$$

is a state feedback optimal control and the value function is

$$V(s, y) = \frac{1}{2}\delta^2\big(1 - e^{\frac{s-T}{\delta^2}}\big)y^2, \quad \forall (s, y) \in [0, T] \times \mathbf{R}.$$

On the other hand, the second-order adjoint equation is

$$\begin{cases} dP(t) = \Big(\frac{1}{\lambda} - \frac{1-\lambda}{\lambda}e^{\frac{t-T}{\delta^2}}\Big)dt + \int_{\mathbf{E}} R(t, e)\tilde{N}(dedt), & t \in [s, T], \\ P(T) = 0, \end{cases}$$

which implies

$$P(t) = \frac{1-\lambda}{\lambda}\delta^2\big(1 - e^{\frac{s-T}{\delta^2}}\big) + \frac{1}{\lambda}(t - T), \quad \forall t \in [s, T].$$

Then the decreasing property of function $\rho(t) \equiv V_{xx}(t, \bar{x}^{s,y;\bar{u}}(t)) + P(t)$ (noting $\rho(T) = 0$) results in

$$V_{xx}(t, \bar{x}^{s,y;\bar{u}}(t)) = \delta^2\big(1 - e^{\frac{t-T}{\delta^2}}\big) < -\frac{1-\lambda}{\lambda}\delta^2\big(1 - e^{\frac{s-T}{\delta^2}}\big) - \frac{1}{\lambda}(t - T) = -P(t), \quad \forall t \in [0, T).$$

We proceed to study the super-subjets of the value function in the time variable t along an optimal trajectory. Different from its deterministic counterpart (see [18]) and similar to but more complicated than its diffusion (without jumps) counterpart (see [18] or [19]), we observe that it is *not* the generalized Hamiltonian G that is to be maximized in the stochastic MP unless V is sufficiently smooth. Instead, it is the following $\mathcal{G}$-function which contains an additional term than the $\mathcal{H}$-function in the stochastic MP (Theorem 2.3). Associated with an optimal pair $(\bar{x}^{s,y;\bar{u}}(\cdot), \bar{u}(\cdot))$, its corresponding adjoint processes $(p(\cdot), q(\cdot), \gamma(\cdot,\cdot))$ and $(P(\cdot), Q(\cdot), R(\cdot,\cdot))$ satisfying (20), we define a $\mathcal{G}$-function $\mathcal{G} : [0, T] \times \mathbf{R}^n \times \mathbf{U} \to \mathbf{R}$ as

$$
\begin{aligned}
&\mathcal{G}(t,x,u)\\
&:= \mathcal{H}(t,x,u) + \frac{1}{2}\mathrm{tr}\Big\{R(t,e)\bar{c}(t,e)\bar{c}(t,e)^\top\Big\}\pi(de)\\
&\equiv \langle p(t), b(t,x,u)\rangle + \mathrm{tr}\{q(t)^\top \sigma(t,x,u)\} + \int_{\mathbf{E}} \langle \gamma(t,e), c(t,x,u,e)\rangle \pi(de) - f(t,x,u)\\
&\quad + \frac{1}{2}\mathrm{tr}\Big\{P(t)\sigma(t,x,u)\sigma(t,x,u)^\top - 2P(t)\sigma(t,x,u)\bar{\sigma}(t)^\top\Big\}\\
&\quad + \frac{1}{2}\mathrm{tr}\Big\{\int_{\mathbf{E}} \big[P(t)c(t,x,u,e)c(t,x,u,e)^\top - 2P(t)c(t,x,u,e)\bar{c}(t,e)^\top\big]\pi(de)\Big\}\\
&\quad - \frac{1}{2}\mathrm{tr}\Big\{\int_{\mathbf{E}} \big[R(t,e)c(t,x,u,e)c(t,x,u,e)^\top - 2R(t,e)c(t,x,u,e)\bar{c}(t,e)^\top\big]\pi(de)\Big\}.
\end{aligned}
\tag{51}
$$

Remark 3.3 Recall definitions of the Hamiltonian function H (17) and the generalized Hamiltonian function G (16), we can easily verify that they have the following relations to the above $\mathcal{G}$-function. For a $\Psi \in C([0,T] \times \mathbf{R}^n))$, we have

$$
\begin{aligned}
\mathcal{G}(t,x,u) &:= G(t,x,u,\Psi(t,x),p(t),P(t)) + \mathrm{tr}\{\sigma(t,x,u)^\top [q(t) - P(t)\bar{\sigma}(t)]\}\\
&\quad - \frac{1}{2}\mathrm{tr}\Big\{P(t)\int_{\mathbf{E}} \big[\bar{c}(t,e)\bar{c}(t,e)^\top + \Delta c(t,e;u)\Delta c(t,e;u)^\top\big]\pi(de)\Big\}\\
&\quad + \frac{1}{2}\mathrm{tr}\Big\{\int_{\mathbf{E}} R(t,e)\Delta c(t,e;u)\Delta c(t,e;u)^\top \pi(de)\Big\}\\
&\quad + \int_{\mathbf{E}} \big[\Psi(t,x+c(t,x,u,e)) - \Psi(t,x) + \langle p(t)+\gamma(t,e), c(t,x,u,e)\rangle\big]\pi(de)\\
&= H(t,x,u,p(t),q(t),\gamma(t,\cdot)) - \frac{1}{2}\mathrm{tr}\Big\{P(t)\Big[\bar{\sigma}(t)\bar{\sigma}(t)^\top + \int_{\mathbf{E}} \bar{c}(t,e)\bar{c}(t,e)^\top \pi(de)\Big]\Big\}\\
&\quad + \frac{1}{2}\mathrm{tr}\Big\{P(t)\Big[\Delta\sigma(t;u)\Delta\sigma(t;u)^\top + \int_{\mathbf{E}} \Delta c(t,e;u)\Delta c(t,e;u)^\top \pi(de)\Big]\Big\}\\
&\quad + \frac{1}{2}\mathrm{tr}\Big\{\int_{\mathbf{E}} R(t,e)\Delta c(t,e;u)\Delta c(t,e;u)^\top \pi(de)\Big\}.
\end{aligned}
$$

Note that, unlike the definition of generalized Hamiltonian function G, the $\mathcal{G}$-function can be defined associated with only $\Psi \in C([0,T] \times \mathbf{R}^n))$.

We first recall a few results on *right Lesbesgue points* for functions with values in abstract spaces (see also in pp. 2013-2014 of [8]).

Definition 3.3 *Let Z be a Banach space and let $z : [a,b] \to Z$ be a measurable function that is Bochner integrable. We say that t is a right Lesbesgue point of z if*

$$
\lim_{h\to 0^+} \frac{1}{h}\int_t^{t+h} |z(r) - z(t)|_Z dr = 0.
$$

Lemma 3.1 *Let $z : [a,b] \to Z$ be as in Definition 3.3. Then the set of right Lesbesgue points of z is of full measure in $[a,b]$.*

The second main result in this subsection is the following.

Theorem 3.3 *Suppose that* **(H1)**~**(H3)** *hold and let* $(s,y) \in [0,T) \times \mathbf{R}^n$ *be fixed. Let* $(\bar{x}^{s,y;\bar{u}}(\cdot), \bar{u}(\cdot))$ *be an optimal pair of our stochastic control problem. Let* $(p(\cdot), q(\cdot), \gamma(\cdot,\cdot))$ *and* $(P(\cdot), Q(\cdot), R(\cdot,\cdot))$ *are first-order and second-order adjoint processes, respectively. Then we have*

$$\mathcal{G}(t, \bar{x}^{s,y;\bar{u}}(t), \bar{u}(t)) \in \mathcal{P}^{1,+}_{t+} V(t, \bar{x}^{s,y;\bar{u}}(t)), \quad a.e. t \in [s,T], \mathbf{P}\text{-}a.s., \tag{52}$$

where the $\mathcal{G}$*-function is defined by (51).*

Proof For any $t \in (s,T)$, take $\tau \in (t,T]$. Denote by $x_\tau(\cdot)$ the solution of the following SDEP

$$\begin{aligned} x_\tau(r) = \bar{x}^{s,y;\bar{u}}(t) &+ \int_\tau^r b(\theta, x_\tau(\theta), \bar{u}(\theta))d\theta + \int_\tau^r \sigma(\theta, x_\tau(\theta), \bar{u}(\theta))dW(\theta) \\ &+ \int_{\mathbf{E}}\int_\tau^r c(\theta, x_\tau(\theta-), \bar{u}(\theta), e)\tilde{N}(ded\theta), \quad r \in [\tau, T]. \end{aligned} \tag{53}$$

Set $\xi_\tau(r) := x_\tau(r) - \bar{x}^{s,y;\bar{u}}(r), \tau \le r \le T$. Working under the new probability measure $\mathbf{P}(\cdot|\mathcal{F}^s_\tau)(\omega)$, we have the following estimate by (10):

$$\mathbb{E}[\sup_{\tau\le r\le T} |\xi_\tau(r)|^2 | \mathcal{F}^s_\tau](\omega) \le C|\bar{x}^{s,y;\bar{u}}(\tau,\omega) - \bar{x}^{s,y;\bar{u}}(t,\omega)|^2, \quad \mathbf{P}\text{-}a.s.\omega.$$

Taking $\mathbb{E}(\cdot|\mathcal{F}^s_t)(\omega)$ on both sides and noting that $\mathcal{F}^s_t \subseteq \mathcal{F}^s_\tau$, by (11), we obtain

$$\mathbb{E}[\sup_{\tau\le r\le T} |\xi_\tau(r)|^2 | \mathcal{F}^s_t](\omega) \le C|\tau - t|, \quad \mathbf{P}\text{-}a.s.\omega. \tag{54}$$

The process $\xi_\tau(\cdot)$ satisfies the following variational equations:

$$\begin{cases} d\xi_\tau(r) = \bar{b}_x(r)\xi_\tau(r)dr + \bar{\sigma}_x(r)\xi_\tau(r)dW(r) + \int_{\mathbf{E}} \bar{c}_x(r,e)\xi_\tau(r-)\tilde{N}(dedr) \\ \qquad\qquad + \varepsilon_{\tau1}(r)dr + \varepsilon_{\tau2}(r)dW(r) + \int_{\mathbf{E}} \varepsilon_{\tau3}(r,e)\tilde{N}(dedr), \quad r \in [t,T], \\ \xi_\tau(\tau) = -\int_t^\tau \bar{b}(r)dr - \int_t^\tau \bar{\sigma}(r)dW(r) - \int_{\mathbf{E}}\int_t^\tau \bar{c}(r,e)\tilde{N}(dedr), \end{cases} \tag{55}$$

and

$$\begin{cases} d\xi_\tau(r) = \left\{\bar{b}_x(r)\xi_\tau(r) + \frac{1}{2}\xi_\tau(r)^\top \bar{b}_{xx}(r)\xi_\tau(r)\right\}dr \\ \qquad\qquad + \left\{\bar{\sigma}_x(r)\xi_\tau(r) + \frac{1}{2}\xi_\tau(r)^\top \bar{\sigma}_{xx}(r)\xi_\tau(r)\right\}dW(r) \\ \qquad\qquad + \int_{\mathbf{E}} \left\{\bar{c}_x(r,e)\xi_\tau(r-) + \frac{1}{2}\xi_\tau(r-)^\top \bar{c}_{xx}(r,e)\xi_\tau(r-)\right\}\tilde{N}(dedr) \\ \qquad\qquad + \varepsilon_{\tau4}(r)dr + \varepsilon_{\tau5}(r)dW(r) + \int_{\mathbf{E}} \varepsilon_{\tau6}(r,e)\tilde{N}(dedr), \quad r \in [t,T], \\ \xi_\tau(\tau) = -\int_t^\tau \bar{b}(r)dr - \int_t^\tau \bar{\sigma}(r)dW(r) - \int_{\mathbf{E}}\int_t^\tau \bar{c}(r,e)\tilde{N}(dedr), \end{cases} \tag{56}$$

where

$$\begin{cases} \varepsilon_{\tau 1}(r) := \int_0^1 \big[b_x(r, \bar{x}^{s,y;\bar{u}}(r) + \theta\xi_\tau(r), \bar{u}(r)) - \bar{b}_x(r)\big]\xi_\tau(r)d\theta, \\ \varepsilon_{\tau 2}(r) := \int_0^1 \big[\sigma_x(r, \bar{x}^{s,y;\bar{u}}(r) + \theta\xi_\tau(r), \bar{u}(r)) - \bar{\sigma}_x(r)\big]\xi_\tau(r)d\theta, \\ \varepsilon_{\tau 3}(r,\cdot) := \int_0^1 \big[c_x(r, \bar{x}^{s,y;\bar{u}}(r-) + \theta\xi_\tau(r-), \bar{u}(r), \cdot) - \bar{c}_x(r,\cdot)\big]\xi_\tau(r-)d\theta, \\ \varepsilon_{\tau 4}(r) := \int_0^1 (1-\theta)\xi_\tau(r)^\top \big[b_{xx}(r, \bar{x}^{s,y;\bar{u}}(r) + \theta\xi_\tau(r), \bar{u}(r)) - \bar{b}_{xx}(r)\big]\xi_\tau(r)d\theta, \\ \varepsilon_{\tau 5}(r) := \int_0^1 (1-\theta)\xi_\tau(r)^\top \big[\sigma_{xx}(r, \bar{x}^{s,y;\bar{u}}(r) + \theta\xi_\tau(r), \bar{u}(r)) - \bar{\sigma}_{xx}(r)\big]\xi_\tau(r)d\theta, \\ \varepsilon_{\tau 6}(r,\cdot) := \int_0^1 (1-\theta)\xi_\tau(r-)^\top \big[c_{xx}(r, \bar{x}^{s,y;\bar{u}}(r-) + \theta\xi_\tau(r-), \bar{u}(r), \cdot) - \bar{c}_{xx}(r,\cdot)\big]\xi_\tau(r-)d\theta. \end{cases}$$

Similar to the proof of (41) and (42), there exists a deterministic continuous and increasing function $\delta : [0, \infty) \to [0, \infty)$, with $\frac{\delta(r)}{r} \to 0$ as $r \to 0$, such that

$$\begin{cases} \mathbb{E}\Big[\int_\tau^T |\varepsilon_{\tau 1}(r)|^2 dr \big| \mathcal{F}_t^s\Big](\omega) \le \delta(|\tau - t|), \quad \mathbf{P}\text{-}a.s.\omega, \\ \mathbb{E}\Big[\int_\tau^T |\varepsilon_{\tau 2}(r)|^2 dr \big| \mathcal{F}_t^s\Big](\omega) \le \delta(|\tau - t|), \quad \mathbf{P}\text{-}a.s.\omega, \\ \mathbb{E}\Big[\int_\tau^T ||\varepsilon_{\tau 3}(r,\cdot)||_{\mathcal{L}^2}^2 dr \big| \mathcal{F}_t^s\Big](\omega) \le \delta(|\tau - t|), \quad \mathbf{P}\text{-}a.s.\omega, \\ \mathbb{E}\Big[\int_\tau^T |\varepsilon_{\tau 4}(r)|^2 dr \big| \mathcal{F}_t^s\Big](\omega) \le \delta(|\tau - t|^2), \quad \mathbf{P}\text{-}a.s.\omega, \\ \mathbb{E}\Big[\int_\tau^T |\varepsilon_{\tau 5}(r)|^2 dr \big| \mathcal{F}_t^s\Big](\omega) \le \delta(|\tau - t|^2), \quad \mathbf{P}\text{-}a.s.\omega, \\ \mathbb{E}\Big[\int_\tau^T ||\varepsilon_{\tau 6}(r,\cdot)||_{\mathcal{L}^2}^2 dr \big| \mathcal{F}_t^s\Big](\omega) \le \delta(|\tau - t|^2), \quad \mathbf{P}\text{-}a.s.\omega. \end{cases} \tag{57}$$

Note that $\bar{u}(\cdot)|_{[\tau,T]} \in \mathcal{U}[\tau, T]$, $\mathbf{P}$-$a.s.$ Thus by the definition of the value function V, we have

$$V(\tau, \bar{x}^{s,y;\bar{u}}(t)) \le \mathbb{E}\Big[\int_\tau^T f(r, x_\tau(r), \bar{u}(r))dr + h(x_\tau(T)) \big| \mathcal{F}_\tau^s\Big], \quad \forall \tau \in (t, T], \mathbf{P}\text{-}a.s.$$

Taking $\mathbb{E}(\cdot|\mathcal{F}_t^s)$ on both sides and noting that $\mathcal{F}_t^s \subseteq \mathcal{F}_\tau^s$, we conclude that

$$V(\tau, \bar{x}^{s,y;\bar{u}}(t)) \le \mathbb{E}\Big[\int_\tau^T f(r, x_\tau(r), \bar{u}(r))dr + h(x_\tau(T)) \big| \mathcal{F}_t^s\Big], \quad \forall \tau \in (t, T], \mathbf{P}\text{-}a.s. \tag{58}$$

Choose a common subset $\Omega_0 \subseteq \Omega$ with $P(\Omega_0) = 1$ such that for any $\omega_0 \in \Omega_0$,

$$\begin{cases} V(t, \bar{x}^{s,y;\bar{u}}(t, \omega_0)) = \mathbb{E}\Big[\int_s^T \bar{f}(r)dr + h(\bar{x}^{s,y;\bar{u}}(T)) \big| \mathcal{F}_t^s\Big](\omega_0), \\ (54), (57) \text{ are satisfied for any rational } \tau > t, \text{ and } \bar{u}(\cdot)|_{[\tau,T]} \in \mathcal{U}[\tau, T]. \end{cases}$$

Let $\omega_0 \in \Omega_0$ be fixed, then for any rational $\tau > t$, we have (noting (58))

$$\begin{aligned}
& V(\tau, \bar{x}^{s,y;\bar{u}}(t,\omega_0)) - V(t, \bar{x}^{s,y;\bar{u}}(t,\omega_0)) \\
\le\ & \mathbb{E}\Big\{ -\int_t^\tau \bar{f}(r)dr + \int_\tau^T [f(r, x_\tau(r), \bar{u}(r)) - \bar{f}(r)]dr + h(x_\tau(T)) - h(\bar{x}^{s,y;\bar{u}}(T)) \big| \mathcal{F}_t^s \Big\}(\omega_0) \\
=\ & \mathbb{E}\Big\{ -\int_t^\tau \bar{f}(r)dr + \int_\tau^T \langle \bar{f}_x(r), \xi_\tau(r)\rangle dr + \langle h_x(\bar{x}^{s,y;\bar{u}}(T)), \xi_\tau(T)\rangle \\
& + \frac{1}{2}\int_\tau^T \xi_\tau(r)^\top \bar{f}_{xx}(r)\xi_\tau(r)dr + \frac{1}{2}\xi_\tau(T)^\top h_{xx}(\bar{x}^{s,y;\bar{u}}(T))\xi_\tau(T) \big| \mathcal{F}_t^s \Big\}(\omega_0) + o(|\tau - t|).
\end{aligned}$$

As in (48) (using the duality technique), we have

$$\begin{aligned}
& V(\tau, \bar{x}^{s,y;\bar{u}}(t,\omega_0)) - V(t, \bar{x}^{s,y;\bar{u}}(t,\omega_0)) \\
\le\ & -\mathbb{E}\Big\{ -\int_t^\tau \bar{f}(r)dr \big| \mathcal{F}_t^s \Big\}(\omega_0) - \mathbb{E}\Big\{ \langle p(\tau), \xi_\tau(\tau)\rangle + \frac{1}{2}\xi_\tau(\tau)^\top P(\tau)\xi_\tau(\tau) \big| \mathcal{F}_t^s \Big\}(\omega_0) \\
& + o(|\tau - t|).
\end{aligned} \tag{59}$$

Now let us estimate the terms on the right-hand side of (59). To this end, we first note that for any $\varphi(\cdot), \hat{\varphi}(\cdot) \in L^2_{\mathcal{F}}([0,T]; \mathbf{R}^n)$, $\psi(\cdot) \in L^2_{\mathcal{F},p}([0,T]; \mathbf{R}^{n\times d})$, $\Phi(\cdot,\cdot)) \in F^2_p([0,T]; \mathbf{R}^n)$, we have the following three estimates:

$$\begin{aligned}
& \mathbb{E}\Big\{ \Big\langle \int_t^\tau \varphi(r)dr, \int_t^\tau \hat{\varphi}(r)dr \Big\rangle \big| \mathcal{F}_t^s \Big\}(\omega_0) \\
& \quad \le \Big\{ \mathbb{E}\Big[\big|\int_t^\tau \varphi(r)dr\big|^2 \big| \mathcal{F}_t^s \Big](\omega_0) \Big\}^{\frac{1}{2}} \Big\{ \mathbb{E}\Big[\big|\int_t^\tau \hat{\varphi}(r)dr\big|^2 \big| \mathcal{F}_t^s \Big](\omega_0) \Big\}^{\frac{1}{2}} \\
& \quad \le (\tau - t)\Big\{ \int_t^\tau \mathbb{E}\big[|\varphi(r)|^2 | \mathcal{F}_t^s\big](\omega_0)dr \int_t^\tau \mathbb{E}\big[|\hat{\varphi}(r)|^2 | \mathcal{F}_t^s\big](\omega_0)dr \Big\}^{\frac{1}{2}} \\
& \quad = o(|\tau - t|), \quad \text{as } \tau \downarrow t, \quad \forall t \in [s,T),
\end{aligned}$$

$$\begin{aligned}
& \mathbb{E}\Big\{ \Big\langle \int_t^\tau \varphi(r)dr, \int_t^\tau \psi(r)dW(r) \Big\rangle \big| \mathcal{F}_t^s \Big\}(\omega_0) \\
& \quad \le \Big\{ \mathbb{E}\Big[\big|\int_t^\tau \varphi(r)dr\big|^2 \big| \mathcal{F}_t^s \Big](\omega_0) \Big\}^{\frac{1}{2}} \Big\{ \mathbb{E}\Big[\big|\int_t^\tau \psi(r)dW(r)\big|^2 \big| \mathcal{F}_t^s \Big](\omega_0) \Big\}^{\frac{1}{2}} \\
& \quad \le (\tau - t)^{\frac{1}{2}}\Big\{ \int_t^\tau \mathbb{E}\big[|\varphi(r)|^2 | \mathcal{F}_t^s\big](\omega_0)dr \int_t^\tau \mathbb{E}\big[|\psi(r)|^2 | \mathcal{F}_t^s\big](\omega_0)dr \Big\}^{\frac{1}{2}} \\
& \quad = o(|\tau - t|), \quad \text{as } \tau \downarrow t, \quad a.e. t \in [s,T),
\end{aligned}$$

$$\begin{aligned}
& \mathbb{E}\Big\{ \Big\langle \int_t^\tau \varphi(r)dr, \int_{\mathbf{E}}\int_t^\tau \Phi(r,e)\tilde{N}(dedr) \Big\rangle \big| \mathcal{F}_t^s \Big\}(\omega_0) \\
& \le \Big\{ \mathbb{E}\Big[\big|\int_t^\tau \varphi(r)dr\big|^2 \big| \mathcal{F}_t^s \Big](\omega_0) \Big\}^{\frac{1}{2}} \Big\{ \mathbb{E}\Big[\big|\int_{\mathbf{E}}\int_t^\tau \Phi(r,e)\tilde{N}(dedr)\big|^2 \big| \mathcal{F}_t^s \Big](\omega_0) \Big\}^{\frac{1}{2}} \\
& \le (\tau - t)^{\frac{1}{2}}\Big\{ \int_t^\tau \mathbb{E}\big[|\varphi(r)|^2 | \mathcal{F}_t^s\big](\omega_0)dr \int_{\mathbf{E}}\int_t^\tau \mathbb{E}\big[|\Phi(r,e)|^2 | \mathcal{F}_t^s\big](\omega_0)\pi(de)dr \Big\}^{\frac{1}{2}} \\
& = o(|\tau - t|), \quad \text{as } \tau \downarrow t, \quad a.e. t \in [s,T).
\end{aligned}$$

All the last equalities in the above three inequalities is due to the fact that the sets of right Lebesgue points have full Lebesgue measures for integrable functions by Lemma 3.1 and $t \mapsto \mathcal{F}_t^s$ is right continuous in t. Thus applying Itô's formula to $\langle p(\cdot), \xi_\tau(\cdot)\rangle$, by (55), (18) we have

$$\begin{aligned}
&\mathbb{E}\Big\{\langle p(\tau), \xi_\tau(\tau)\rangle \big| \mathcal{F}_t^s\Big\}(\omega_0)\\
&= \mathbb{E}\Big\{\langle p(t), \xi_\tau(\tau)\rangle + \langle p(\tau) - p(t), \xi_\tau(\tau)\rangle \big| \mathcal{F}_t^s\Big\}(\omega_0)\\
&= \mathbb{E}\Big\{\Big\langle p(t), -\int_t^\tau \bar{b}(r)dr - \int_t^\tau \bar{\sigma}(r)dW(r) - \int_{\mathbf{E}}\int_t^\tau \bar{c}(r,e)\tilde{N}(dedr)\Big\rangle\\
&\quad + \Big\langle -\int_t^\tau \Big(\bar{b}_x(r)p(r) + \bar{\sigma}_x(r)q(r) + \int_{\mathbf{E}} \bar{c}_x(r,e)\gamma(r,e)\pi(de)\Big)dr\\
&\quad + \int_t^\tau q(r)dW(r) + \int_{\mathbf{E}}\int_t^\tau \gamma(r,e)\tilde{N}(dedr), -\int_t^\tau \bar{b}(r)dr\\
&\quad - \int_t^\tau \bar{\sigma}(r)dW(r) - \int_{\mathbf{E}}\int_t^\tau \bar{c}(r,e)\tilde{N}(dedr)\Big\rangle \big| \mathcal{F}_t^s\Big\}(\omega_0)\\
&= \mathbb{E}\Big\{-\langle p(t), \int_t^\tau \bar{b}(r)dr\rangle - \int_t^\tau \mathrm{tr}\{q(r)^\top \bar{\sigma}(r)\}dr\\
&\quad - \int_{\mathbf{E}}\int_t^\tau \langle \gamma(r,e), \bar{c}(r,e)\rangle \pi(de)dr \big| \mathcal{F}_t^s\Big\}(\omega_0) + o(|\tau - t|), \quad \text{as } \tau \downarrow t,\ a.e. t \in [s,T).
\end{aligned} \tag{60}$$

Similarly, applying Itô's formula to $\xi_\tau(\cdot)^\top P(\cdot)\xi_\tau(\cdot)\rangle$, by (56) and (19), we have

$$\begin{aligned}
&\mathbb{E}\Big\{\xi_\tau(\tau)^\top P(\tau)\xi_\tau(\tau) \big| \mathcal{F}_t^s\Big\}(\omega_0)\\
&= \mathbb{E}\Big\{\int_t^\tau \mathrm{tr}\{\bar{\sigma}(r)^\top P(t)\bar{\sigma}(r)\}dr + \int_{\mathbf{E}}\int_t^\tau \mathrm{tr}\{\bar{c}(r,e)^\top P(t)\bar{c}(r,e)\}\pi(de)dr \big| \mathcal{F}_t^s\Big\}(\omega_0)\\
&\quad + o(|\tau - t|), \quad \text{as } \tau \downarrow t, \quad a.e. t \in [s,T).
\end{aligned} \tag{61}$$

It follows from (59), (60) and (61) that for any rational $\tau \in (t, T]$,

$$\begin{aligned}
&V(\tau, \bar{x}^{s,y;\bar{u}}(t,\omega_0)) - V(t, \bar{x}^{s,y;\bar{u}}(t,\omega_0))\\
&\le \mathbb{E}\Big\{-\int_t^\tau \bar{f}(r)dr + \langle p(t), \int_t^\tau \bar{b}(r)dr\rangle + \int_t^\tau \mathrm{tr}\{q(r)^\top \bar{\sigma}(r)\}dr\\
&\quad - \frac{1}{2}\int_t^\tau \mathrm{tr}[\bar{\sigma}(r)^\top P(t)\bar{\sigma}(r)]dr + \int_{\mathbf{E}}\int_t^\tau \langle \gamma(r,e)^\top \bar{c}(r,e)\rangle \pi(de)dr\\
&\quad - \frac{1}{2}\int_{\mathbf{E}}\int_t^\tau \mathrm{tr}\{\bar{c}(r,e)^\top P(t)\bar{c}(r,e)\}\pi(de)dr \big| \mathcal{F}_t^s\Big\}(\omega_0) + o(|\tau - t|)\\
&\le (\tau - t)\Big\{\mathcal{H}(t, \bar{x}^{s,y;\bar{u}}(t,\omega_0), \bar{u}(t)) + \frac{1}{2}\mathrm{tr}\{R(t,e)\bar{c}(t,e)\bar{c}(t,e)^\top\}\pi(de)\Big\}\\
&= (\tau - t)\mathcal{G}(t, \bar{x}^{s,y;\bar{u}}(t,\omega_0), \bar{u}(t)) + o(|\tau - t|), \quad \text{as } \tau \downarrow t, \quad a.e. t \in [s,T).
\end{aligned} \tag{62}$$

By the same argument as in the paragraph following (49), we conclude that (62) holds for any $\tau > t$. By definition (30), then (52) holds. The proof is complete. □

Remark 3.4 As aforementioned, it is worth to point out that by comparing the result of Theorem 3.3 with those analogue in the diffusion case (see Theorem 4.7 on pp. 263 of [18]), we observe that an additional term $-\frac{1}{2}\mathrm{tr}\{R(t,e)\bar{c}(t,e)\bar{c}(t,e)^\top\}\pi(de)$ – containing the integrand with respect to the compensated martingale measure $\tilde{N}$ in the second-order adjoint equation appears in the $\mathcal{G}$-function of (52). This is different from the continuous diffusion case where this $\mathcal{G}$-function coincides with the $\mathcal{H}$-function appearing in the maximum principle.

Now, let us combine Theorem 3.2 and 3.3 to get the following result.

Theorem 3.4 *Suppose that* **(H1)**$\sim$**(H3)** *hold and let* $(s,y) \in [0,T) \times \mathbf{R}^n$ *be fixed. Let* $(\bar{x}^{s,y;\bar{u}}(\cdot),\bar{u}(\cdot))$ *be the optimal pair of our stochastic control problem. Let* $(p(\cdot),q(\cdot),\gamma(\cdot,\cdot))$ *and* $(P(\cdot),Q(\cdot),R(\cdot,\cdot))$ *are first-order and second-order adjoint processes, respectively. Then we have*

$$[\mathcal{G}(t,\bar{x}^{s,y;\bar{u}}(t),\bar{u}(t)),\infty) \times \{-p(t)\} \times [-P(t),\infty) \subseteq \mathcal{P}^{1,2,+}_{t+,x} V(t,\bar{x}^{s,y;\bar{u}}(t)), \ a.e. t \in [s,T], \mathbf{P}\text{-}a.s., \tag{63}$$

$$\mathcal{P}^{1,2,-}_{t+,x} V(t,\bar{x}^{s,y;\bar{u}}(t)) \subseteq (-\infty,\mathcal{G}(t,\bar{x}^{s,y;\bar{u}}(t),\bar{u}(t))] \times \{-p(t)\} \times (-\infty,P(t)], \ a.e. t \in [s,T], \mathbf{P}\text{-}a.s. \tag{64}$$

Proof The first conclusion (63) can be proved by combining the proofs of (34) and (52) and making use of (3.1). We now show (64). For any $q \in \mathcal{P}^{1,-}_{t+} V(t,\bar{x}^{s,y;\bar{u}}(t))$, by definition (31) and (62) we have

$$\begin{aligned} 0 &\le V(\tau,\bar{x}^{s,y;\bar{u}}(t)) - V(t,\bar{x}^{s,y;\bar{u}}(t)) - q(\tau-t) \\ &= \big(\mathcal{G}(t,\bar{x}^{s,y;\bar{u}}(t),\bar{u}(t)) - q\big)(\tau-t) + o(|\tau-t|), \quad a.e. t \in [s,T], \mathbf{P}\text{-}a.s. \end{aligned}$$

Then it is necessary that

$$q \le \mathcal{G}(t,\bar{x}^{s,y;\bar{u}}(t),\bar{u}(t)), \quad a.e. t \in [s,T], \mathbf{P}\text{-}a.s.$$

From this and (35), we have (64). The proof is complete. □

Theorem 3.4 is a generalization of the classical result on the relationship between stochastic MP and DPP (see Theorem 2.1 of [7]). On the other hand, we do have a simple example showing that both the set inclusions in (63) and (64) may be strict.

Example 3.2 Consider the following linear stochastic control system with Poisson jumps ($n = d = 1$):

$$\begin{cases} dx^{s,y;u}(t) = x^{s,y;u}(t)u(t)dt + x^{s,y;u}(t)dW(t) + x^{s,y;u}(t-)u(t)\tilde{N}(dt), \quad t \in [s,T], \\ x^{s,y;u}(s) = y. \end{cases}$$

Here N is a Poisson process with the intensity λdt and $\tilde{N}(dt) := N(dt) - \lambda dt (\lambda > 0)$ is the compensated martingale measure. The control domain is $\mathbf{U} = [0,1]$ and the cost functional is

$$J(s,y;u(\cdot)) = -\mathbb{E}x^{s,y;u}(T).$$

The corresponding HJB equation (15) now reads

$$\begin{cases} -v_t(t,x) - \lambda v(t,x) - \frac{1}{2}x^2 v_{xx}(t,x) \\ \quad + \sup_{0\le u\le 1} [(1-\lambda)v_x(t,x)xu + \lambda v(t,x+xu)] = 0, \quad t \in [s,T], \\ v(T,x) = -x. \end{cases}$$

It is not difficulty to verify that the following function is a viscosity solution of (3.2):

$$V(t,x) = \begin{cases} -x, & \text{if } x \le 0, \\ -xe^{T-t}, & \text{if } x > 0, \end{cases}$$

which clearly satisfy (12) and (13). Thus, by the uniqueness of the viscosity solution, V coincides with the value function of the optimal control problem. However, it is *not* in $C^{1,2}([0,T] \times \mathbf{R})$, since $V_x(t,x)$ has a jump at $(t,0)$ for all $0 \le t < T$. On the other hand, we have (noting (18))

$$\begin{cases} -dp(t) = [\bar{u}(t)p(t) + q(t) + \lambda\gamma(t)\bar{u}(t)]dt - q(t)dW(t) - \gamma(t)\tilde{N}(dt), \quad t \in [s,T], \\ p(T) = 1. \end{cases}$$

We can solve that

$$(p(t), q(t), \gamma(t)) = (e^{\int_t^T \bar{u}(s)ds}, 0, 0).$$

Then with

$$\mathcal{H}(t, \bar{x}^{s,y;\bar{u}}(t), u) = e^{\int_t^T \bar{u}(s)ds}\bar{x}^{s,y;\bar{u}}(t)u - \lambda\bar{u}(t)\bar{x}^{s,y;\bar{u}}(t)^2 u + \frac{1}{2}\lambda\bar{x}^{s,y;\bar{u}}(t)^2 u^2,$$

by Theorem 2.3, we have for $t \in [s,T]$,

$$\bar{u}(t) = \begin{cases} 1, & \bar{x}^{s,y;\bar{u}}(t) > 0, \\ 0, & \bar{x}^{s,y;\bar{u}}(t) \le 0. \end{cases}$$

Now consider $s = 0, y = 0$. Clearly, $(\bar{x}^{0,0;\bar{u}}(\cdot), \bar{u}(\cdot)) \equiv (0,0)$ is an optimal control. Theorem 2.1 of [7] does not apply, because $V_x(t,x)$ does not exist along the *whole* trajectory $\bar{x}^{0,0;\bar{u}}(t), t \in [0,T]$. However, one can show that

$$\begin{cases} \mathcal{P}^{1,2,+}_{t+,x} V(t, \bar{x}^{0,0;\bar{u}}(t)) = \mathcal{P}^{1,2,+}_{t+,x} V(t,0) = [0,\infty) \times [-e^{T-t}, -1] \times [0,\infty), \quad t \in [0,T], \\ \mathcal{P}^{1,2,-}_{t+,x} V(t, \bar{x}^{0,0;\bar{u}}(t)) = \mathcal{P}^{1,2,-}_{t+,x} V(t,0) = [-\infty, 0] \times \varnothing \times \varnothing, \quad t \in [0,T], \\ p(t) = 1, \quad t \in [0,T], \qquad P(t) = 0, \quad t \in [0,T], \\ \mathcal{G}(t, \bar{x}^{0,0;\bar{u}}(t), \bar{u}(t)) = \mathcal{G}(t,0,0) = 0, \quad t \in [0,T]. \end{cases} \tag{65}$$

Thus, it is clear that both the set inclusions in (63) and (64) of Theorem 3.4 are strict for $t \in [0,T)$.

The following result is the special case when we assume the value function is enough smooth.

Corollary 3.1 *Suppose that* **(H1)**$\sim$**(H3)** *hold and let* $(s,y) \in [0,T) \times \mathbf{R}^n$ *be fixed. Let* $(\bar{x}^{s,y;\bar{u}}(\cdot), \bar{u}(\cdot))$ *be an optimal pair of our stochastic optimal control problem and* $(p(\cdot), q(\cdot), \gamma(\cdot,\cdot))$ *be the first-order adjoint processes. Suppose the value function* $V \in C^{1,2}([0,T] \times \mathbf{R}^n)$. *Then*

$$\begin{aligned} V_t(t, \bar{x}^{s,y;\bar{u}}(t)) &= G(t, \bar{x}^{s,y;\bar{u}}(t), \bar{u}(t), -V(t, \bar{x}^{s,y;\bar{u}}(t)), -V_x(t, \bar{x}^{s,y;\bar{u}}(t)), -V_{xx}(t, \bar{x}^{s,y;\bar{u}}(t))) \\ &= \sup_{u \in \mathbf{U}} G(t, \bar{x}^{s,y;\bar{u}}(t), u, -V(t, \bar{x}^{s,y;\bar{u}}(t)), -V_x(t, \bar{x}^{s,y;\bar{u}}(t)), -V_{xx}(t, \bar{x}^{s,y;\bar{u}}(t))), \\ & \qquad a.e. t \in [s,T], \mathbf{P}\text{-}a.s., \end{aligned} \tag{66}$$

where G is defined by (16). Moreover, if $V \in C^{1,3}([0,T] \times \mathbf{R}^n)$ *and* V_{tx} *is also continuous, then*

$$\begin{cases} V_x(t, \bar{x}^{s,y;\bar{u}}(t)) = -p(t), \qquad \forall t \in [s,T], \mathbf{P}\text{-}a.s., \\ V_{xx}(t, \bar{x}^{s,y;\bar{u}}(t))\bar{\sigma}(t) = -q(t), \qquad a.e. t \in [s,T], \mathbf{P}\text{-}a.s., \\ V_x(t, \bar{x}^{s,y;\bar{u}}(t-) + \bar{c}(t,\cdot)) - V_x(t, \bar{x}^{s,y;\bar{u}}(t-)) = -\gamma(t,\cdot), \qquad a.e. t \in [s,T], \mathbf{P}\text{-}a.s. \end{cases} \tag{67}$$

By martingale representation theorem (see Lemma 2.3, [16]) and Itô's formula (see [10]), the proof technique is quite similar to Theorem 4.1, Chapter 4 of [18]. So we omit the detail. In fact, the relationship in (67) also can be seen in Theorem 2.1 of [7]. See also (3) in Introduction of this chapter.

Remark 3.5 (i) On the assumption that the value function V is smooth, the first equality in (66) show us the relationship between the derivative of V with respective to the time variable and the generalized Hamiltonian function G defined by (16) along an optimal state trajectory.

(ii) It is interesting to note that the second equality in (66) may be regard as a "maximum principle" in terms of the value function and its derivatives. It is different from the stochastic MP aforementioned (Theorem 2.3), where no value function or its derivatives is involved.

(iii) The three equalities in (67) show us the relationship between the derivative of V with respective to the state variable and the adjoint processes $p, q, \gamma(\cdot)$. More precisely, the three adjoint processes $p, q, \gamma(\cdot)$ can be expressed in terms of the derivatives of V with respective to the state variable along an optimal state trajectory. It is also interesting to note that from the third equality in (67), we observe that the jump amplitude of $V_x(t, \bar{x}^{s,y;\bar{u}}(t))$ equal to $-\gamma(t,\cdot)$ which is just that $V_x(t, \bar{x}^{s,y;\bar{u}}(t)) = -p(t)$ tell us by the first-order adjoint equation (18).

Remark 3.6 By Remark 3.2 and Example 3.1, it can be seen that though the first classical relation in (67) of Corollary 3.1 is recovered from Theorem 3.2 when the value function V is smooth enough, the nonsmooth version of the second classical relation in (67), i.e., $q(t) = P(t)\bar{\sigma}(t)$, does not hold in general. We are also interested to the question that to what extent the third classical relation in (67) can be generalized when V is not smooth. However, it seems that Theorem 3.2 tells us nothing in this context while the following result gives the general relationship among $p, q, \gamma(\cdot), P, \bar{\sigma}$ and $\bar{c}(\cdot)$.

Proposition 3.3 *Under the assumption of Theorem 3.4, we have*

$$\begin{aligned} 0 \le\ & tr\{\bar{\sigma}(t)^\top [q(t) - P(t)\bar{\sigma}(t)]\} - \frac{1}{2} tr\Big\{P(t) \int_{\mathbf{E}} \bar{c}(t,e)\bar{c}(t,e)^\top \pi(de)\Big\} \\ & + \int_{\mathbf{E}} \big[\psi_1(t, x + \bar{c}(t,e)) - \psi_1(t,x) + \langle p(t) + \gamma(t,e), \bar{c}(t,e)\rangle\big]\pi(de), \quad a.e. t \in [s,T], \mathbf{P}\text{-}a.s., \end{aligned} \tag{68}$$

or, equivalently,

$$\mathcal{G}(t, \bar{x}^{s,y;\bar{u}}(t), \bar{u}(t)) \ge G(t, \bar{x}^{s,y;\bar{u}}(t), \bar{u}(t), -\psi_1(t, \bar{x}^{s,y;\bar{u}}(t)), p(t), P(t)), \quad a.e. t \in [s,T], \mathbf{P}\text{-}a.s., \tag{69}$$

where $\psi_1 \in C^{1,2}([0,T] \times \mathbf{R}^n)$, *such that* $\psi_1(t',x') > V(t',x'), \forall (t',x') \neq (t,x) \in [s,T] \times \mathbf{R}^n$.

Proof By (63) and the fact that V is a viscosity solution of (15), we have

$$\begin{aligned}
0 \geq & -\mathcal{G}(t, \bar{x}^{s,y;\bar{u}}(t), \bar{u}(t)) + \sup_{u \in \mathbf{U}} G(t, \bar{x}^{s,y;\bar{u}}(t), u, -\psi_1(t, \bar{x}^{s,y;\bar{u}}(t)), p(t), P(t)) \\
\geq & -G(t, \bar{x}^{s,y;\bar{u}}(t), \bar{u}(t), -\psi_1(t, \bar{x}^{s,y;\bar{u}}(t)), p(t), P(t)) \\
& + \sup_{u \in \mathbf{U}} G(t, \bar{x}^{s,y;\bar{u}}(t), u, -\psi_1(t, \bar{x}^{s,y;\bar{u}}(t)), p(t), P(t)) \\
& - \operatorname{tr}\{\bar{\sigma}(t)^\top [q(t) - P(t)\bar{\sigma}(t)]\} + \frac{1}{2}\operatorname{tr}\Big\{P(t) \int_{\mathbf{E}} \bar{c}(t,e)\bar{c}(t,e)^\top \pi(de)\Big\} \\
& - \int_{\mathbf{E}} [\psi_1(t, x + \bar{c}(t,e)) - \psi_1(t,x) + \langle p(t) + \gamma(t,e), \bar{c}(t,e)\rangle] \pi(de).
\end{aligned}$$

Then, (68) or (69) follows. The proof is complete. □

4. Conclusion

In this chapter, we have derived the relationship between the maximum principle and dynamic programming principle for the stochastic optimal control problem of jump diffusions. Without involving any derivatives of the value function, relations among the adjoint processes, the generalized Hamiltonian and the value function are derived in the language of viscosity solutions and the associated super-subjets. The conditions under which the above results are valid are very mild and reasonable. The results in this chapter bridge an important gap in the literature.

Author details

Jingtao Shi
Shandong University, P. R. China

5. References

[1] Alvarez, D. & Tourin, A. (1996). Viscosity solutions of nonlinear integral-differential equations. *Ann. Inst. H. Poincaré Anal. Non Linéaire*, Vol. 13, No. 3, 293–317.

[2] Barles, G., Buckdahn, R. & Pardoux, E. (1997). Backward stochastic differential equations and integral-partial differential equations. *Stochastics & Stochastics Reports*, Vol. 60, 57–83.

[3] Barles, G. & Imbert, C. (2008). Second-order elliptic integral-differential equations: Viscosity solutions' theory revisited. *Ann. Inst. H. Poincaré Anal. Non Linéaire*, Vol. 25, No. 3, 567–585.

[4] Bensoussan, A. (1981). Lectures on stochastic control. *Lecture Notes in Mathematics*, Vol. 972, Springer–Verlag, Berlin.

[5] Bismut, J.M. (1978). An introductory approach to duality in optimal stochastic control. *SIAM Journal on Control and Optimization*, Vol. 20, No. 1, 62–78.

[6] Biswas, I.H., Jakobsen, E.R. & Karlsen, K.H. (2010). Viscosity solutions for a system of integro-PDEs and connections to optimal switching and control of jump-diffusion processes. *Applied Mathematics and Optimization*, Vol. 62, No. 1, 47–80.

[7] Framstad, N.C., Øksendal, B. & Sulem, A. (2004). A sufficient stochastic maximum principle for optimal control of jump diffusions and applications to finance. *Journal of Optimization Theory and Applications*, Vol. 121, No. 1, 77–98. (Errata (2005), this journal, Vol. 124, No. 2, 511–512.)

[8] Gozzi, F., Swiech, A. & Zhou, X.Y. (2005). A corrected proof of the stochastic verification theorem within the framework of viscosity solutions. *SIAM Journal on Control and Optimization*, Vol. 43, No. 6, 2009–2019.

[9] Haussmann, U.G. (1981). On the adjoint process for optimal control of diffusion processes. *SIAM Journal on Control and Optimization*, Vol. 19, No. 2, 221–243.

[10] Ikeda, N. & Watanabe, S. (1989). *Stochastic Differential Equations and Diffusion Processes.* North-Holland, Kodansha.

[11] Li, J. & Peng, S.G. (2009). Stochastic optimization theory of backward stochastic differential equations with jumps and viscosity solutions of Hamilton-Jacobi-Bellman equations. *Nonlinear Analysis*, Vol. 70, 1776–1796.

[12] Øksendal, B. & Sulem, A. (2005). *Applied Stochastic Control of Jump Diffusions.* Springer–Verlag, Berlin.

[13] Pardoux, E. & Peng, S.G. (1990). Adapted solution of a backward stochastic differential equation. *Systems & Control Letters*, Vol. 14, 55–61.

[14] Pham, H. (1998). Optimal stopping of controlled jump diffusion processes: A viscosity solution approach. *Journal of Mathematical Systems, Estimations and Control*, Vol. 8, No. 1, 1–27.

[15] Situ, R. (2005). *Theory of Stochastic Differential Equations with Jumps and Applications.* Springer, New York.

[16] Tang, S.J. & Li, X.J. (1994). Necessary conditions for optimal control of stochastic systems with random jumps. *SIAM Journal on Control and Optimization*, Vol. 32, No. 5, 1447–1475.

[17] Yong, J.M. (1992). *Dynamic Programming Method and the Hamilton-Jacobi-Bellman Equations* (in Chinese). Shanghai Science and Technology Press, Shanghai.

[18] Yong, J.M. & Zhou, X.Y. (1999). *Stochastic Controls: Hamiltonian Systems and HJB Equations.* Springer–Verlag, New York.

[19] Zhou, X.Y. (1990). The connection between the maximum principle and dynamic programming in stochastic control. *Stochastics & Stochastics Reports*, Vol. 31, 1–13.

Chapter 7

Singular Stochastic Control in Option Hedging with Transaction Costs

Tze Leung Lai and Tiong Wee Lim

Additional information is available at the end of the chapter

1. Introduction

An option written on an underlying asset (e.g., stock) confers on its holder the right to receive a certain payoff before or on a certain (expiration) date T. The payoff $f(\cdot)$ is a function of the price of the underlying asset at the time of exercise (i.e., claiming the payoff), or more generally, a functional of the asset price path up to the time of exercise. We focus here on European options, for which exercise is allowable only at T, which are different from American options, for which early exercise at any time before T is also allowed. For example, the holder of a European call (resp. put) option has the right to buy (resp. sell) the underlying asset at T at a certain (strike) price K. Denoting by S_T the asset price at T, the payoff of the option is $f(S_T)$, with $f(S) = (S-K)^+$ and $(K-S)^+$ for a call and put, respectively.

Black & Scholes [1] made seminal contributions to the theory of option pricing and hedging by modeling the asset price as a geometric Brownian motion and assuming that (i) the market has a risk-free asset with constant rate of return r, (ii) no transaction costs are imposed on the sale or purchase of assets, (iii) there are no limits on short selling, and (iv) trading occurs continuously. Specifically, the asset price S_t at time t satisfies the stochastic differential equation

$$dS_t = \alpha S_t \, dt + \sigma S_t \, dW_t, \quad S_0 > 0, \tag{1}$$

where $\alpha \in \mathbb{R}$ and $\sigma > 0$ are the mean and standard deviation (or volatility) of the asset's return, and $\{W_t, \ t \geq 0\}$ is a standard Brownian motion (with $W_0 = 0$) on some filtered probability space $(\Omega, \mathcal{F}, \{\mathcal{F}_t, \ t \geq 0\}, \mathbb{P})$. The absence of transaction fees permits the construction of a continuously rebalanced portfolio consisting of $\mp\Delta$ unit of the asset for every ± 1 unit of the European option such that its rate of return equals the risk-free rate r, where $\Delta = \partial c/\partial S$ (resp. $\partial p/\partial S$) for a call (resp. put) whose price is c (resp. p). By requiring this portfolio to be *self-financing* (in the sense that all subsequent rebalancing is financed entirely by the initial capital and, if necessary, by short selling the risk-free asset) and to perfectly replicate the outcome of the European option at expiration, Black & Scholes [1] have shown that the "fair" value of the option in the absence of arbitrage opportunities is the initial amount of capital $\hat{E}\{e^{-rT}f(S_T)\}$, where $\hat{E}$ denotes expectation under the equivalent martingale measure (with respect to which S_t has drift $\alpha = r$). Instead of considering geometric Brownian motion $S_t = S_0 \exp\{(r - \sigma^2/2)t + \sigma W_t\}$, it is convenient to work directly with Brownian motion W_t. The fact that σW_t and $W_{\sigma^2 t}$ have the same distribution suggests the change of variables

$$s = \sigma^2(t - T), \quad z = \log(S/K) - (\rho - 1/2)s \tag{2}$$

with $\rho = r/\sigma^2$. The Black-Scholes option pricing formulas $\hat{E}\{e^{-r(T-t)}f(S_T) \,|\, S_t = S\}$ are given explicitly by $c(s,z) = Ke^{\rho s}[e^{z-s/2}\Phi(z/\sqrt{-s} - \sqrt{-s}) - \Phi(z/\sqrt{-s})]$ for the call and $p(s,z) = Ke^{\rho s}[\Phi(-z/\sqrt{-s}) - e^{z-s/2}\Phi(-\{z/\sqrt{-s} - \sqrt{-s}\})]$ for the put, where Φ is the standard normal distribution function. Correspondingly, the option deltas are $\Delta(s,z) = \pm\Phi(\pm\{z/\sqrt{-s} - \sqrt{-s}\})$ with + for the call and − for the put.

In the presence of transaction costs, perfect hedging of a European option is not possible (since it results in an infinite turnover of the underlying asset and is, therefore, ruinously expensive) and trading in an option involves an essential element of risk. This hedging risk can be characterized as the difference between the realized cash flow from a hedging strategy which uses the initial option premium to trade in the underlying asset and bond, and the desired option payoff at maturity. By embedding option hedging within the framework of portfolio selection introduced by Magill & Constantinides [12] and Davis & Norman [13], Hodges & Neuberger [15] used a risk-averse utility function to assess this shortfall (or "replication error") and formulated the option hedging problem as one of maximizing the investor's expected utility of terminal wealth. This involves the value functions of two singular stochastic control problems, for trading in the market with and without a (short or long) position in the option, and the optimal hedge is given by the difference in the trading strategies corresponding to these two problems. The nature of the optimal hedge is that an investor with an option position should rebalance his portfolio only when the number of units of the asset falls "too far" out of line. For the negative exponential utility function, Davis et al. [14], Clewlow & Hodges [11], and Zakamouline [20] have developed numerical methods to compute the optimal hedge by making use of discrete-time dynamic programming on an approximating binomial tree for the asset price; see Kushner & Dupuis [3] for the general theory of Markov chain approximations for continuous-time processes and their use in the numerical solution of optimal stopping and control problems. More recently, Lai & Lim [18] introduced a new numerical method for solving the singular stochastic control problems associated with utility maximization, yielding a much simpler algorithm to compute the buy and sell boundaries and value functions in the utility-based approach.

The new method is motivated by the equivalence between singular stochastic control and optimal stopping, which was first observed in the pioneering work of Bather & Chernoff [10] on the problem of controlling the motion of a spaceship relative to its target on a finite horizon with an infinite amount of fuel and has since been established for the general class of bounded variation follower problems by Karatzas & Shreve [7, 8], Karatzas & Wang [9] and Boetius [2]. By transforming the original singular stochastic control problem to an optimal stopping problem associated with a Dynkin game, the solution can be computed by applying standard backward induction to an approximating Bernoulli walk. Lai & Lim [18] showed how this backward induction algorithm can be modified, by making use of finite difference methods if necessary, for the more general singular stochastic control problem of option hedging even though it is not reducible to an equivalent optimal stopping problem because of the presence of additional value functions.

In Section 2, we review the equivalence theory between singular stochastic control and optimal stopping. We also outline the development of the computational schemes of Lai & Lim [18] to solve stochastic control problems that are equivalent to optimal stopping. In Section 3, we introduce the utility-based option hedging problem, outline how the algorithm in Section 2 can be modified to solve stochastic control problems for which equivalence does not exist, and provide numerical examples to illustrate the use of the coupled backward induction algorithm to compute the optimal buy and sell boundaries of a short European call option. We conclude in Section 4.

2. Singular stochastic control and optimal stopping

Bather & Chernoff [10] pioneered the study of singular stochastic control in their analysis of the problem of controlling the motion of a spaceship relative to its target on a finite horizon with an infinite amount of fuel. A key idea in their analysis is the reduction of the stochastic control problem to an optimal stopping problem via a change of variables. This spaceship control problem is an example of a *bounded variation follower problem* and the equivalence between singular stochastic control and optimal stopping has since been established for a general class of bounded variation follower problems by Karatzas & Shreve [7, 8], Karatzas & Wang [9] and Boetius [2]. In this section, we review this equivalence for a particular formulation of the bounded variation follower problem in which the control $\xi^+ - \xi^-$ is not applied additively to the Brownian motion $\{Z_u\}$ and outline the backward induction algorithm for solving the equivalent Dynkin game.

2.1. A bounded variation follower problem and its equivalent optimal stopping problem

Suppose that the state process $\mathbf{S} = \{S_t,\ t \geq 0\}$ representing the underlying stochastic environment (in the absence of control) is given by (1). In our subsequent application to option hedging, $\mathbf{S}$ represents the price of the asset underlying the option, whereas in other applications such as to the problem of reversible investment by Guo & Tomecek [21], $\mathbf{S}$ represents an economic indicator reflecting demand for a certain commodity. A singular control process is given by a pair (ξ^+, ξ^-) of adapted, nondecreasing, LCRL processes such that $d\xi^+$ and $d\xi^-$ are supported on disjoint subsets. We are interested in problems with a finite time horizon, i.e., we consider the time interval $[0, T]$ for some terminal time $T \in (0, \infty)$. Given any times $0 \leq s \leq t \leq T$, $\xi^+_{t+} - \xi^+_s$ and $\xi^-_{t+} - \xi^-_s$ represent the cumulative increase and decrease, respectively, in control level resulting from the controller's decisions over the time interval $[s, t]$, with $\xi^+_0 = \xi^-_0 = 0$. The total control value is therefore given by the finite variation process

$$x_t = x_0 + \xi^+_t - \xi^-_t. \tag{3}$$

A pair (ξ^+, ξ^-) is an *admissible* singular control if, in addition to the above requirements, $x_t \in \overline{\mathcal{I}}$ for all $t \in [0, T]$, where $\mathcal{I}$ is an open, possibly unbounded interval of $\mathbb{R}$ and $\overline{\mathcal{I}}$ is its closure.

Let $F(t,S,x)$, $\kappa^{\pm}(t,S)$ and $G(S,x)$ be sufficiently smooth functions, with F and G representing the running and terminal reward, respectively, and $\kappa^{\pm}$ the costs of exerting control. The goal of the controller is to maximize an objective function of the form:

$$J_{t,S,x}(\xi^+,\xi^-) = E_{t,S,x}\left\{ \int_t^T e^{-r(u-t)} F(u,S_u,x_u)\,du - \int_{[t,T)} e^{-r(u-t)}\kappa^+(u,S_u)\,d\xi_u^+ \right.$$
$$\left. - \int_{[t,T)} e^{-r(u-t)}\kappa^-(u,S_u)\,d\xi_u^- + e^{-r(T-t)}G(S_T,x_T) \right\},$$

where $E_{t,S,x}$ denotes conditional expectation given $S_t = S$ and $x_t = x$. The value function of the stochastic control problem is

$$V(t,S,x) = \sup_{(\xi^+,\xi^-)\in\mathcal{A}_{t,x}} J_{t,S,x}(\xi^+,\xi^-), \quad (t,S,x) \in [0,T]\times(0,\infty)\times\overline{\mathcal{I}}, \tag{4}$$

where $\mathcal{A}_{t,x}$ denotes the set of all admissible controls which satisfy $x_t = x$.

We derive formally the Hamilton-Jacobi-Bellman equation associated with the stochastic control problem (4), which provides key insights into the nature of the optimal control. Consider, for now, a smaller set $\mathcal{A}_{t,x}^k$ of admissible controls such that $\xi^{\pm}$ are absolutely continuous processes, i.e., $d\xi_t^{\pm} = \ell_t^{\pm}\,dt$ with $0 \le \ell_t^+, \ell_t^- \le k < \infty$. Under this restriction, the value function (4) becomes

$$V^k(t,S,x) = \sup_{(\ell^+,\ell^-)\in\mathcal{A}_{t,x}^k} J_{t,S,x}^k(\ell^+,\ell^-), \quad (t,S,x) \in [0,T]\times(0,\infty)\times\overline{\mathcal{I}},$$

where

$$J_{t,S,x}^k(\ell^+,\ell^-)$$
$$= E_{t,S,x}\left\{ \int_t^T e^{-r(u-t)}\left[F(u,S_u,x_u) - \kappa^+(u,S_u)\ell_u^+ - \kappa^-(u,S_u)\ell_u^-\right]du + e^{-r(T-t)}G(S_T,x_T)\right\}.$$

Since the infinitesimal generator of the stochastic system comprising (1) and $dx_t = (\ell_t^+ - \ell_t^-)\,dt$ (corresponding to (3) for absolutely continuous $\xi^{\pm}$) is

$$\alpha S\frac{\partial}{\partial S} + \frac{\sigma^2 S^2}{2}\frac{\partial^2}{\partial S^2} + (\ell^+ - \ell^-)\frac{\partial}{\partial x},$$

the Bellman equation for $V^k(t,S,x)$ is

$$\max_{0\le \ell^+,\ell^- \le k}\left\{ \left[\mathcal{L}_{t,S} + (\ell^+ - \ell^-)\frac{\partial}{\partial x}\right]V^k(t,S,x) + F(t,S,x) - \kappa^+(t,S)\ell^+ - \kappa^-(t,S)\ell^- \right\} = 0,$$

where

$$\mathcal{L}_{t,S} = \frac{\partial}{\partial t} + \alpha S\frac{\partial}{\partial S} + \frac{\sigma^2 S^2}{2}\frac{\partial^2}{\partial S^2} - r,$$

or equivalently,

$$\max_{0\le \ell^+,\ell^- \le k}\left\{\left[\frac{\partial V^k}{\partial x}(t,S,x)-\kappa^+(t,S)\right]\ell^+ - \left[\frac{\partial V^k}{\partial x}(t,S,x)+\kappa^-(t,S)\right]\ell^-\right\}$$

$$+\mathcal{L}_{t,S}V^k(t,S,x)+F(t,S,x)=0,\quad (t,S,x)\in[0,T)\times(0,\infty)\times\overline{\mathcal{I}}.$$

Assuming the value function to be an increasing function of x, the optimal control is obtained by considering the following three possible cases (all the other permutations of inequalities are impossible):

(i) If $\partial V^k(t,S,x)/\partial x-\kappa^+(t,S)\ge 0$ and $\partial V^k(t,S,x)/\partial x+\kappa^-(t,S)>0$, then the maximum is achieved by $\ell^-=0$ and exerting control ξ^+ at the maximum possible rate $\ell^+=k$.

(ii) If $\partial V^k(t,S,x)/\partial x-\kappa^+(t,S)<0$ and $\partial V^k(t,S,x)/\partial x+\kappa^-(t,S)\le 0$, then the maximum is achieved by $\ell^+=0$ and exerting control ξ^- at the maximum possible rate $\ell^-=k$.

(iii) If $\partial V^k(t,S,x)/\partial x-\kappa^+(t,S)\le 0$ and $\partial V^k(t,S,x)/\partial x+\kappa^-(t,S)\ge 0$, then the maximum is achieved by not exerting any control, i.e., $\ell^+=\ell^-=0$, and $V^k(t,S,x)$ satisfies the partial differential equation (PDE) $\mathcal{L}_{t,S}V^k(t,S,x)+F(t,S,x)=0$.

Thus, the state space $[0,T]\times(0,\infty)\times\overline{\mathcal{I}}$ is partitioned into three regions, which we denote by $\mathcal{N}$ (corresponding to no control), $\mathcal{B}$ (control ξ^+ is exerted), and $\mathcal{S}$ (control ξ^- is exerted). The boundaries between the no-control region $\mathcal{N}$ and the regions $\mathcal{B}$ and $\mathcal{S}$ are denoted by $\partial\mathcal{B}$ and $\partial\mathcal{S}$.

As $k\to\infty$, the set $\mathcal{A}^k_{t,x}$ of admissible controls becomes the set $\mathcal{A}_{t,x}$ of problem (4) and the state space remains partitioned into the regions $\mathcal{N}$, $\mathcal{B}$ and $\mathcal{S}$. If $(t,S,x)\in\mathcal{B}$ (resp. $\mathcal{S}$), then the control ξ^+ (resp. ξ^-) must be instantaneously exerted to bring the state to the boundary $\partial\mathcal{B}$ (resp. $\partial\mathcal{S}$). Thus, besides an initial jump from $\mathcal{B}$ or $\mathcal{S}$ to the boundary $\partial\mathcal{B}$ or $\partial\mathcal{S}$ (if necessary), the optimal control process acts thereafter only when $(t,S,x)\in\partial\mathcal{B}$ or $\partial\mathcal{S}$ so as to keep the state in $\mathcal{N}\cup\partial\mathcal{B}\cup\partial\mathcal{S}$. Because the optimal process behaves like the *local time* of the (optimally controlled) state process at the boundaries, such a control is termed *singular*. In $\mathcal{B}$, since the optimal control is to increase x by a positive amount δx (up to that required to take the state to $\partial\mathcal{B}$) at the cost of $\kappa^+(t,S)$ per unit increase, the value function satisfies the equation

$$V(t,S,x)=V(t,S,x+\delta x)-\kappa^+(t,S)\delta x\quad(\text{in }\mathcal{B}).$$

Similarly, since the optimal control in $\mathcal{S}$ is to decrease x by a positive amount δx (up to that required to take the state to $\partial\mathcal{S}$) at the cost of $\kappa^-(t,S)$ per unit decrease, the value function satisfies the equation

$$V(t,S,x)=V(t,S,x-\delta x)-\kappa^-(t,S)\delta x\quad(\text{in }\mathcal{S}).$$

Letting $\delta x\to 0$ leads to gradient constraints for the value function in $\mathcal{B}$ and $\mathcal{S}$. In $\mathcal{N}$, $V(t,S,x)$ continues to satisfy the PDE given in (iii) above. From these observations, we obtain the following free boundary problem (FBP) for the value function $V(t,S,x)$:

$$\mathcal{L}_{t,S}V(t,S,x)+F(t,S,x)=0 \qquad \text{in } \mathcal{N}, \tag{5a}$$

$$\frac{\partial V}{\partial x}(t,S,x)=\kappa^{+}(t,S) \qquad \text{in } \mathcal{B}, \tag{5b}$$

$$\frac{\partial V}{\partial x}(t,S,x)=-\kappa^{-}(t,S) \qquad \text{in } \mathcal{S}, \tag{5c}$$

$$V(T,S,x)=G(S,x). \tag{5d}$$

It also follows that the Hamilton-Jacobi-Bellman equation associated with (4) is the following variational inequality with gradient constraints:

$$\max\left\{\mathcal{L}_{t,S}V(t,S,x)+F(t,S,x),\ \frac{\partial V}{\partial x}(t,S,x)-\kappa^{+}(t,S),\ -\frac{\partial V}{\partial x}(t,S,x)-\kappa^{-}(t,S)\right\}=0, \tag{6}$$

$(t,S,x)\in[0,T)\times(0,\infty)\times\overline{\mathcal{I}}$.

With $\rho=r/\sigma^2$ and $\beta=\alpha/\sigma^2$, a more parsimonious parameterization of (5) can be obtained by considering the change of variables (2) (without K and with ρ replaced by β here since the state process $\mathbf{S}$ has rate of return α under the "physical" measure rather than r under the risk-neutral measure) and $v(s,z,x)=e^{-\rho s}V(t,S,x)$. Applying the chain rule of differentiation yields

$$\frac{\partial V}{\partial S}=\frac{e^{\rho s}}{S}\frac{\partial v}{\partial z},\quad \frac{\partial^2 V}{\partial S^2}=\frac{e^{\rho s}}{S^2}\left(\frac{\partial^2 v}{\partial z^2}-\frac{\partial v}{\partial z}\right),\quad \frac{\partial V}{\partial t}-rV=e^{\rho s}\left[\sigma^2\frac{\partial v}{\partial s}-\left(\alpha-\frac{\sigma^2}{2}\right)\frac{\partial v}{\partial z}\right].$$

We also define $\tilde{F}(s,z,x)=e^{r(T-t)}F(t,S,x)/\sigma^2$ and $\tilde{\kappa}^{\pm}(s,z)=e^{-\rho s}\kappa^{\pm}(t,S)$. Upon substitution into (5), we arrive at the FBP

$$\left\{\frac{\partial}{\partial s}+\frac{1}{2}\frac{\partial^2}{\partial z^2}\right\}v(s,z,x)+\tilde{F}(s,z,x)=0 \qquad \text{in } \mathcal{N}, \tag{7a}$$

$$\frac{\partial v}{\partial x}(s,z,x)=\tilde{\kappa}^{+}(s,z) \qquad \text{in } \mathcal{B}, \tag{7b}$$

$$\frac{\partial v}{\partial x}(s,z,x)=-\tilde{\kappa}^{-}(s,z) \qquad \text{in } \mathcal{S}, \tag{7c}$$

$$v(0,z,x)=G(e^z,x), \tag{7d}$$

Note that $v(s,z,x)$ is the value function of the corresponding singular stochastic control problem for the Brownian motion $\{Z_u,\ u\le 0\}$:

$$v(s,z,x)=\sup_{(\xi^+,\xi^-)\in\mathcal{A}_{s,x}} E_{s,z,x}\left\{\int_s^0 \tilde{F}(u,Z_u,x_u)\,du-\int_{[s,0)}\tilde{\kappa}^{+}(u,Z_u)\,d\xi_u^{+}\right.$$
$$\left.-\int_{[s,0)}\tilde{\kappa}^{-}(u,Z_u)\,d\xi_u^{-}+G(e^{Z_0},x_0)\right\}, \tag{8}$$

where $E_{s,z,x}$ denotes conditional expectation given $Z_s=z$ and $x_s=x$.

We now introduce the change of variables

$$w(s,z,x) = \frac{\partial v}{\partial x}(s,z,x), \quad (s,z,x) \in [-\sigma^2 T, 0] \times \mathbb{R} \times \overline{\mathcal{I}}. \tag{9}$$

From (7) it follows that w solves the FBP

$$\left\{\frac{\partial}{\partial s} + \frac{1}{2}\frac{\partial^2}{\partial z^2}\right\} w(s,z,x) + \phi(s,z,x) = 0 \qquad \text{in } \mathcal{N}, \tag{10a}$$

$$w(s,z,x) = \tilde{\kappa}^+(s,z) \qquad \text{in } \mathcal{B}, \tag{10b}$$

$$w(s,z,x) = -\tilde{\kappa}^-(s,z) \qquad \text{in } \mathcal{S}, \tag{10c}$$

$$w(0,z,x) = g(e^z, x), \tag{10d}$$

where $\phi(s,z,x) = \partial\tilde{F}(s,z,x)/\partial x$ and $g(\cdot,x) = \partial G(\cdot,x)/\partial x$. The FBP (10) can be restated as an optimal stopping problem associated with a Dynkin game, for which $w(s,z,x)$ is the value function. Specifically,

$$\begin{aligned} w(s,z,x) = \underline{w}(s,z,x) &:= \sup_{\tau^- \in \mathcal{T}(s,0)} \inf_{\tau^+ \in \mathcal{T}(s,0)} I_{s,z,x}(\tau^+,\tau^-) \\ = \overline{w}(s,z,x) &:= \inf_{\tau^+ \in \mathcal{T}(s,0)} \sup_{\tau^- \in \mathcal{T}(s,0)} I_{s,z,x}(\tau^+,\tau^-), \end{aligned} \tag{11}$$

where $\mathcal{T}(a,b)$ denotes the set of stopping times taking values between a and b $(> a)$, and

$$\begin{aligned} I_{s,z,x}(\tau^+,\tau^-) = E_{s,z,x}\Bigg\{ \int_s^{\tau^+\wedge\tau^-} \phi(u, Z_u, x_u)\, du + \tilde{\kappa}^+(\tau^+, Z_{\tau^+}) I_{\{\tau^+<\tau^-<0\}} \\ - \tilde{\kappa}^-(\tau^-, Z_{\tau^-}) I_{\{\tau^-<\tau^+<0\}} + g(e^{Z_0}, x_0) I_{\{\tau^-=\tau^+=0\}} \Bigg\}. \end{aligned}$$

The Dynkin game is a "stochastic game of timing" in which there are two players P and M, each of whom chooses a stopping time (τ^+ and τ^-, respectively) in $\mathcal{T}(s,0)$. As long as the game is in progress, P keeps paying M at the rate $\phi(s,z,x)$ per unit of time. The game terminates as soon as one of the players decides to stop, i.e., at $\tau^+ \wedge \tau^-$. If player M stops first, he pays P the amount $\tilde{\kappa}^-(\tau^-, Z_{\tau^-})$. If player P stops first, he pays M the amount $\tilde{\kappa}^+(\tau^+, Z_{\tau^+})$ (resp. $g(e^{Z_0}, x_0)$) when the game terminates before (resp. at) the end of the time horizon 0. The objective of P is to minimize his expected total payment to M whereas the objective of M is to maximize this quantity.

In addition to the relationship (9) between the value functions v and w, the optimal continuation region of the Dynkin game (11) coincides with the no-control region of the singular stochastic control problem (8) in the sense that if $(\xi^{+,*}, \xi^{-,*})$ is an optimal control of (8) and we define the stopping times $\tau^{+,*} = \inf\{u \in [s,0) : \xi^{+,*}(u) > 0\}$ and $\tau^{-,*} = \inf\{u \in [s,0) : \xi^{-,*}(u) > 0\}$ ($\inf\emptyset = 0$), then $(\tau^{+,*}, \tau^{-,*})$ is a *saddlepoint* of the game with the property that $w(s,z,x) = I_{s,z,x}(\tau^{+,*}, \tau^{-,*})$.

2.2. Example: Reversible investment

Before we outline the computational algorithm for solving the Dynkin game (11), we give an example in mathematical economics of a stochastic control problem which has the form (4). In the notation of (4), the problem of reversible investment is one in which a company, by adjusting its production capacity x_t through expansion ξ_t^+ and contraction ξ_t^- according to market fluctuations S_t, wishes to maximize its overall expected net profit $E_{t,S,x}J_{t,S,x}(\xi^+,\xi^-)$ over a finite horizon. The net profit of such an investment depends on the running production function $F(t,S,x)$ of the actual capacity, the benefits of contraction $\kappa^-(t,S) \equiv K^- < 0$, and the cost of expansion $\kappa^+(t,S) \equiv K^+ > 0$, with $K^+ + K^- > 0$. The economic uncertainty about S_t (such as the price or demand for the product) is modeled by geometric Brownian motion (1).

Guo & Tomecek [21] studied the infinite-horizon ($T = \infty$) reversible investment problem and provided an explicit solution to the problem with the so-called Cobb-Douglas production function $F(t,S,x) = S^\lambda x^\mu$, where $\lambda \in (0,n)$, $n = [-(\alpha - \sigma^2/2) + \sqrt{(\alpha-\sigma^2/2)^2 - 2\sigma^2 r}]/\sigma^2 > 0$ and $\mu \in (0,1]$. The optimal strategy is for the company to increase (resp. decrease) capacity when (S,x) belongs to the investment (resp. disinvestment) region $\mathcal{B}$ (resp. $\mathcal{S}$). Here, $\mathcal{B} = \{(S,x) : x \le X_b(S)\}$ and $\mathcal{S} = \{(S,x) : x \ge X_s(S)\}$, where $X_i(S) = (S/\nu_i)^{\lambda/(1-\alpha)}$, $i = b, s$, ν_b and ν_s are unique solutions to

$$\frac{\alpha}{\lambda - m}(\nu_b^{\lambda-m} - \nu_s^{\lambda-m}) = -\frac{r}{m}(K^+\nu_b^{-m} + K^-\nu_s^{-m}),$$
$$\frac{\alpha}{n - \lambda}(\nu_b^{\lambda-n} - \nu_s^{\lambda-n}) = \frac{r}{n}(K^+\nu_b^{-n} + K^-\nu_s^{-n}),$$

and $m = [-(\alpha - \sigma^2/2) - \sqrt{(\alpha-\sigma^2/2)^2 - 2\sigma^2 r}]/\sigma^2 < 0$.

In the case of finite horizon ($T < \infty$), the investment and disinvestment regions have similar forms but are not stationary in time, i.e., $\mathcal{B} = \{(t,S,x) : x \le X_b(t,S)\}$ and $\mathcal{S} = \{(t,S,x) : x \ge X_s(t,S)\}$. It is not possible to express the boundaries $X_i(t,S)$ explicitly. We can solve for them (after applying the change of variables $(t,S,x) \mapsto (s,z,x)$ given by (2)) by making use of the backward induction algorithm described in the next section; for details and numerical results, see Lai et al. [16].

2.3. Computational algorithm for solving the Dynkin game

In view of the equivalence between the stochastic control problem (8) and the Dynkin game (11), which is an optimal stopping problem with (disjoint) stopping regions $\mathcal{B}$ and $\mathcal{S}$ coinciding with the control regions of (8) as well as continuation region $\mathcal{N}$ coinciding with the no-control region of (8), it suffices to solve (11) rather than (8) directly. The backward induction algorithm outlined below is similar to the algorithms studied by Lai et al. [17], for which convergence properties have been established.

Suppose the buy and sell boundaries can be expressed as functions $X_b(s,z)$ and $X_s(s,z)$ such that $\mathcal{B} = \{(s,z,x) : x \le X_b(s,z)\}$ and $\mathcal{S} = \{(s,z,x) : x \ge X_s(s,z)\}$. Whereas $w(s,z,x)$ is given by (10b) and (10c) in the buy and sell regions (i.e., the stopping region in the Dynkin game), the continuation value of the Dynkin game is a solution of the partial differential equation (10a) and can be computed using backward induction on a symmetric Bernoulli random walk which approximates standard Brownian motion. Specifically, let $T_{\max}$ denote the largest terminal date of interest, take small positive δ and ϵ such that $N := \sigma^2 T_{\max}/\delta$ is an integer, and let $\mathbf{Z}_\delta = \{0, \pm\sqrt{\delta}, \pm 2\sqrt{\delta}, \dots\}$ and $\mathbf{X}_\epsilon = \{0, \pm\epsilon, \pm 2\epsilon, \dots\}$. For $i = 1, 2, \dots, N$, with $s_0 = 0$, $s_i = s_{i-1} - \delta$, $z \in \mathbf{Z}_\delta$, $x \in \mathbf{X}_\epsilon$, the continuation value at s_i can be computed using

$$\tilde{w}(s_i, z, x) = \delta\phi(s_i, z, x) + [w(s_i + \delta, z + \sqrt{\delta}, x) + w(s_i + \delta, z - \sqrt{\delta}, x)]/2 \tag{12}$$

with $w(0, z, x) = g(e^z, x)$. The following algorithm allows us to solve for $X_b(s_i, z)$ and $X_s(s_i, z)$ for $z \in \mathbf{Z}_\delta$.

Algorithm 1. Let $w(0, z, x) = g(e^z, x)$ for $z \in \mathbf{Z}_\delta$ and $x \in \mathbf{X}_\epsilon$. For $i = 1, 2, \dots, N$ and $z \in \mathbf{Z}_\delta$:

(i) Starting at $x_0 \in \mathbf{X}_\epsilon$ with $\tilde{w}(s_i, z, x_0) < \tilde{\kappa}^+(s_i, z)$, search for the first $j \in \{1, 2, \dots\}$ (denoted by j^*) for which $\tilde{w}(s_i, z, x_0 + j\epsilon) \geq \tilde{\kappa}^+(s_i, z)$ and set $X_b(s_i, z) = x_0 + j^*\epsilon$.

(ii) For $j \in \{1, 2, \dots\}$, let $x_j = X_b(s_i, z) + j\epsilon$. Compute, and store for use at s_{i+1}, $w(s_i, z, x_j) = \tilde{w}(s_i, z, x_j)$ as defined by (12). Search for the first j (denoted by j^*) for which $\tilde{w}(s_i, z, x_j) \geq -\tilde{\kappa}^-(s_i, z)$ and set $X_s(s_i, z) = X_b(s_i, z) + j^*\epsilon$.

(iii) For $x \in \mathbf{X}_\epsilon$ outside the interval $[X_b(s_i, z), X_s(s_i, z)]$, set $w(s_i, z, x) = \tilde{\kappa}^+(s_i, z)$ or $-\tilde{\kappa}^-(s_i, z)$ according to whether $x \leq X_b(s_i, z)$ or $x \geq X_s(s_i, z)$.

The following backward induction equation summarizes this algorithm:

$$w(s_i, z, x) = \begin{cases} \tilde{\kappa}^+(s_i, z) & \text{if } \tilde{w}(s_i, z, x) < \tilde{\kappa}^+(s_i, z), \\ -\tilde{\kappa}^-(s_i, z) & \text{if } \tilde{w}(s_i, z, x) > -\tilde{\kappa}^-(s_i, z), \\ \tilde{w}(s_i, z, x) & \text{otherwise.} \end{cases} \tag{13}$$

Lai et al. [17] have shown that under suitable conditions discrete-time random walk approximations to continuous-time optimal stopping problems can approximate the value function with an error of the order $O(\delta)$ and the stopping boundary with an error of the order $o(\sqrt{\delta})$, where δ is the interval width in discretizing time for the approximating random walk. To prove this result, they approximate the underlying optimal stopping problem by a recursively defined family of "canonical" optimal stopping problems which depend on δ and for which the continuation and stopping regions can be completely characterized, and use an induction argument to provide bounds on the absolute difference between the boundaries of the continuous-time and discrete-time stopping problems as well as that between the value functions of the two problems. Since the Dynkin game is also an optimal stopping problem (with two stopping boundaries), their result can be extended to the present setting to establish that (13) is able to approximate the value function (11) with an error of the order $O(\delta)$ and Algorithm 1 is able to approximate the stopping boundaries $Z_i(s, x) := X_i^{-1}(s, z)$, $i = \text{b}, \text{s}$, corresponding to the optimal stopping problem (11) with an error of the order $o(\sqrt{\delta})$. Further refinements to the random walk approximations can be made by correcting for the excess over the boundary when stopping cocurs in the discrete-time problem; for details and other applications, see Chernoff [4], Chernoff & Petkau [5, 6] and Lai et al. [17].

3. Utility-based option theory in the presence of transaction costs

Consider now a risk-averse investor who trades in a risky asset whose price is given by the geometric Brownian motion (1) and a bond which pays a fixed risk-free rate $r > 0$ with the objective of maximizing the expected utility of his terminal wealth Ω_T^0. The number of units x_t the investor holds in the asset is given by (3), where x_0 denotes the initial asset position and ξ_t^+ (resp. ξ_t^-) represents the cumulative number of units of the asset bought (resp. sold) within the time interval $[0,t]$, $0 \le t \le T$. If the investor pays fractions $0 < \lambda < 1$ and $0 < \mu < 1$ of the dollar value transacted on purchase and sale of the asset, the dollar value y_t of his investment in bond is given by

$$dy_t = ry_t\,dt - aS_t\,d\xi_t^+ + bS_t\,d\xi_t^-,$$

with $a = 1+\lambda$ and $b = 1-\mu$, or more explicitly,

$$y_T = y_t e^{r(T-t)} - a\int_{[t,T)} e^{r(T-u)} S_u\,d\xi_u^+ + b\int_{[t,T)} e^{r(T-u)} S_u\,d\xi_u^-.$$

Let $U : \mathbb{R} \to \mathbb{R}$ be a concave and increasing (hence risk-averse) utility function. We can express the investor's problem in terms of the value function

$$V^0(t,S,x,y) = \sup_{(\xi^+,\xi^-)\in\mathcal{A}_{t,x,y}} E_{t,S,x,y}\left[U\left(\Omega_T^0\right)\right], \tag{14}$$

where $\mathcal{A}_{t,x,y}$ denotes the set of all admissible controls which satisfy $x_t = x$ and $y_t = y$, and $E_{t,S,x,y}$ denotes conditional expectation given $S_t = S$, $x_t = x$ and $y_t = y$. For the special case of the negative exponential utility function $U(z) = 1 - e^{-\gamma z}$, which has constant absolute risk aversion (CARA) γ, we can reduce the number of state variables by one by working with

$$H^0(t,S,x) = 1 - V^0(t,S,x,0)$$
$$= \inf_{(\xi^+,\xi^-)\in\mathcal{A}_{t,x}} E_{t,S,x}\left[\exp\left\{\gamma\left(\int_{[t,T)} e^{r(T-u)} S_u(a\,d\xi_u^+ - b\,d\xi_u^-) - Z^0(S_T,x_T)\right)\right\}\right], \tag{15}$$

where

$$Z^0(S,x) = xS(aI_{\{x<0\}} + bI_{\{x\ge 0\}}) \tag{16}$$

denotes the liquidated value of the asset by trading x units of the asset at price S to zero unit.

If the investor is presented with an opportunity to enter into a position in a European call option written on the given asset, with strike price K and expiration date T, the problem can be formulated in the same way as (14) but with Ω_T^0 replaced by Ω_T^i with $i = \mathrm{s}$ indicating a short call position and $i = \mathrm{b}$ indicating a long call position. The corresponding value functions $V^i(t,S,x,y)$ also admit reductions in dimensionality via $H^i(t,S,x) = 1 - V^i(t,S,x,0)$, but with $Z^0(S_T,x_T)$ in (15) replaced by $Z^i(S_T,x_T)$. If the option is *asset settled*, then the writer delivers one unit of the asset in return for a payment of K when the buyer exercises the option at maturity T, so

$$Z^i(S,x) = Z^0(S, x - \Delta^i(S)) + K\Delta^i(S), \quad i = \mathrm{s},\mathrm{b}, \tag{17}$$

where $\Delta^{\mathrm{s}}(S) = I_{\{S>K\}}$ (short call) and $\Delta^{\mathrm{b}}(S) = -I_{\{S>K\}}$ (long call). In the case of a *cash settled* option, the writer delivers $(S_T - K)^+$ in cash at T, so

$$Z^i(S,x) = Z^0(S,x) - (S-K)\Delta^i(S), \quad i = \mathrm{s},\mathrm{b}. \tag{18}$$

As in Section 2.1, we apply the change of variables (2) (with ρ replaced by $\beta = \alpha/\sigma^2$) to (15) and work with the resulting value function $h^i(s,z,x) = H^i(t,S,x)$. Corresponding to the definitions (16)–(18) of terminal settlement value are

$$A^0(z,x) = xe^z(aI_{\{x<0\}} + bI_{\{x\geq 0\}}),$$
$$A^i(z,x) = A^0(z,x-D^i(z)) + D^i(z) \quad \text{for asset settlement,} \quad i = \mathrm{s},\mathrm{b},$$
$$A^i(z,x) = A^0(z,x) - (e^z-1)D^i(z) \quad \text{for cash settlement,} \quad i = \mathrm{s},\mathrm{b},$$

with $D^{\mathrm{s}}(z) = I_{\{z>0\}}$ (short call) and $D^{\mathrm{b}}(z) = -I_{\{z>0\}}$ (long call). For use in (22d) below, we define $B^i(z,x) := \partial A^i(z,x)/\partial x$, which are given explicitly by $B^0(z,x) = A^0(z,x)/x$ and, for $i = \mathrm{s},\mathrm{b}$, $B^i(z,x) = B^0(z,x-D^i(z))$ under asset settlement and $B^i(z,x) = B^0(z,x)$ under cash settlement. In the sequel, we fix $i = 0,\mathrm{s},\mathrm{b}$ and drop the superscript i in the value functions (and associated quantities).

3.1. Associated free boundary problems and their solutions

The formulation of the option (pricing and) hedging problem as two stochastic control problems of the form (14) goes back to Hodges & Neuberger [15]. Davis et al. [14] derived the Hamilton-Jacobi-Bellman (HJB) equations associated with the control problems $V(t,S,x,y)$ and $H(t,S,x) = 1 - V(t,S,x,0)$ in the same way as we did in Section 2.1. By applying the transformation $h(s,z,x) = H(t,S,x)$ to their HJB equations, we obtain the following free boundary problem (FBP) for $h(s,z,x)$:

$$\frac{\partial h}{\partial s} + \frac{1}{2}\frac{\partial^2 h}{\partial z^2} = 0, \qquad x \in [X_{\mathrm{b}}(s,z), X_{\mathrm{s}}(s,z)], \tag{19a}$$
$$\frac{\partial h}{\partial x}(s,z,x) = w_{\mathrm{b}}(s,z,x), \qquad x \leq X_{\mathrm{b}}(s,z), \tag{19b}$$
$$\frac{\partial h}{\partial x}(s,z,x) = w_{\mathrm{s}}(s,z,x), \qquad x \geq X_{\mathrm{s}}(s,z), \tag{19c}$$
$$h(0,z,x) = \exp\{-\gamma K A(z,x)\}, \tag{19d}$$

where

$$w_{\mathrm{b}}(s,z,x) = -a\gamma K e^{z+(\beta-\rho-1/2)s} h(s,z,x), \tag{20a}$$
$$w_{\mathrm{s}}(s,z,x) = -b\gamma K e^{z+(\beta-\rho-1/2)s} h(s,z,x). \tag{20b}$$

Associated with FBP (19) are three regions: $\mathcal{B} = \{(s,z,x) : x \leq X_{\mathrm{b}}(s,z)\}$ where it is optimal to buy the (risky) asset, $\mathcal{S} = \{(s,z,x) : x \geq X_{\mathrm{s}}(s,z)\}$ where it is optimal to sell the asset, and $\mathcal{N} = [-\sigma^2 T, 0] \times \mathbb{R} \times \mathbb{R} \setminus (\mathcal{B} \cup \mathcal{S})$ where it is optimal to not transact. Since $\partial/\partial s + (1/2)\partial^2/\partial z^2$ is the infinitesimal generator of space-time Brownian motion, this means that while (s, Z_s, x_s) is inside the no-transaction region, the dynamics of $h(s, Z_s, x_s)$ is driven by the standard Brownian motion $\{Z_s,\ s \leq 0\}$. In the buy and sell regions, it follows from (19b) and (19c) that

$$h(s,z,x) = \exp\left\{-a\gamma K e^{z+(\beta-\rho-1/2)s}[x - X_{\mathrm{b}}(s,z)]\right\} h(s,z,X_{\mathrm{b}}(s,z)), \quad x \leq X_{\mathrm{b}}(s,z), \tag{21a}$$

$$h(s,z,x) = \exp\left\{-b\gamma K e^{z+(\beta-\rho-1/2)s}[x - X_{\mathrm{s}}(s,z)]\right\} h(s,z,X_{\mathrm{s}}(s,z)), \quad x \geq X_{\mathrm{s}}(s,z). \tag{21b}$$

Finally, if we let $w(s,z,x) = \partial h(s,z,x)/\partial x$, then $w(s,z,x)$ satisfies the FBP

$$\frac{\partial w}{\partial s} + \frac{1}{2}\frac{\partial^2 w}{\partial z^2} = 0, \qquad x \in [X_b(s,z), X_s(s,z)], \tag{22a}$$

$$w(s,z,x) = w_b(s,z,x), \qquad x \le X_b(s,z), \tag{22b}$$

$$w(s,z,x) = w_s(s,z,x), \qquad x \ge X_s(s,z), \tag{22c}$$

$$w(0,z,x) = -\gamma K B(z,x) h(0,z,x). \tag{22d}$$

If the function $h(s,z,x)$ is known, then by analogy to (10), the FBP (22) is an optimal stopping problem associated with a Dynkin game, and its solution can be computed using the following analog of the backward induction equation (13):

$$w(s_i,z,x) = \begin{cases} w_b(s_i,z,x) & \text{if } \tilde{w}(s_i,z,x) < w_b(s_i,z,x), \\ w_s(s_i,z,x) & \text{if } \tilde{w}(s_i,z,x) > w_s(s_i,z,x), \\ \tilde{w}(s_i,z,x) & \text{otherwise,} \end{cases} \tag{23}$$

where $\tilde{w}(s,z,x)$ is given by (12) with $\phi \equiv 0$. On the other hand, if the boundaries $X_b(s,z)$ and $X_s(s,z)$ are given, then the FBP (19) can also be solved by backward induction: For $z \in \mathbf{Z}_\delta$, compute $h(s_i,z,x)$ using (21) (with s replaced by s_i) if $x \in \mathbf{X}_\epsilon$ is outside the interval $[X_b(s_i,z), X_s(s_i,z)]$, and if $x \in \mathbf{X}_\epsilon \cap [X_b(s_i,z), X_s(s_i,z)]$, let $h(s_i,z,x) = \tilde{h}(s_i,z,x)$ with

$$\tilde{h}(s,z,x) = [h(s+\delta, z+\sqrt{\delta}, x) + h(s+\delta, z-\sqrt{\delta}, x)]/2. \tag{24}$$

By replacing the unknown h in (20a) and (20b) by $\tilde{h}$ and redefining them as $\tilde{w}_b$ and $\tilde{w}_s$, Lai & Lim [18] have developed the coupled backward induction algorithm described below to solve for $X_b(s_i,z)$ and $X_s(s_i,z)$, as well as to compute values of $h(s_i,z,x)$ for $x \in \mathbf{X}_\epsilon \cap [X_b(s_i,z), X_s(s_i,z)]$.

Algorithm 2. Let $h(0,z,x) = \exp\{-\gamma K A(z,x)\}$ and $w(0,z,x) = -\gamma K B(z,x) h(0,z,x)$ for $z \in \mathbf{Z}_\delta$ and $x \in \mathbf{X}_\epsilon$. For $i = 1,2,\ldots,N$ and $z \in \mathbf{Z}_\delta$:

(i) Starting at $x_0 \in \mathbf{X}_\epsilon$ with $\tilde{w}(s_i,z,x_0) < \tilde{w}_b(s_i,z,x_0)$, search for the first $j \in \{1,2,\ldots\}$ (denoted by j^*) for which $\tilde{w}(s_i,z,x_0+j\epsilon) \ge \tilde{w}_b(s_i,z,x_0+j\epsilon)$ and set $X_b(s_i,z) = x_i + j^*\epsilon$.

(ii) For $j \in \{1,2,\ldots\}$, let $x_j = X_b(s_i,z) + j\epsilon$. Compute, and store for use at s_{i+1}, $w(s_i,z,x_j) = \tilde{w}(s_i,z,x_j)$ as defined by (12) with $\phi \equiv 0$ and $h(s_i,z,x_j) = \tilde{h}(s_i,z,x_j)$ by (24). Search for the first j (denoted by j^*) for which $\tilde{w}(s_i,z,x_j) \ge \tilde{w}_s(s_i,z,x_j)$ and set $X_s(s_i,z) = X_b(s_i,z) + j^*\epsilon$.

(iii) For $x \in \mathbf{X}_\epsilon$ outside the interval $[X_b(s_i,z), X_s(s_i,z)]$, compute $h(s_i,z,x)$ using (21) and set $w(s_i,z,x) = w_b(s_i,z,x)$ or $w_s(s_i,z,x)$ as defined by (20) according to whether $x \le X_b(s_i,z)$ or $x \ge X_s(s_i,z)$.

It can be established that the convergence property of this algorithm is similar to that of Algorithm 1 even though (22) is not a stopping problem like (10) is. Specifically, the backward inductions (23) as well as (24) and (21) applied to $\{s_N = -\sigma^2 T, s_{N-1}, \ldots, s_1, s_0 = 0\} \times \mathbf{Z}_\delta \times \mathbf{X}_\epsilon$ are able to approximate the value functions $w(s,z,x)$ and $h(s,z,x)$ of the corresponding continuous-time control problems with an error of the order $O(\delta)$, and Algorithm 2 is able to approximate the buy and sell boundaries $Z_i(s,x) := X_i^{-1}(s,z)$, $i = \mathrm{b},\mathrm{s}$, corresponding to these control problems with an error of the order $o(\sqrt{\delta})$; see Lai et al. [19] for details as well as an extension to the problem of optimal investment and consumption.

3.2. Numerical results

Associated with each of the problems $V^i(t,S,x,y)$, $i = 0$ given by (14) corresponding to an investor with no option position and $i = \mathrm{s},\mathrm{b}$ being analogs of (14) corresponding to an investor with a short or long call position, respectively, is an optimal trading strategy x_t^i ($i = 0,\mathrm{s},\mathrm{b}$) of the form

$$x_t = \begin{cases} X_\mathrm{b}(t,S_t) & \text{if } x_{t-} < X_\mathrm{b}(t,S_t), \\ X_\mathrm{s}(t,S_t) & \text{if } x_{t-} > X_\mathrm{s}(t,S_t), \\ x_{t-} & \text{if } X_\mathrm{b}(t,S_t) \le x_{t-} \le X_\mathrm{s}(t,S_t). \end{cases}$$

The optimal *hedging* strategies for the option writer and buyer are then given by $x_t^\mathrm{s} - x_t^0$ and $x_t^\mathrm{b} - x_t^0$, respectively. In the case of no transaction costs ($\lambda = \mu = 0$), it can be shown that

$$x_t^0 = X_0(t,S) := \frac{e^{-r(T-t)}}{\gamma S}\frac{\alpha - r}{\sigma^2}, \quad x_t^i = \Delta^i(t,S) + X_0(t,S), \quad i = \mathrm{s},\mathrm{b},$$

where $\Delta^\mathrm{b}(t,S)$ (resp. $\Delta^\mathrm{s}(t,S) = -\Delta^\mathrm{b}(t,S)$) denotes the Black-Scholes delta for a long (resp. short) call option, given explicitly (as a function of (s,z) after applying the change of variable (2)) in Section 1. In the case of $\alpha = r$ (risk-neutrality), it can be shown that $x_t^0 \equiv 0$ whether or not there are transaction costs. In particular, if $\alpha = r$ and $\lambda = \mu = 0$, the optimal hedging strategy is to hold Δ shares of stock at all times (see Section 1). Thus, the Black-Scholes option theory is a special case of the more general utility-based option theory.

Whereas Clewlow & Hodges [11] and Zakamouline [20] made use of discrete-time dynamic programming on an approximating binomial tree for the asset price to solve the control problems directly for the optimal hedge, Lai & Lim [18] made use of the simpler Algorithm 2 outlined in Section 3.1. They provided extensive numerical results for the CARA utility function with $\alpha = r$, for which only one pair of boundaries need to be computed (since it is then optimal not to trade in the risky asset when the investor does not have an option position). As an illustration of Algorithm 2, we compute and show in Fig. 1 the optimal buy (lower) and sell (upper) boundaries for a short asset-settled call (solid black lines) with strike price $K = 20$ and for the case of no option (solid red lines) at four different times before expiration ($T - t = 1.5$, 0.5, 0.25, 0.1) when proportional transaction costs are incurred at the rate of $\lambda = \mu = 0.5\%$; the dashed lines correspond to $X_0(t,\cdot)$ and $\Delta^\mathrm{s}(t,\cdot) + X_0(t,\cdot)$ for the case of no transaction costs. Other parameters are: absolute risk aversion $\gamma = 2.0$, risk-free rate $r = 8.5\%$, asset return rate $\alpha = 10\%$ and asset volatility $\sigma = 5\%$. Note that the red boundaries are consistent with the intuitive notion of "buy at the low and sell at the high" when investing only in a risky asset (and bond). However, unlike the case of $\alpha = r$ in which the buy and sell boundaries corresponding to a short asset-settled call always lie between 0 and 1, the black boundaries in this case (where $\alpha \neq r$) do not necessarily take values in the interval $[0,1]$.

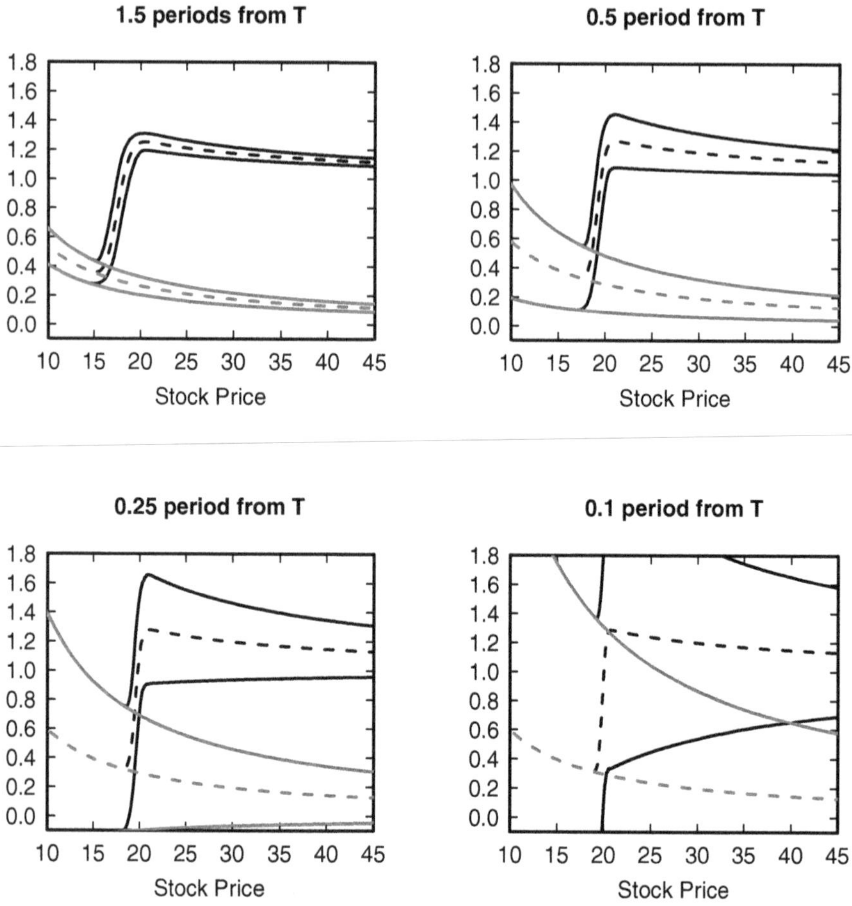

Figure 1. Optimal buy (lower) and sell (upper) boundaries from negative exponential (CARA) utility maximization for a short asset-settled call with strike price $K = 20$ (solid black lines) and for the case of no option (solid red lines), with proportional transaction costs incurred at the rate of $\lambda = \mu = 0.5\%$, absolute risk aversion $\gamma = 2.0$, risk-free rate $r = 8.5\%$, asset return rate $\alpha = 10\%$ and asset volatility $\sigma = 5\%$, at 1.5, 0.5, 0.25 and 0.1 period(s) from expiration T. For each pair of boundaries, the "buy asset" region is below the buy boundary and the "sell asset" region is above the sell boundary; the no-transaction region is between the two boundaries. The dashed lines correspond to the case of no transaction costs.

4. Conclusion

For the so-called bounded variation follower problems, the equivalence between singular stochastic control and optimal stopping can be harnessed to provide a much simpler solution to the control problem by solving the corresponding Dynkin game. This approach can be used on certain control problems for which there does not exist an equivalent stopping problem. We show how the "standard" algorithm can be modified to provide a coupled backward induction algorithm for solving the utility-based option hedging problem and provide numerical illustrations on the vanilla call option.

Author details

Tze Leung Lai
Department of Statistics
Stanford University, U.S.A.

Tiong Wee Lim
Department of Statistics and Applied Probability
National University of Singapore, Republic of Singapore

5. References

[1] F. Black and M. Scholes. The pricing of options and corporate liabilities. *Journal of Political Economy*, 81:637–654, 1973.
[2] F. Boetius. Bounded variation singular stochastic control and Dynkin game. *SIAM Journal on Control and Optimization*, 44:1289–1321, 2005.
[3] H.J. Kushner and P.G. Dupuis. *Numerical Methods for Stochastic Control in Continuous Time*. Springer-Verlag, New York, 1992.
[4] H. Chernoff. Sequential tests for the mean of a normal distribution. *Annals of Mathematical Statistics*, 36:55–68, 1965.
[5] H. Chernoff and A.J. Petkau. An optimal stopping problem for sums of dichotomous random variables. *Annals of Probability*, 4:875–889, 1976.
[6] H. Chernoff and A.J. Petkau. Numerical solutions for Bayes sequential decision problems. *SIAM Journal on Scientific and Statistical Computing*, 7:46–59, 1986.
[7] I. Karatzas and S.E. Shreve. Connections between optimal stopping and singular stochastic control I. *SIAM Journal on Control and Optimization*, 22:856–877, 1984.
[8] I. Karatzas and S.E. Shreve. Connections between optimal stopping and singular stochastic control II. *SIAM Journal on Control and Optimization*, 23:433–451, 1985.
[9] I. Karatzas and H. Wang. Connections between bounded-variation control and Dynkin games. In J.L. Menaldi, E. Rofman, and A. Sulem, editors, *Optimal Control and Partial Differential Equations—Innovations and Applications*, pages 353–362. IOS Press, Amsterdam, 2001.
[10] J.A. Bather and H. Chernoff. Sequential decisions in the control of a spaceship. In *Proceedings of the Fifth Berkeley Symposium on Mathematical Statistics and Probability, Volume 3—Physical Sciences*, pages 181–207, Berkeley, 1967. University of California Press.
[11] L. Clewlow and S.D. Hodges. Optimal delta-hedging under transaction costs. *Journal of Economic Dynamics and Control*, 21:1353–1376, 1997.
[12] M.J.P. Magill and G.M. Constantinides. Portfolio selection with transaction costs. *Journal of Economic Theory*, 13:245–263, 1976.

[13] M.H.A. Davis and A.R. Norman. Portfolio selection with transaction costs. *Mathematics of Operations Research*, 15:676–713, 1990.

[14] M.H.A. Davis, V.G. Panas, and T. Zariphopoulou. European option pricing with transaction costs. *SIAM Journal on Control and Optimization*, 31:470–493, 1993.

[15] S.D. Hodges and A. Neuberger. Optimal replication of contingent claims under transaction costs. *Review of Futures Markets*, 8:222–239, 1989.

[16] T.L. Lai, T.W. Lim, and K. Ross. A new approach to solving multi-dimensional singular stochastic control problems with applications to investment theories and queueing networks. Working paper, Stanford University, 2012.

[17] T.L. Lai, Y.-C. Yao, and F. AitSahlia. Corrected random wall approximations to free boundary problems in optimal stopping. *Advances in Applied Probability*, 39:753–775, 2007.

[18] T.L. Lai and T.W. Lim. Option hedging theory under transaction costs. *Journal of Economic Dynamics and Control*, 33:1945–1961, 2009.

[19] T.L. Lai, T.W. Lim, and W. Zhou. Backward induction algorithms for singular stochastic control problems associated with transaction costs. Working paper, Stanford University, 2012.

[20] V.I. Zakamouline. European option pricing and hedging with both fixed and proportional transaction costs. *Journal of Economic Dynamics and Control*, 30:1–25, 2006.

[21] X. Guo and P. Tomecek. Connections between singular control and optimal switching. *SIAM Journal on Control and Optimization*, 47:421–443, 2008.

Chapter 8

Stochastic Observation Optimization on the Basis of the Generalized Probabilistic Criteria

Sergey V. Sokolov

Additional information is available at the end of the chapter

1. Introduction

Till now the synthesis problem of the optimum control of the observation process has been considered and solved satisfactorily basically for the linear stochastic objects and observers by optimization of the *quadratic* criterion of quality expressed, as a rule, through the a posteriori dispersion matrix [1-4]. At the same time, the statement of the synthesis problem for the optimum observation control in a more general case assumes, first, a nonlinear character of the object and observer, and, second, the application of the non-quadratic criteria of quality, which, basically, can provide the potentially large estimation accuracy[3-6].

In connection with the fact that the solution of the given problem in such a statement generalizing the existing approaches, represents the obvious interest, we formulate it more particularly as follows.

2. Description of the task

Let the Markovian vector process ξ_t, described generally by the nonlinear stochastic differential equation in the symmetrized form

$$\dot{\xi}_t = f(\xi,t) + f_0(\xi,t)n_t, \qquad \xi(t_0) = \xi_0, \tag{1}$$

where f, f_0 are known N – dimensional vector and $N \times M$ – dimensional matrix nonlinear functions;

n_t is the white Gaussian normalized M – dimensional vector - noise; be observed by means of the vector nonlinear observer of form:

$$Z = H(\xi,t) + W_t,$$

where $Z - L \leq N$ – dimensional vector of the output signals of the meter;

$h(\xi,t)$ – a known nonlinear L- dimension vector - function of observation;

W_t – a white Gaussian L- dimension vector - noise of measurement with the zero average and the matrix of intensity D_W.

The function of the a posteriori probability density (APD) of process $\rho(\xi,t) = \rho(\xi,t|Z_\tau,\ \tau \in [t_0,t])$ is described by the known integro-differential equation in partial derivatives (Stratonovich equation), the right-hand part of which explicitly depends on the observation function h:

$$\frac{\partial\rho(\xi,t)}{\partial t} = L\{\rho(\xi,t)\} + [Q - Q_0]\rho(\xi,t),$$

where $L\{\rho(\xi,t)\} = -div\left\{\left[f + \frac{1}{2}\frac{\partial f_0}{\partial \xi}\left(f_0^T\right)^{(V)}\right]\rho\right\} + \frac{1}{2}div\left\{\overline{div}\left[f_0 f_0^T \rho\right]\right\}$ – the Focker-Plank-operator,

$(A)^{(V)}$ is the operation for transforming the $n \times m$ matrix **A** into vector $(A)^{(V)}$ formed from its elements as follows:

$$A^{(V)} = \left|a_{11}a_{21}\ldots a_{m1}a_{12}a_{22}\ldots a_{m2}\ldots a_{1n}a_{2n}\ldots a_{mn}\right|^T,$$

$\overline{div}$ is the symbol for the operation of divergence of the matrix row,

$$Q = Q(\xi,t) = -\frac{1}{2}\left[Z - H(\xi,t)\right]^T D_W^{-1}\left[Z - H(\xi,t)\right],$$

$$Q_0 = \int_{-\infty}^{\infty} Q(\xi,t)\rho(\xi,t)d\xi.$$

As the main problem of the a posteriori analysis of the observable process ξ_t is the obtaining of the maximum reliable information about it, then the synthesis problem of the optimum observer would be natural to formulate as the definition of the form of the functional dependence $h(\xi,t)$, providing the maximum of the a posteriori probability (MAP) of signal ξ_t on the given interval of occurrence of its values $\xi_* = [\xi_{\min}, \xi_{\max}]$ during the required interval of time $T = [t_0, t_k]$, i.e. in view of the positive definiteness $\rho(\xi,t)$

$$\max\left\{J = \int_T \int_{\xi_*} \rho(\xi,t)d\xi\, dt\right\}$$

or

$$\min\left\{-\int_T\int_{\xi_*}\rho(\xi,t)\,d\xi\,dt\right\}.$$

Generally instead of criterion MAP one can use, for example, the criterion of the minimum of the a posteriori entropy on interval $\xi_* = \left[\xi_{\min}, \xi_{\max}\right]$ or the criterion of the minimum of the integrated deviation of the a posteriori density from the density of the given form etc., that results in the need for representing the criterion of optimality J in the more generalized form:

$$J = \int_T\int_{\xi_*}\Phi\left[\rho(\xi,t)\right]d\xi\,dt,$$

where $\mathbf{\Phi}$ is the known nonlinear function which takes into account generally the feasible analytical restrictions on the vector ξ_t;

$T = [t_0, t_k]$ is a time interval of optimization;

ξ_* is some bounded set of the state parameters ξ_t.

In the final forming of structure of the criterion of optimality J it is necessary to take into account the limited opportunities of the practical realization of the function of observation $h(\xi,t)$, as well, that results, in its turn, in the additional restriction on the choice of functional dependence $h(\xi,t)$. The formalization of the given restriction, for example, in the form of the requirement of the minimization of the integrated deviation of function H from the given form H_0 on interval ξ_* during time interval T allows to write down analytically the form of the minimized criterion J as follows:

$$J = \int_T\int_{\xi_*}\Phi\left[\rho(\xi,t)\right]d\xi\,dt + \int_T\int_{\xi_*}\left[H(\xi,t) - H_0(\xi,t)\right]^T\left[H(\xi,t) - H_0(\xi,t)\right]d\xi\,dt = \int_T W_*(t)\,dt. \quad (2)$$

Thus, the final statement of the synthesis problem of the optimum observer in view of the above mentioned reasoning consists in defining function $h(\xi,t)$, giving the minimum to functional (2).

3. Synthesis of observations optimal control

Function APD, included in it, is described explicitly by the integro-differential Stratonovich equation with the right-hand part dependent on $h(\xi,t)$. The analysis of the experience of the instrument realization of the meters shows, that their synthesis consists, in essence, in defining the parameters of some functional series, approximating the output characteristic of the device projected with the given degree of accuracy. As such a series one uses, as a rule, the final expansion of the nonlinear components of vector $h(\xi,t)$ in some given system of the multidimensional functions: power, orthogonal etc.

Having designated vector of the multidimensional functions as $\left|\psi_1...\psi_S\right|^T = \psi$, we present the approximation of vector $h(\xi,t)$ as

$$H(\xi,t) = \left(E \otimes \psi^T\right) h = \psi_E h, \tag{3}$$

$$h = \left|h_{11}...h_{1S}h_{21}...h_{2S}...h_{N1}...h_{NS}\right|^T,$$

where $h_i(\xi,t) = \sum_{j=1}^{S} h_{ij}(t)\psi_j(\xi)$ is the i-th component of vector h, the factors of which define the concrete technical characteristics of the device,

$\otimes$ is the symbol of the Kronecker product.

For the subsequent analytical synthesis of optimum vector - function $h(\xi,t)$ in form of (3) we rewrite the equation of the APD $\rho(\xi,t)$ in the appropriate form

$$\frac{\partial \rho}{\partial t} = L[\rho] + h^T H_1[\rho] - h^T H_2[\rho] h, \tag{4}$$

where

$$H_1[\rho] = \left[\psi_E^T - \int_{\xi_*} \psi_E^T(\xi)\rho(\xi,t)d\xi\right] D_W^{-1} Z \rho(\xi,t),$$

$$H_2[\rho] = \frac{\rho(\xi,t)}{2}\left[\psi_E^T D_W^{-1}\psi_E - \int_{\xi_*} \psi_E^T(\xi) D_W^{-1}\psi_E(\xi)\rho(\xi,t)d\xi\right].$$

The constructions carried out the problem of search of optimum vector $h(\xi,t)$ is reduced to the synthesis of the optimum in-the- sense -of-(2) control h of the process with the distributed parameters described by Stratonovich equation (in view of representing vector $H_0(\xi,t)$ in the form similar to (3)

$$H_0(\xi,t) = \psi_E h_0).$$

The optimum control of process $\rho(\xi,t)$ will be searched in the class of the limited piecewise-continuous functions with the values from the open area H_*. For its construction we use the method of the dynamic programming, according to which the problem is reduced to the minimization of the known functional [1]

$$\min_{h \in H_*}\left\{\frac{dV}{dt} + W_*\right\} = 0 \tag{5}$$

under the final condition $V(t_k) = 0$ with respect to the optimum functional $V = V(\rho,t)$, parametrically dependent on $t \in [t_0, t_k]$ and determined on the set of functions satisfying (4).

For the processes, described by the linear equations in partial derivative, and criteria of the form of the above-stated ones, functional V is found in the form of the integrated quadratic form [1], therefore in this case we have:

$$V = \int_{\xi_*} v(\xi,t)\rho^2(\xi,t)d\xi.$$

Calculating derivative $\frac{dV}{dt}$

$$\frac{dV}{dt} = \int_{\xi_*}\left(\frac{dv}{dt}\rho^2 + 2v\rho\frac{d\rho}{dt}\right)d\xi = \int_{\xi_*}\left(\frac{dv}{dt}\rho^2 + 2v\rho L[\rho] + 2v\rho h^T H_1[\rho] - 2v\rho h^T H_2[\rho]h\right)d\xi,$$

the functional equation for v is obtained in the following form:

$$\min_{h\in H_*}\int_{\xi_*}\left(\frac{dv}{dt}\rho^2 + 2v\rho L[\rho] + 2v\rho\left(h^T H_1[\rho] - h^T H_2[\rho]h\right) + \Phi[\rho]\right)d\xi +$$

$$+(h-h_0)^T\int_{\xi_*}\psi_E^T(\xi)\psi_E(\xi)d\xi(h-h_0) = 0,$$

whence we have optimum vector h_{opt}:

$$h_{opt} = \left\{\int_{\xi_*}\left[\psi_E^T(\xi)\psi_E(\xi) - v\rho\left(H_2 + H_2^T\right)\right]d\xi\right\}^{-1}\int_{\xi_*}\left(\psi_E^T(\xi)\psi_E(\xi)h_0 - v\rho H_1\right)d\xi =$$

$$= B(v,\rho)\int_{\xi_*}\left(\psi_1(\xi)h_0 - v\rho H_1\right)d\xi.$$

Using condition $\left\{\frac{dV}{dt} + W_*\right\}_{h=h_{opt}} = 0$, for $v(\xi,t)$ we have the following equation:

$$\frac{dv}{dt} = -2v\rho^{-1}L[\rho] - \rho^{-2}\left(h_0^T\psi_{1\xi}B^T - \int_{\xi_*}v\rho H_1^T d\xi\, B^T\right)\left(2v\rho H_1 - \psi_1 h_0\right) +$$

$$+\rho^{-2}h_0^T\psi_1\left(B\psi_{1\xi}h_0 - B\int_{\xi_*}v\rho H_1 d\xi\right) + \rho^{-2}\left(h_0^T\psi_{1\xi}B^T - \int_{\xi_*}v\rho H_1^T d\xi\, B^T\right)\times$$

$$\times\left(2v\rho H_2-\psi_1\right)\left(B\psi_{1\xi}h_0-B\int_{\xi_*}v\rho H_1 d\xi\right)-\rho^{-2}h_0^T\psi_1 h_0-\rho^{-2}\Phi\left[\rho\right], \tag{6}$$

where

$$\psi_{1\xi}=\int_{\xi_*}\psi_1\left(\xi\right)d\xi,$$

which is connected with the equation of the APD, having after substitution into it expression h_{opt} the following form:

$$\frac{d\rho}{dt}=L\left[\rho\right]+\left(h_0^T\psi_{1\xi}B^T-\int_{\xi_*}v\rho H_1^T d\xi B^T\right)H_1-$$

$$-\left(h_0^T\psi_{1\xi}B^T-\int_{\xi_*}v\rho H_1^T d\xi B^T\right)H_2\left(B\psi_{1\xi}h_0-B\int_{\xi_*}v\rho H_1 d\xi\right). \tag{7}$$

4. Observations suboptimal control

The solution of the obtained equations (6), (7) exhausts completely the problem stated, allowing to generate the required optimum vector - function h of form (3). On the other hand, the solution problem of system (6), (7) is the point-to-point boundary-value problem for integrating the system of the integro-differential equations in partial derivatives, general methods of the exact analytical solution of which , as it is known, does not exist now. Not considering the numerous approximated methods of the solution of the given problem oriented on the trade-off of accuracy against volume of the computing expenses, then as one of the solution methods for this problem we use the method based on the expansion of function **v**, ***p*** in series by some system of the orthonormal functions of the vector argument :

$$V\left(\xi,t\right)=\sum_{\mu}\alpha_{\mu}\left(t\right)\phi_{\mu}\left(\xi\right)=\phi^T\alpha,$$

$$\rho\left(\xi,t\right)=\sum_{\mu}\beta_{\mu}\left(t\right)\phi_{\mu}\left(\xi\right)=\phi^T\beta,$$

where μ is the index running a set of values from (0,...,0) to (M,...,M) [2];

φ is the vector of the orthonormal functions of argument ξ;

α, β are vectors of factors of the appropriate expansions.

In this case the solution is reduced to the solution of the point-to-point boundary-value problem for integrating the system of the following equations, already ordinary ones:

$$\dot{\beta}=\int_{\xi_*}\phi L\left[\phi^T\beta\right]d\xi+\int_{\xi_*}\phi\left[h_0\psi_{1\xi}B^T(\alpha,\beta,\phi)-\int_{\xi_*}\phi^T\alpha\phi^T\beta H_1^T\left(\phi^T\beta\right)d\xi B^T(\alpha,\beta,\phi)\right]H_1\left(\phi^T\beta\right)d\xi-$$

$$-\int_{\xi_*}\phi\left[h_0^T\psi_{1\xi}B^T(\alpha,\beta,\phi)-\int_{\xi_*}\phi^T\alpha\phi^T\beta H_1^T\left(\phi^T\beta\right)d\xi B^T(\alpha,\beta,\phi)\right]H_2\left(\phi^T\beta\right)-$$

$$-\left(B(\alpha,\beta,\phi)\psi_{1\xi}h_0-B(\alpha,\beta,\phi)\int_{\xi_*}\phi^T\alpha\phi^T\beta H_1\left(\phi^T\beta\right)d\xi\right)d\xi,$$

$$\dot{\alpha}=\int_{\xi_*}\phi^T\left\{-2\phi^T\alpha\left(\phi^T\beta\right)^{-1}L\left[\phi^T\beta\right]-\left(\phi^T\beta\right)^{-2}\left[h_0^T\psi_{1\xi}B^T(\alpha,\beta,\phi)-\right.\right.$$

$$\left.-\int_{\xi_*}\phi^T\alpha\phi^T\beta H_1^T\left(\phi^T\beta\right)d\xi B^T(\alpha,\beta,\phi)\right]\left(2\phi^T\alpha\phi^T\beta H_1\left(\phi^T\beta\right)-\psi_1h_0\right)+$$

$$+\left(\phi^T\beta\right)^{-2}h_0^T\psi_1\left(B(\alpha,\beta,\phi)\psi_{1\xi}h_0-B(\alpha,\beta,\phi)\int_{\xi_*}\phi^T\alpha\phi^T\beta H_1\left(\phi^T\beta\right)d\xi\right)+\tag{8}$$

$$+\left(\phi^T\beta\right)^{-2}\left[h_0^T\psi_{1\xi}B^T(\alpha,\beta,\phi)-\int_{\xi_*}\phi^T\alpha\phi^T\beta H_1^T\left(\phi^T\beta\right)d\xi B^T(\alpha,\beta,\phi)\right]\times$$

$$\times\left(2\phi^T\alpha\phi^T\beta H_2\left(\phi^T\beta\right)-\psi_1\right)\left(B(\alpha,\beta,\phi)\psi_{1\xi}h_0-\right.$$

$$\left.\left.-B(\alpha,\beta,\phi)\int_{\xi_*}\phi^T\alpha\phi^T\beta H_1\left(\phi^T\beta\right)d\xi\right)-\left(\phi^T\beta\right)^{-2}h_0^T\psi_1h_0-\left(\phi^T\beta\right)^{-2}\Phi\left[\phi^T\beta\right]\right\}d\xi$$

under boundary value conditions $\alpha(T)=0,\ \beta(t_0)=\beta_0$, where the values of the components are defined from the expansion of function $\rho(\xi,t_0)=\rho_0$.

From the point of view of the practical realization the integration of system (8) under the boundary-value conditions appears to be more simple than integration (6), (7), but from the point of view of organization of the estimation process in the real time it is still hindered: first, the volume of the necessary temporary and computing expenses is great, secondly the feasibility of the adjustment of the vector of factors h in the real time of arrival of the signal of measurement Z - is excluded, the prior simulation of realizations Z appears to be necessary (in this case in the course of the instrument realization, as a rule, one fails to maintain the precisely given values h all the same). Thus, the use of the approximated methods of the problem solution (8) is quite proved in this case, then as one of which we consider the method of the invariant imbedding [3], used above and providing the required approximated solution in the real time.

As the application of the given method assumes the specifying of all the components of the required approximately estimated vector in the differential form, then for the realization of the feasibility of the synthesis of vector h through the given method in the real time we introduce a dummy variable v, allowing to take into account from here on expression h_{opt} as the differential equation

$$\dot{v} = h_{\text{opt}}\left(\phi^T \alpha, \phi^T \beta\right),$$

forming with equations (8) a unified system. The application of the method of the invariant imbedding results in this case in the following system of equations:

$$\begin{vmatrix} \dot{v}_0 \\ \dot{\beta}_0 \end{vmatrix} = \begin{vmatrix} h_0 \\ \int\limits_{\xi_*} \phi\left[\phi^T \beta_0\right] + h_0^T\left(H_1\left[\phi^T \beta_0\right] - H_2\left[\phi^T \beta\right] h_0\right) d\xi \end{vmatrix} - D \int\limits_{\xi_*} \phi\left(\phi^T \beta_0\right)^{-2} \Phi\left[\phi^T \beta_0\right] d\xi,$$

$$\dot{D} = 2 \int\limits_{\xi_*} \phi\left(\frac{\partial}{\partial \beta_0}\left\{L\left[\phi^T \beta_0\right]\right\} + h_0^T \frac{\partial H_1}{\partial \beta_0}\left[\phi^T \beta_0\right] - h_0^T \frac{\partial H_2}{\partial \beta_0}\left[\phi^T \beta_0\right] h_0\right) d\xi\, D +$$

$$+2D \int\limits_{\xi_*} \phi\left(\phi^T \beta_0\right)^{-1}\left(\phi^T L\left[\phi^T \beta_0\right] - h_0^T\left(\phi^T \otimes H_2\left[\phi^T \beta_0\right]\right) h_0\right) d\xi -$$

$$-\int\limits_{\xi_*} \phi\left(H_1^T\left[\phi^T \beta_0\right] - 2 h_0^T H_2\left[\phi^T \beta_0\right]\right) d\xi\, \mu +$$

$$+2D \int\limits_{\xi_*} \phi\left(\phi^T \beta_0\right)^{-2}\left(2\left(\phi^T \beta_0\right)^{-1} \phi^T \Phi\left[\phi^T \beta_0\right] - \frac{\partial}{\partial \beta_0} \Phi\left[\phi^T \beta_0\right]\right) d\xi\, D,$$

$$\mu = 2 \psi_{1\xi}^{-1} \int\limits_{\xi_*}\left(\phi^T \beta_0\right)\left[\left(\phi^T \otimes H_2\right) h_0 - \frac{1}{2} H_1 \phi^T\right] d\xi.$$

By virtue of the fact that matrix D in the method of the invariant imbedding plays the role of the weight matrix at the deviation of the vector of the approximated solution from the optimum one, in this case for variables β_{i0} the appropriate components D characterize the degree of their deviation from the factors of expansion of the true APD (components D_0 - are deviations of the parameters at the initial moment). The essential advantage of the approach considered, despite the formation of the approximated solution, is the feasibility of the synthesis of the optimum observation function in the real time, i.e. in the course of arrival of the measuring information.

5. Example

For the illustration of the feasibility of the practical use of the suggested method the numerical simulation of the process of forming vector $h = \left|h_1 h_2\right|^T$ of factors of the observer

$Z = h_1\xi + h_2\xi^2 + W_t$ for target $\dot{\xi} = -\xi^3 + n_t$ was carried out the normalized Gaussian white noises of the target and meter. As the criterion of optimization the criterion of the maximum of the a posteriori probability of the existence of the observable process on interval ξ_* = = [-2.5, 2.5] was chosen that provided the additional restriction in the form of the requirement of the minimal deviation of vector h from the given vector $h_0 = |0.95, 0.3|^T$ that allows to write down the minimized functional as

$$J = -\int_T\int_{\xi_*} \rho(\xi,t)\,d\xi\,dt + \int_T (h-h_0)^T D_H (h-h_0)\,dt,$$

where

$$D_H = \int_{\xi_*} \begin{vmatrix} \xi \\ \xi^2 \end{vmatrix} \begin{vmatrix} \xi & \xi^2 \end{vmatrix} d\xi = \begin{vmatrix} 10.4 & \cdots & 0 \\ \vdots & & \vdots \\ 0 & \cdots & 39.1 \end{vmatrix}, \quad T = |0;\, 600|.$$

In this case the equation of the APD has the form

$$\frac{\partial\rho}{\partial t} = \frac{\partial}{\partial\xi}\left(\xi^3\rho\right) + \frac{1}{2}\frac{\partial^2\rho}{\partial\xi^2} + h^T H_1 - h^T H_2 h,$$

where

$$H_1 = \left(\begin{vmatrix} \xi \\ \xi^2 \end{vmatrix} - \int_{\xi_*} \begin{vmatrix} \xi \\ \xi^2 \end{vmatrix} \rho(\xi,t)\,d\xi \right) Z\rho,$$

$$H_2 = \frac{\rho}{2}\left(\begin{vmatrix} \xi^2 & \xi^3 \\ \xi^3 & \xi^4 \end{vmatrix} - \int_{\xi_*} \begin{vmatrix} \xi^2 & \xi^3 \\ \xi^3 & \xi^4 \end{vmatrix} \rho(\xi,t)\,d\xi \right).$$

The optimum vector h is defined from expression h_{opt} as

$$h_{opt} = \left[D_H - \int_{\xi_*} V\rho^2 \left(\begin{vmatrix} \xi^2 & \xi^3 \\ \xi^3 & \xi^4 \end{vmatrix} - \int_{\xi_*} \begin{vmatrix} \xi^2 & \xi^3 \\ \xi^3 & \xi^4 \end{vmatrix} \rho(\xi,t)\,d\xi \right) d\xi \right]^{-1} \times$$

$$\times \left(D_H h_0 - Z \int_{\xi_*} V\rho^2 \left[\begin{vmatrix} \xi \\ \xi^2 \end{vmatrix} - \int_{\xi_*} \begin{vmatrix} \xi \\ \xi^2 \end{vmatrix} \rho(\xi,t)\,d\xi \right] d\xi \right).$$

Using the Fourier expansion up to the 3-rd order for the approximated representation of functions V, ρ

$$V(\xi,t)=\frac{1}{2}\alpha_0+\sum_{k=1}^{2}\alpha_{1k}\cos k\omega_0\xi+\alpha_{2k}\sin k\omega_0\xi\ ,$$

$$\rho(\xi,t)=\sum_{k=1}^{2}\beta_{1k}\cos k\omega_0\xi+\beta_{2k}\sin k\omega_0\xi\ ,$$

$$\omega_0=\frac{2\pi}{5},$$

(then for $\upsilon\rho^2$ the following representation holds true

$$\upsilon\rho^2(\xi,t)=\gamma_0+\sum_{k=1}^{6}\gamma_{1k}\cos k\omega_0\xi+\gamma_{2k}\sin k\omega_0\xi\ ,$$

$\gamma_0,\gamma_{ik}=\gamma(\alpha,\beta)$ are functions linearly dependent on factors α_{ik} and quadratically - on β_{ik}) and introducing designations

$$\zeta_C(k,n)=\int_{\xi_*}\xi^n\cos k\omega_0\xi\,d\xi=2\sum_{i=1,3}^{n}i!C_i^n\frac{(2.5)^{n-i}}{(k\omega_0)^{i+1}}\sin\left(k\pi+\frac{i\pi}{2}\right),\quad n=2;4;$$

$$\zeta_S(k,m)=\int_{\xi_*}\xi^m\sin k\omega_0\xi\,d\xi=-2\sum_{i=0,2}^{m}i!C_i^m\frac{(2.5)^{m-i}}{(k\omega_0)^{i+1}}\cos\left(k\pi+\frac{i\pi}{2}\right),\quad m=1;3;$$

vector h_{opt} of the factors of the observer we write down as follows:

$$h_{\text{opt}}=\left(D_H+\left|\begin{matrix}\gamma_0\left(5\sum_{k=1}^{2}\beta_{1k}\zeta_C(k,2)-10{,}4\right)-\sum_{k=1}^{6}\gamma_{1k}\zeta_C(k,2) & \cdots & 5\gamma_0\sum_{k=1}^{2}\beta_{2k}\zeta_S(k,3)-\sum_{k=1}^{6}\gamma_{2k}\zeta_S(k,3)\\ \vdots & & \vdots\\ 5\gamma_0\sum_{k=1}^{2}\beta_{2k}\zeta_S(k,3)-\sum_{k=1}^{6}\gamma_{2k}\zeta_S(k,3) & \cdots & \gamma_0\left(5\sum_{k=1}^{2}\beta_{1k}\zeta_C(k,4)-39{,}1\right)-\sum_{k=1}^{6}\gamma_{1k}\zeta_C(k,4)\end{matrix}\right|\right)^{-1}\times$$

$$\times\left(D_Hh_0-Z\left|\begin{matrix}\sum_{k=1}^{6}\gamma_{2k}\zeta_S(k,1)-5\gamma_0\sum_{k=1}^{2}\beta_{2k}\zeta_S(k,1)\\ \vdots\\ \sum_{k=1}^{6}\gamma_{1k}\zeta_C(k,2)-5\gamma_0\sum_{k=1}^{2}\beta_{1k}\zeta_C(k,2)\end{matrix}\right|\right)=h(\alpha,\beta).$$

Then the system of equations for the factors of expansion has the following form:

$$\dot{\beta}_{1(2)i} = \frac{1}{2}\sum_{K=1}^{2} \beta_{1(2)K}\left(3\left[\zeta_C(k-i,2) \pm \zeta_C(k+i,2)\right] \mp\right.$$

$$\left.\mp k\omega_0\left[\zeta_S(k+i,3)+\zeta_S(k-i,3)\right]\right) \mp \frac{(i\omega_0)^2}{2}\beta_{1(2)i} +$$

$$+Zh^T(\alpha,\beta)\begin{vmatrix} \sum_{K=1}^{2}\beta_{2(1)K}\frac{1}{2}\left[\zeta_S(k+i,1)+\zeta_S(k-i,1)\right]-\beta_{2K}\zeta_S(k,1)\beta_{1(2)i} \\ \vdots \\ \sum_{K=1}^{2}\beta_{1(2)K}\frac{1}{2}\left[\zeta_C(k-i,2)\pm\zeta_C(k+i,2)\right]-\beta_{1K}\zeta_C(k,2)\beta_{1(2)i} \end{vmatrix} -$$

$$-\frac{1}{2}h^T(\alpha,\beta)\begin{vmatrix} \sum_{K=1}^{2}\beta_{1(2)K}\frac{1}{2}\left[\zeta_C(k-i,2)\pm\zeta_C(k+i,2)\right]-\beta_{1K}\zeta_C(k,2)\beta_{1(2)i} & \cdots \\ \vdots & \\ \sum_{K=1}^{2}\beta_{2(1)K}\frac{1}{2}\left[\zeta_S(k+i,3)+\zeta_S(k-i,3)\right]-\beta_{2K}\zeta_S(k,3)\beta_{1(2)i} & \cdots \end{vmatrix}$$

$$\begin{vmatrix} \cdots & \sum_{K=1}^{2}\beta_{2(1)K}\frac{1}{2}\left[\zeta_S(k+i,3)+\zeta_S(k-i,3)\right]-\beta_{2K}\zeta_S(k,3)\beta_{1(2)i} \\ & \vdots \\ \cdots & \sum_{K=1}^{2}\beta_{1(2)K}\frac{1}{2}\left[\zeta_C(k-i,4)\pm\zeta_C(k+i,4)\right]-\beta_{1K}\zeta_C(k,4)\beta_{1(2)i} \end{vmatrix} h(\alpha,\beta),$$

$$\beta_{1i}(0)=10^{-2}, \qquad \beta_{2i}(0)=0, \qquad i=1,2;$$

$$\dot{\alpha}_0 = -2\left(3\sum_{K=1}^{2}\alpha_{1K}\zeta_C(k,2)+\alpha_{1K}\left[\chi_C(k,\beta)+\omega_C(k,\beta)\right]+\alpha_{2K}\left[\chi_S(k,\beta)+\omega_S(k,\beta)\right]\right)-$$

$$-2Zh^T(\alpha,\beta)\begin{vmatrix} \sum_{K=1}^{2}\alpha_{2K}\zeta_S(k,1) \\ \sum_{K=1}^{2}\alpha_{1K}\zeta_C(k,2) \end{vmatrix} - h^T(\alpha,\beta)\begin{vmatrix} \sum_{K=1}^{2}\alpha_{1K}\zeta_C(k,2) & \cdots & \sum_{K=1}^{2}\alpha_{2K}\zeta_S(k,3) \\ \vdots & & \vdots \\ \sum_{K=1}^{2}\alpha_{2K}\zeta_S(k,3) & \cdots & \sum_{K=1}^{2}\alpha_{1K}\zeta_C(k,4) \end{vmatrix} \times$$

$$\times h(\alpha,\beta)+\mu(\beta)-(h-h_0)^T\mu_1(\beta)(h-h_0),$$

$$\dot{\alpha}_{1(2)i} = -2\left\{\frac{3}{2}\sum_{K=1}^{2}\alpha_{1(2)K}\left[\zeta_C(k-i,2)\pm\zeta_C(k+i,2)\right]+\right.$$

$$+\alpha_{1K}\left[\chi_{1(2)C}(k,i,\beta)+\omega_{1(2)C}(k,i,\beta)\right]+\alpha_{2K}\left[\chi_{1(2)S}(k,i,\beta)+\omega_{1(2)S}(k,i,\beta)\right]\Big\}-$$

$$-2Zh^T(\alpha,\beta)\left(\begin{vmatrix}\sum_{K=1}^{2}\alpha_{2(1)K}\frac{1}{2}\left[\zeta_S(k+i,1)+\zeta_S(k-i,1)\right]\\ \sum_{K=1}^{2}\alpha_{1(2)K}\frac{1}{2}\left[\zeta_C(k-i,2)\pm\zeta_C(k+i,2)\right]\end{vmatrix}-\alpha_{1(2)i}\begin{vmatrix}\sum_{K=1}^{2}\beta_{2K}\zeta_S(k,1)\\ \sum_{K=1}^{2}\beta_{1K}\zeta_C(k,2)\end{vmatrix}\right)-$$

$$-h^T(\alpha,\beta)\left(\begin{vmatrix}\sum_{K=1}^{2}\alpha_{1(2)K}\frac{1}{2}\left[\zeta_C(k-i,2)\pm\zeta_C(k+i,2)\right] & \cdots\\ \vdots & \\ \sum_{K=1}^{2}\alpha_{2(1)K}\frac{1}{2}\left[\zeta_S(k+i,3)+\zeta_S(k-i,3)\right] & \cdots\end{vmatrix}\right.$$

$$\left.\begin{matrix}\cdots & \sum_{K=1}^{2}\alpha_{2(1)K}\frac{1}{2}\left[\zeta_S(k+i,3)+\zeta_S(k-i,3)\right]\\ & \vdots\\ \cdots & \sum_{K=1}^{2}\alpha_{1(2)K}\frac{1}{2}\left[\zeta_C(k-i,4)\pm\zeta_C(k+i,4)\right]\end{matrix}\right|-$$

$$\left.-\alpha_{1(2)i}\begin{vmatrix}\sum_{K=1}^{2}\beta_{1K}\zeta_C(k,2) & \cdots & \sum_{K=1}^{2}\beta_{2K}\zeta_S(k,3)\\ \vdots & & \vdots\\ \sum_{K=1}^{2}\beta_{2K}\zeta_S(k,3) & \cdots & \sum_{K=1}^{2}\beta_{1K}\zeta_C(k,4)\end{vmatrix}\right)h(\alpha,\beta)+$$

$$+\mu_{C(S)}(i,\beta)-(h-h_0)^T\mu_{1C(S)}(i,\beta)(h-h_0),$$

$$\alpha(t_K)=0,$$

where the expressions of factors χ, ω, μ (determined by the numerical integration in the course of solving) aren't given as complicated. In the reduced form the system obtained can be given as

$$\dot{\beta}=\Phi_\beta\left[\beta,h(\alpha,\beta)\right],$$

$$\dot{\alpha} = G_1(\alpha, h) + G_2(\beta),$$

$$G_2(\beta) = \left|\mu(\beta) \vdots \mu_C(1,\beta) \vdots \mu_C(2,\beta) \vdots \mu_S(1,\beta) \vdots \mu_S(2,\beta)\right|^T.$$

The approximated solving of the given boundary-value problem by the method of the invariant imbedding results in the required system of the equations allowing to carry out simultaneously the definition of vector h_{opt} and formation of vector β in the real time:

$$\begin{vmatrix} \dot{\nu}_0 \\ \dot{\beta}_0 \end{vmatrix} = \begin{vmatrix} h_0 \\ \Phi_\beta(\beta_0, h_0) \end{vmatrix} + DG_2(\beta_0),$$

$$\dot{D} = 2\frac{\partial \Phi_\beta}{\partial \beta}(\beta_0, h_0)D - \frac{\partial \Phi_\beta}{\partial \alpha}(\beta_0, \alpha)\bigg|_{\alpha=0} + 2D\frac{\partial G_2}{\partial \beta}(\beta_0)D - D\frac{\partial G_1}{\partial \alpha}(\alpha, \beta_0, h_0)\bigg|_{\alpha=0}.$$

The integration of the given system was made by the Runge-Kutta method on interval [0; 600] s. with the step equal to 0,05 s.

For the comparison of efficiency of the approach suggested with that of the existing ones the formation of the optimum- by-the- criterion-of-the-MAP estimation $\hat{\xi}$ by two ways was carried out: on the basis of the MAP - filter with the linear observer [4] and by defining the maximum of the function of the APD, approximated by series $\phi^T \beta_0$ (where β_0 is the solution of the last system of the estimation equations), by means of the method of the random draft. The search of the maximum of the APD was carried out on the simulation interval [500; 600] s. for the estimations of vector β_0, taken with interval 1 s. The generated test sample of dimension 100 was the normalized Gaussian sequence.

The calculation of the estimation errors was made by comparing the current values of estimations with the target coordinate and subsequent defining of the average values of the errors on interval [500; 600] s. Upon terminating the simulation interval the value of the average error obtained in this way for the estimation equations [4], using the linear observer, has exceeded the average estimation error carried out by the technique suggested, using the information of the optimum observer, by the factor of ~ 1,52.

Author details

Sergey V. Sokolov
Rostov State University of Means of Communication, Russia

6. References

[1] Sirazetdinov T.K. Optimization of systems with the distributed parameters. M: Science, 1977.

[2] Pugachev V. S., Sinitsyn I.N. Stochastic differential systems. M: Science, 1985.

[3] Pervachev S.V., Perov A.I. Adaptive filtration of messages. - Moscow, Radio and communication, 1991.

[4] Sejdzh E.P., White C.S. Optimum control of systems. M: Radio and communication, 1982.

[5] V.V. Khutortsev, I.V. Baranov. The optimization of observations control in problem of discrete searching for Poisson model of the observed objects flow // Radiotechnika, №3, 2010.

[6] Harisov V. N., Anikin A.L. Observations optimal control in problem of detection of signals // Radio systems, № 81, 2004.

Chapter 9

Iterations for a General Class of Discrete-Time Riccati-Type Equations: A Survey and Comparison

Ivan Ivanov

Additional information is available at the end of the chapter

1. Introduction

The discrete-time linear control systems have been applied in the wide area of applications such as engineering, economics, biology. Such type systems have been intensively considered in the control literature in both the deterministic and the stochastic framework. The stability and optimal control of stochastic differential equations with Markovian switching has recently received a lot of attention, see Freiling and Hochhaus [8], Costa, Fragoso, and Marques [2], Dragan and Morozan [4, 5]. The equilibrium in these discrete-time stochastic systems can be found via the maximal solution of the corresponding set of discrete-time Riccati equations.

We consider a set of discrete-time generalized Riccati equations that arise in quadratic optimal control of discrete-time stochastic systems subjected to both state-dependent noise and Markovian jumps, i.e. the discrete-time Markovian jump linear systems (MJLS). The iterative method to compute the maximal and stabilizing solution of wide class of discrete-time nonlinear equations is derived by Dragan, Morozan and Stoica [6, 7].

We study a problem for computing the maximal symmetric solution to the following set of discrete-time generalized algebraic Riccati equations (DTGAREs):

$$\begin{aligned} X(i) = \mathcal{P}(i,\mathbf{X}) := & \textstyle\sum_{l=0}^{r} A_l(i)^T \mathcal{E}_i(\mathbf{X}) A_l(i) + C^T(i)C(i) \\ & -(\textstyle\sum_{l=0}^{r} A_l(i)^T \mathcal{E}_i(\mathbf{X}) B_l(i) + L(i)) \\ & \times R(i,\mathbf{X})^{-1} (\textstyle\sum_{l=0}^{r} B_l(i)^T \mathcal{E}_i(\mathbf{X}) A_l(i) + L(i)^T), \quad i = 1,\dots,N, \end{aligned} \tag{1}$$

where $R(i,\mathbf{X}) = R(i) + \sum_{l=0}^{r} B_l(i)^T \mathcal{E}_i(\mathbf{X}) B_l(i)$ and $\mathcal{E}(\mathbf{X}) = (\mathcal{E}_1(\mathbf{X}),\dots,\mathcal{E}_N(\mathbf{X}))$ with $\mathbf{X} = (X(1),\dots,X(N))$ and

$$\mathcal{E}_i(\mathbf{X}) = \sum_{j=1}^{N} p_{ij} X(j), \qquad X(j) \text{ is an } n \times n \text{ matrix}, \quad \text{for } i = 1,\dots,N.$$

In addition the operator

$$\Pi(i,\mathbf{X}) = \begin{pmatrix} \sum_{l=0}^{r} A_l(i)^T \mathcal{E}_i(\mathbf{X}) A_l(i) & \sum_{l=0}^{r} A_l(i)^T \mathcal{E}_i(\mathbf{X}) B_l(i) + L(i) \\ \sum_{l=0}^{r} B_l(i)^T \mathcal{E}_i(\mathbf{X}) A_l(i) + L(i)^T & \sum_{l=0}^{r} B_l(i)^T \mathcal{E}_i(\mathbf{X}) B_l(i) \end{pmatrix}$$

is assumed to be linear and positive, i.e. $\mathbf{X} \geq 0$ implies $\Pi(i,\mathbf{X}) \geq 0$ for $i = 1,\ldots,N$. That is a natural assumption (see assumption H_1, [6]). The notation $\mathbf{X} \geq 0$ means that $X(i) \geq 0, i = 1,\ldots,N$.

Such systems of discrete-time Riccati equations $X(i) = \mathcal{P}(i,\mathbf{X}), i = 1,\ldots,N$ are used to determine the solutions of linear-quadratic optimization problems for a discrete-time MJLS [5]. More precisely, these optimization problems are described by controlled systems of the type:

$$x(t+1) = [A_0(\eta_t) + \sum_{l=1}^{r} w_l(t) A_l(\eta_t)]x(t) + [B_0(\eta_t) + \sum_{l=1}^{r} w_l(t) B_l(\eta_t)]u(t) \tag{2}$$

for $x(0) = x_0$ and the output

$$y = C(\eta_t)\, x(t) + D(\eta_t)\, x(t)$$

where $\{\eta_t\}_{t\geq 0}$ is a Markov chain taking values in $\{1,2,\ldots,N\}$ with transition probability matrix $(p_{ij})_{i,j=1}^{N}$. Moreover, $\{w(t)\}_{t\geq 0}$ is a sequence of independent random vectors ($w(t) = (w_1(t),\ldots,w_r(t))^T$), for details see e.g. [5–7].

We define the matrices A_l, B_l such that $A_l = (A_l(1),\ldots,A_l(N))$, $B_l = (B_l(1),\ldots,B_l(N))$ where $A_l(i)$ is an $n \times n$ matrix and $B_l(i)$ is an $n \times k$ matrix $l = 0,1,\ldots,r$ and $i = 1,\ldots,N$, and $\mathbf{A} = (A_0, A_1, A_2, \ldots, A_r)$ and $\mathbf{B} = (B_0, B_1, B_2, \ldots, B_r)$. We present the Definition 4.1 [7] in the form:

Definition 1.1. *We say that the couple* $(\mathbf{A}, \mathbf{B})$ *is stabilizable if for some* $\mathbf{F} = (F(1),\ldots,F(N))$ *the closed loop system:*

$$x(t+1) = [A_0(\eta_t) + B_0(\eta_t)F(\eta_t) + \sum_{l=1}^{r} w_l(t)(A_l(\eta_t) + B_l(\eta_t)F(\eta_t))]x(t)$$

is exponentially stable in mean square (ESMS).

The matrix $\mathbf{F}$ involved in the above definition is called stabilizing feedback gain.

We will investigate the computation of the maximal solution of a set of equations (1). A solution $\tilde{\mathbf{X}}$ of (1) is called maximal if $\tilde{\mathbf{X}} \geq \mathbf{X}$ for any solution $\mathbf{X}$.

We will consider three cases. In the first case the weighting matrices $R(i) = D^T(i)\, D(i), i = 1,\ldots,N$ are assumed to be positive definite and $Q(i) = C^T(i)\, C(i), i = 1,\ldots,N$ are positive semidefinite. Thus, the matrices $R(i,\mathbf{X}) = R(i) + \sum_{l=0}^{r} B_l(i)^T \mathcal{E}_i(\mathbf{X}) B_l(i), i = 1,\ldots,N$ are positive definite. We present an overview of several computational methods [6, 10] to compute the maximal and stabilizing solutions of a considered class of discrete-time Riccati equations. In addition, we apply a new approach, where the variables are changed and an equivalent set of the same type Riccati equations is obtained. A new iteration for the maximal solution to this equivalent set of nonlinear equations is proposed. This is the subject of section 2.

We investigate the applicability of the existing methods for the maximal solution to (1) where the weighting matrices $R(i), Q(i)$ are indefinite in the second case. These weighting matrices are indefinite, but the matrices $R(i, \mathbf{X}), i = 1, \ldots, N$ are still positive definite. Similar investigations have been executed by Rami and Zhou [12, 14] in case in the infinite time horizon. The important tool for finding the maximal solution is a semidefinite programming associated with the linear matrix inequalities (LMI). The method for the maximal solution used an LMI optimization problem is called the LMI approach or the LMI method. Rami and Zhou [14] have described a technics for applying the LMI method to indefinite linear quadratic problem in infinite time horizon. Here we will extend their findings and will modify their technics to indefinite linear quadratic problem of Markovian jump linear systems. We propose a new optimization problem suitable to the occasion. The investigation is accompanied by comparisons of the LMI approach on different numerical examples. This is the subject of section 3.

The third case is considered in section 4. Here, the solution of (1) under the assumption that at least one of matrices $R(i, \mathbf{X}), i = 1, \ldots, N$ is positive semidefinite is analysed. In this case set of equations (1) can be written as

$$X(i) = \sum_{l=0}^{r} A_l(i)^T \mathcal{E}_i(\mathbf{X}) A_l(i) + C^T(i)C(i) - S(i, \mathbf{X}) \left(R(i, \mathbf{X})\right)^{\dagger} S(i, \mathbf{X})^T \quad (3)$$

for $i = 1, \ldots, N$ with the additional conditions

$$R(i, \mathbf{X}) \geq 0, \quad \text{and} \quad \left(I - \left(R(i, \mathbf{X})\right)^{\dagger} R(i, \mathbf{X})\right) S(i, \mathbf{X})^T = 0\,. \quad (4)$$

Such type generalized Riccati equation is introduced in [15]. The notation $Z^{\dagger}$ stands for the Moore-Penrose inverse of a matrix Z. We derive a suitable iteration formula for computing the maximal solution of (3) - (4) and the convergence properties of the induced matrix sequence are proved. In addition, the LMI approach is modified and applied to the case of semidefinite matrices $R(i, \mathbf{X})$ $(i = 1, \ldots, N)$. Numerical simulations for comparison the derived methods are presented in the section.

We are executing some numerical experiments in this investigation. Based on the results from experiments the considered methods are compared in all cases. In the examples we consider a MJLS with three operation modes describing an economic system, adapted from [17] which studies a time-variant macroeconomic model where some of the parameters are allowed to fluctuate in an exogenous form, according to a Markov chain. The operation modes are interpreted as the general situation: "neutral", "bad" or "good" $(N = 3)$. See [17] and references therein for more details. Our experiments are carried out in the MATLAB on a 1,7GHz PENTIUM computer. In order to execute our experiments the suitable MATLAB procedures are used.

The notation $\mathcal{H}^n$ stands for the linear space of symmetric matrices of size n over the field of real numbers. For any $X, Y \in \mathcal{H}^n$, we write $X > Y$ or $X \geq Y$ if $X - Y$ is positive definite or $X - Y$ is positive semidefinite. The notations $\mathbf{X} = (X(1), X(2), \ldots, X(N)) \in \mathcal{H}^n$ and the inequality $\mathbf{Y} \geq \mathbf{Z}$ mean that for $i = 1, \ldots, N$, $X(i) \in \mathcal{H}^n$ and $Y(i) \geq Z(i)$, respectively. The linear space $\mathcal{H}^n$ is a Hilbert space with the Frobenius inner product $< X, Y >= trace(XY)$. Let $\|.\|$ denote the spectral matrix norm.

2. The positive definite case

Let us assume that the weighting matrices $R(i), i = 1, \ldots, N$ are positive definite and $Q(i), i = 1, \ldots, N$ are positive semidefinite. Thus the matrices $R(i, \mathbf{X}), i = 1, \ldots, N$ are positive definite. In this section, we consider set of equations (1) where a matrix $\mathbf{X}$ belongs to the domain:

$$Dom\,\mathcal{P} = \{\mathbf{X} \in \mathcal{H}^n \,|\, R(i, \mathbf{X}) = R(i) + \sum_{l=0}^{r} B_l(i)^T \mathcal{E}_i(\mathbf{X}) B_l(i) > 0,\ i = 1, 2, \ldots, N\}.$$

Note that $\mathbf{X} \in Dom\,\mathcal{P}$ implies $\mathbf{Y} \in Dom\,\mathcal{P}$ for all $\mathbf{Y} \geq \mathbf{X}$ and that $Dom\,\mathcal{P}$ is open and convex. We consider the map $Dom\,\mathcal{P} \to \mathcal{H}^n$. We investigate some iterations for finding the maximal solution to (1). For the matrix function $\mathcal{P}(i, \mathbf{X})$ we introduce notations

$$\begin{aligned}
Q(i, \mathbf{Z}) &= \textstyle\sum_{l=0}^{r} A_l(i)^T \mathcal{E}_i(\mathbf{Z}) A_l(i) + C^T(i) C(i);\\
S(i, \mathbf{Z}) &= \textstyle\sum_{l=0}^{r} A_l(i)^T \mathcal{E}_i(\mathbf{Z}) B_l(i) + L(i);\\
F(i, \mathbf{Z}) &= -R(i, \mathbf{Z})^{-1} S(i, \mathbf{Z})^T, \quad (\text{note that } \mathrm{S(i, Z)} = -\mathrm{F(i, Z)^T R(i, Z)})\\
T(i, \mathbf{Z}) &= C^T(i) C(i) + F(i, \mathbf{Z})^T L(i)^T + L(i)\, F(i, \mathbf{Z}) + F(i, \mathbf{Z})^T R(i)\, F(i, \mathbf{Z})\\
&= \begin{pmatrix} I & F(i, \mathbf{Z})^T \end{pmatrix} \begin{pmatrix} C^T(i) C(i) & L(i) \\ L(i)^T & R(i) \end{pmatrix} \begin{pmatrix} I \\ F(i, \mathbf{Z}) \end{pmatrix}
\end{aligned}$$

and we present set of equations (1) as follows:

$$X(i) = Q(i, \mathbf{X}) - S(i, \mathbf{X})\, R(i, \mathbf{X})^{-1} S(i, \mathbf{X})^T,$$

with $i = 1, \ldots, N$.

Then, for the matrix function $\mathcal{P}(i, \mathbf{X})$ we rewrite

$$\mathcal{P}(i, \mathbf{X}) = Q(i, \mathbf{X}) - F(i, \mathbf{X})^T R(i, \mathbf{X})\, F(i, \mathbf{X}).$$

We will study the system $X(i) = \mathcal{P}(i, \mathbf{X})$ for $i = 1, \ldots, N$. We start by some useful properties to $\mathcal{P}(i, \mathbf{X})$. For briefly we use $\tilde{A}_l(i, \mathbf{Z}) = A_l(i) + B_l(i) F(i, \mathbf{Z})$, $l = 0, 1, \ldots, r$ for some $\mathbf{Z} \in \mathcal{H}^n$.

Lemma 2.1. *[10] Assuming $\mathbf{Y} \in \mathcal{H}^n$ and $\mathbf{Z} \in \mathcal{H}^n$ are symmetric matrices, then the following properties of $\mathcal{P}(i, \mathbf{X}), i = 1, \ldots, N$*

$$\begin{aligned}
\mathcal{P}_{\mathbf{Z}}(i, \mathbf{Y}) &= \sum_{l=0}^{r} \tilde{A}_l(i, \mathbf{Z})^T\, \mathcal{E}_i(\mathbf{Y}) \tilde{A}_l(i, \mathbf{Z}) + T(i, \mathbf{Z})\\
&\quad - \left(F(i, \mathbf{Y})^T - F(i, \mathbf{Z})^T\right) R(i, \mathbf{Y})\, (F(i, \mathbf{Y}) - F(i, \mathbf{Z}))
\end{aligned}$$

$$\begin{aligned}
\mathcal{P}_{\mathbf{Z}}(i, \mathbf{Z}) - \mathcal{P}_{\mathbf{Z}}(i, \mathbf{Y}) &= \sum_{l=0}^{r} \tilde{A}_l(i, \mathbf{Z})^T\, \mathcal{E}_i(\mathbf{Z} - \mathbf{Y}) \tilde{A}_l(i, \mathbf{Z})\\
&\quad + \left(F(i, \mathbf{Y})^T - F(i, \mathbf{Z})^T\right) R(i, \mathbf{Y})\, (F(i, \mathbf{Y}) - F(i, \mathbf{Z}))
\end{aligned}$$

hold.

Dragan, Morozan and Stoica [6] have been proposed an iterative procedure for computing the maximal solution of a set of nonlinear equations (1). The proposed iteration [6, iteration (4.7)] is :

$$\begin{aligned} X(i)^{(k)} &= \mathcal{P}_{\mathbf{X}^{(k-1)}}(i,\mathbf{X}^{(k-1)}) + \tfrac{\varepsilon}{k} I_n \\ &= \textstyle\sum_{l=0}^{r} \left[\tilde{A}_l(i,\mathbf{X}^{(k-1)})\right]^T \mathcal{E}_i(\mathbf{X}^{(k-1)}) \left[\tilde{A}_l(i,\mathbf{X}^{(k-1)})\right] \\ &\quad + T(i,\mathbf{X}^{(k-1)}) + \tfrac{\varepsilon}{k} I_n , \\ \text{where} \quad & \tilde{A}_l(i,\mathbf{X}^{(k-1)}) = A_l(i) + B_l(i)F(i,\mathbf{X}^{(k-1)}) , \end{aligned} \tag{5}$$

$k = 1,2,3\ldots$, and ε is a small positive number. Note that iteration (5) is a special case of the general iterative method given in [6, Theorem 3.3]. Based on the Gauss-Seidel technique the following modification is observed by Ivanov [10]:

$$\begin{aligned} X(i)^{(k)} &= \textstyle\sum_{l=0}^{r} \left[\tilde{A}_l(i,\mathbf{X}^{(k-1)})\right]^T \\ &\quad \times \left(\mathcal{E}_{i1}(\mathbf{X}^{(k)}) + p_{ii}X(i)^{(k-1)} + \mathcal{E}_{i2}(\mathbf{X}^{(k-1)})\right) \\ &\quad \times \left[\tilde{A}_l(i,\mathbf{X}^{(k-1)})\right] + T(i,\mathbf{X}^{(k-1)}) , \\ &\quad i = 1,2,\ldots,N, \;\; k = 1,2,3\ldots \end{aligned} \tag{6}$$

where

$$\mathcal{E}_{i1}(\mathbf{Z}) = \sum_{j=1}^{i-1} p_{ij}\, Z(j), \quad \text{and} \quad \mathcal{E}_{i2}(\mathbf{Z}) = \sum_{j=i+1}^{N} p_{ij}\, Z(j) .$$

The convergence properties of matrix sequences defined by (5) and (6) are derived in the corresponding papers.

The method can be applied under the assumption that the matrix inequalities $\mathcal{P}(i, \mathbf{Z}) \geq Z(i)$ and $\mathcal{P}(i, \mathbf{Z}) \leq Z(i)$, $(i = 1,\ldots,N)$ are solvable. Under these conditions the convergence of (6) takes place if the algorithm starts at any suitable initial point $\mathbf{X}^{(0)}$. The new iteration (6) can be considered as an accelerated modification to iteration (5). The convergence result is given by the following theorem:

Theorem 2.1. *[10] Letting there are symmetric matrices* $\hat{\mathbf{X}} = (\hat{X}_1,\ldots,\hat{X}_N) \in Dom\,\mathcal{P}$ *and* $\mathbf{X}^{(0)} = (X_1^{(0)},\ldots,X_N^{(0)})$ *such that* (a) $\mathcal{P}(i, \hat{\mathbf{X}}) \geq \hat{X}(i)$; (b) $\mathbf{X}^{(0)} \geq \hat{\mathbf{X}}$; (c) $\mathcal{P}(i, \mathbf{X}^{(0)}) \leq X(i)^{(0)}$ *for* $i = 1,\ldots,N$. *Then for the matrix sequences* $\{X(1)^{(k)}\}_{k=1}^{\infty},\ldots,\{X(N)^{(k)}\}_{k=1}^{\infty}$ *defined by (6) the following properties are satisfied:*

(i) *We have* $\mathbf{X}^{(k)} \geq \hat{\mathbf{X}}$, $\;\mathbf{X}^{(k)} \geq \mathbf{X}^{(k+1)}$ *and*

$$\mathcal{P}(i,\mathbf{X}^{(k)}) = X(i)^{(k+1)} + \sum_{l=0}^{r} \tilde{A}_l(i,\mathbf{X}^{(k)})^T \mathcal{E}_{i1}(\mathbf{X}^{(k)} - \mathbf{X}^{(k+1)}) \tilde{A}_l(i,\mathbf{X}^{(k)}),$$

where $i = 1,2,\ldots,N$, $\;k = 0,1,2,\ldots$;

(ii) *the sequences* $\{X(1)^{(k)}\},\dots,\{X(N)^{(k)}\}$ *converge to the maximal solution* $\tilde{\mathbf{X}}$ *of the set of equations* $X(i) = \mathcal{P}(i,\mathbf{X})$ *and* $\tilde{\mathbf{X}} \geq \hat{\mathbf{X}}$.

In this section, we are proving that iteration (5) has a linear rate of convergence.

Theorem 2.2. *Assume that conditions a),b),c) of theorem 2.1 are fulfilled for a symmetric solution* $\hat{\mathbf{X}} \in Dom\,\mathcal{P}$ *of set of equations (1). Then, the sequence* $\{\mathbf{X}^{(k)}\}_{k=1}^{\infty}$ *defined by (5) converges to the maximal solution* $\tilde{\mathbf{X}}$. *If*

$$\max_{1\leq i\leq N} \|X(i)^{(0)} - \tilde{X}(i)\| < \frac{2-\sum_{l=0}^{r}\|\tilde{A}_l(i,\tilde{\mathbf{X}})\|^2}{a} = \frac{2-b}{a}$$

where

$$a = \|R(i,\mathbf{X}^{(0)})\|\,\|R(i,\tilde{\mathbf{X}})^{-1}\|^2 \left(\|R(i,\tilde{\mathbf{X}})^{-1}\|\|S(i,\tilde{\mathbf{X}})\|\sum_{l=0}^{r}\|B_l(i)\|^2 + \sum_{l=0}^{r}\|B_l(i)\|\,\|A_l(i)\|\right)^2,$$

then

$$\max_{1\leq i\leq N} \|X(i)^{(k)} - \tilde{X}(i)\| < \max_{1\leq i\leq N} \|X(i)^{(k-1)} - \tilde{X}(i)\|.$$

Proof. Following the course of the proof of theorem 2.1 it has been proved that $\mathbf{X}^{(k)} \geq \hat{\mathbf{X}}$ for all k. Therefore, for $i = 1,\dots,N$ we conclude $R(i,\mathbf{X}^{(k)}) \geq R(i,\hat{\mathbf{X}}) > 0$. It follows that

$$\lim_{k\to\infty} R(i,\mathbf{X}^{(k)}) = R(i,\tilde{\mathbf{X}})$$

and then, the limit $F(i,\tilde{\mathbf{X}}) = \lim_{k\to\infty} F(i,\mathbf{X}^{(k)})$ exists and

$$F(i,\tilde{\mathbf{X}}) = -R(i,\tilde{\mathbf{X}})^{-1} S(i,\tilde{\mathbf{X}})^T.$$

Based on the proof of theorem 2.1 and the properties of lemma 2.1 the following equality is established

$$X(i)^{(k)} = \mathcal{P}_{\mathbf{X}^{(k-1)}}(i,\mathbf{X}^{(k-1)}) + \frac{\varepsilon}{k} I_n \quad \text{and} \quad \tilde{X}(i) = \mathcal{P}_{\tilde{\mathbf{X}}}(i,\tilde{\mathbf{X}}).$$

Moreover

$$\begin{aligned}\mathcal{P}(i,\mathbf{X}^{(k-1)}) &= \sum_{l=0}^{r} \tilde{A}_l(i,\tilde{\mathbf{X}})^T \left(\mathcal{E}_i(\mathbf{X}^{(k-1)})\right) \tilde{A}_l(i,\tilde{\mathbf{X}}) + T(i,\mathbf{X}^{(k-1)})\\ &\quad - \left(F(i,\mathbf{X}^{(k-1)})^T - F(i,\tilde{\mathbf{X}})^T\right) R(i,\mathbf{X}^{(k-1)}) \left(F(i,\mathbf{X}^{(k-1)}) - F(i,\tilde{\mathbf{X}})\right),\end{aligned}$$

and

$$X(i)^{(k)} - \tilde{X}(i) = \mathcal{P}_{\mathbf{X}^{(k-1)}}(i,\mathbf{X}^{(k-1)}) - \mathcal{P}_{\tilde{\mathbf{X}}}(i,\tilde{\mathbf{X}}) + \frac{\varepsilon}{k} I_n$$

$$\begin{aligned}X(i)^{(k)} - \tilde{X}(i) &= \sum_{l=0}^{r} \tilde{A}_l(i,\tilde{\mathbf{X}})^T \left(\mathcal{E}_i(\mathbf{X}^{(k-1)} - \tilde{\mathbf{X}})\right) \tilde{A}_l(i,\tilde{\mathbf{X}}) + \frac{\varepsilon}{k} I_n\\ &\quad - \left(F(i,\mathbf{X}^{(k-1)})^T - F(i,\tilde{\mathbf{X}})^T\right) R(i,\mathbf{X}^{(k-1)}) \left(F(i,\mathbf{X}^{(k-1)}) - F(i,\tilde{\mathbf{X}})\right).\end{aligned}$$

Consider the difference

$$
\begin{aligned}
& F(i,\mathbf{X}^{(k-1)}) - F(i,\tilde{\mathbf{X}}) \\
&= -R(i,\mathbf{X}^{(k-1)})^{-1} S(i,\mathbf{X}^{(k-1)})^T + R(i,\tilde{\mathbf{X}})^{-1} S(i,\tilde{\mathbf{X}})^T \\
&= -R(i,\mathbf{X}^{(k-1)})^{-1} S(i,\mathbf{X}^{(k-1)} \pm \tilde{\mathbf{X}})^T + R(i,\tilde{\mathbf{X}})^{-1} S(i,\tilde{\mathbf{X}})^T \\
&= \left[R(i,\tilde{\mathbf{X}})^{-1} - R(i,\mathbf{X}^{(k-1)})^{-1}\right] S(i,\tilde{\mathbf{X}})^T - R(i,\mathbf{X}^{(k-1)})^{-1} \sum_{l=0}^{r} B_l(i)^T \mathcal{E}_i(\mathbf{X}^{(k-1)} - \tilde{\mathbf{X}}) A_l(i) \\
&= R(i,\tilde{\mathbf{X}})^{-1} \left[R(i,\mathbf{X}^{(k-1)}) - R(i,\tilde{\mathbf{X}})\right] R(i,\mathbf{X}^{(k-1)})^{-1} S(i,\tilde{\mathbf{X}})^T \\
& \quad -R(i,\mathbf{X}^{(k-1)})^{-1} \sum_{l=0}^{r} B_l(i)^T \mathcal{E}_i(\mathbf{X}^{(k-1)} - \tilde{\mathbf{X}}) A_l(i) .
\end{aligned}
$$

Then

$$
\begin{aligned}
& F(i,\mathbf{X}^{(k-1)}) - F(i,\tilde{\mathbf{X}}) \\
&= R(i,\tilde{\mathbf{X}})^{-1} \left[\sum_{l=0}^{r} B_l(i)^T \mathcal{E}_i(\mathbf{X}^{(k-1)} - \tilde{\mathbf{X}}) B_l(i)\right] R(i,\mathbf{X}^{(k-1)})^{-1} S(i,\tilde{\mathbf{X}})^T \\
& \quad -R(i,\mathbf{X}^{(k-1)})^{-1} \sum_{l=0}^{r} B_l(i)^T \mathcal{E}_i(\mathbf{X}^{(k-1)} - \tilde{\mathbf{X}}) A_l(i) .
\end{aligned}
$$

Thus

$$
\begin{aligned}
\|F(i,\mathbf{X}^{(k-1)}) - F(i,\tilde{\mathbf{X}})\| \le{}& \|R(i,\tilde{\mathbf{X}})^{-1}\|^2 \, \|S(i,\tilde{\mathbf{X}})\| \sum_{l=0}^{r} \|B_l(i)\|^2 \, \|\mathcal{E}_i(\mathbf{X}^{(k-1)} - \tilde{\mathbf{X}})\| \\
& +\|R(i,\tilde{\mathbf{X}})^{-1}\| \sum_{l=0}^{r} \|B_l(i)\| \, \|A_l(i)\| \, \|\mathcal{E}_i(\mathbf{X}^{(k-1)} - \tilde{\mathbf{X}})\| ,
\end{aligned}
$$

$$
\|F(i,\mathbf{X}^{(k-1)}) - F(i,\tilde{\mathbf{X}})\| \le \tau_{i,2} \, \|\mathcal{E}_i(\mathbf{X}^{(k-1)} - \tilde{\mathbf{X}})\| ,
$$

where

$$
\tau_{i,2} = \|R(i,\tilde{\mathbf{X}})^{-1}\| \left(\|R(i,\tilde{\mathbf{X}})^{-1}\| \|S(i,\tilde{\mathbf{X}})\| \sum_{l=0}^{r} \|B_l(i)\|^2 + \sum_{l=0}^{r} \|B_l(i)\| \, \|A_l(i)\| \right) .
$$

In addition, using $\mathbf{X}^{(0)} \ge \mathbf{X}^{(k)}$ we in fact have

$$
\|R(i,\mathbf{X}^{(k-1)})\| \le \|R(i,\mathbf{X}^{(0)})\| , \quad i = 1,\ldots,N .
$$

Furthermore, we estimate for $i = 1, \ldots, N$

$$\|\tilde{X}(i) - X(i)^{(k)}\| \leq \sum_{l=0}^{r} \|\tilde{A}_l(i, \tilde{\mathbf{X}})\|^2 \, \|\mathcal{E}_i(\mathbf{X}^{(k-1)} - \tilde{\mathbf{X}})\| + (\tau_{i,2})^2 \, \|\mathcal{E}_i(\mathbf{X}^{(k-1)} - \tilde{\mathbf{X}})\|^2 \, \|R(i, \mathbf{X}^{(0)})\| \,.$$

Note that

$$\|\mathcal{E}_i(\mathbf{X}^{(k-1)} - \tilde{\mathbf{X}})\| \leq \sum_{j=1}^{N} p_{ij} \, \|X^{(k-1)}(j) - \tilde{X}(j)\| \leq \max_{1 \leq j \leq N} \|X(j)^{(k-1)} - \tilde{X}(j)\| \,, \, \forall \, i.$$

Further on,

$$\max_{1 \leq i \leq N} \|\tilde{X}(i) - X(i)^{(k)}\| \leq b \left(\max_{1 \leq j \leq N} \|X(j)^{(k-1)} - \tilde{X}(j)\| \right) + a \left(\max_{1 \leq j \leq N} \|X(j)^{(k-1)} - \tilde{X}(j)\| \right)^2 .$$

Now, assuming that the inequality

$$\max_{1 \leq i \leq N} \|X(i)^{(s)} - \tilde{X}(i)\| < \frac{2 - b}{a}$$

holds for $s = 0, \ldots, k-1$. Then

$$\max_{1 \leq i \leq N} \|\tilde{X}(i) - X(i)^{(k)}\| \leq \max_{1 \leq j \leq N} \|X(j)^{(k-1)} - \tilde{X}(j)\|$$
$$\times \left(a \max_{1 \leq j \leq N} \|X(j)^{(k-1)} - \tilde{X}(j)\| + b - 1 \right) < \max_{1 \leq j \leq N} \|X(j)^{(k-1)} - \tilde{X}(j)\| \,.$$

Thus, the proof of the theorem is complete. □

Let us consider the following example in order to compare iterations (5) and (6).

Example 2.1. *We take the following weighting matrices:*

$$R(1) = \text{diag}\,(0.0126,\ 0.024)\,,$$
$$R(2) = \text{diag}\,(0.09,\ 0.012)\,, \quad R(3) = \text{diag}\,(0.12,\ 0.105)\,,$$
$$Q(1) = 0.75 * eye(n,n)\,, \quad Q(2) = 0.25 * eye(n,n)\,, \quad Q(3) = 0.05 * eye(n,n)\,.$$

The coefficient matrices $A_0(i), A_1(i), B_0(i), B_1(i), L(i), i = 1,2,3$ for system (1) are given through formulas (using the MATLAB *notations):*

$$A_0(1) = randn(n,n)/6; \;\; A_0(2) = randn(n,n)/6; \;\; A_0(3) = randn(n,n)/6;$$
$$A_1(1) = randn(n,n)/7; \;\; A_1(2) = randn(n,n)/7; \;\; A_1(3) = randn(n,n)/7;$$

$$B_0(1) = randn(n,2)/8; \;\; B_0(2) = randn(n,2)/8; \;\; B_0(3) = randn(n,2)/8;$$
$$B_1(1) = randn(n,2)/8; \;\; B_1(2) = randn(n,2)/8; \;\; B_1(3) = randn(n,2)/8;$$

$$L(1) = randn(n,2)/8; \;\; L(2) = randn(n,2)/8; \;\; L(3) = randn(n,2)/8,$$

and the following transition probability matrix

$$\left(p_{ij}\right) = \begin{pmatrix} 0.67 \; 0.17 \; 0.16 \\ 0.30 \; 0.47 \; 0.23 \\ 0.26 \; 0.30 \; 0.44 \end{pmatrix} .$$

We have executed hundred examples of each value of n. We report the maximal number of iteration steps mIt and the average number of iteration steps $avIt$ of each size for all examples needed for achieving the accuracy. The used accuracy equals to $1.e-10$. The results are listed in Table 1. The average number of iteration steps for method (6) smaller than the corresponding average number for method (5). The last column of Table 1 shows how much is the acceleration of method (6).

	method (5)		method (6)		speed up
n	mIt	$avIt$	mIt	$avIt$	
15	212	79.2	166	60.1	0.75
16	141	87.1	112	65.3	0.75
17	216	104.6	165	77.7	0.74
18	235	132.9	177	97.0	0.73
19	389	195.8	288	143.5	0.73
20	1882	311.8	900	221.5	0.71

Table 1. Results for Example 2.1. Comparison between iterations for 100 runs.

Further on, we execute some matrix manipulations on system (1) to derive new recurrence equations. We are going to prove the convergence properties to the proposed new iteration under new assumptions. Following the substitution

$$\mathbf{Y} = (Y(1), \ldots, Y(N)), \text{ where } Y(i) = \mathcal{E}_i(\mathbf{X}) \text{ for } i = 1, \ldots, N,$$

the equivalent system of equations is derived

$$Y(i) = \mathcal{P}(i, \mathbf{Y}), \tag{7}$$

where

$$\begin{aligned} \mathcal{P}(i,\mathbf{Y}) = {} & \textstyle\sum_{l=0}^{r} \hat{A}_l(i)^T Y(i) \hat{A}_l(i) + \hat{C}^T(i)\hat{C}(i) - \hat{S}(i, Y(i)) \\ & \times [R(i) + \textstyle\sum_{l=0}^{r} B_l(i)^T Y(i) B_l(i)]^{-1} \hat{S}(i, Y(i))^T + \textstyle\sum_{j=1}^{N} \gamma_{ij} Y(j), \end{aligned} \tag{8}$$

with appropriate transformations on the matrix coefficients $\hat{A}_l(i), \hat{C}(i), \hat{L}(i)$ and

$$\hat{S}(i, Y(i)) = \sum_{l=0}^{r} \hat{A}_l(i)^T Y(i) B_l(i) + \hat{L}(i),$$

$$\hat{A}_l(i) = \sqrt{\frac{p_{ii}}{1-\delta_{ii}}} A_l(i), \quad \hat{C}(i) = \sqrt{\frac{p_{ii}}{1-\delta_{ii}}} C(i), \qquad l = 0, \ldots, r,$$

$$\hat{L}(i) = \sqrt{\frac{p_{ii}}{1-\delta_{ii}}}\,L(i)\,, \quad \hat{S}(i,Y(i)) = \sum_{l=0}^{r} \hat{A}_l(i)^T\,Y(i)\,B_l(i) + \hat{L}(i),$$

for $i = 1,\ldots,N$, and

$$\Gamma = (\gamma_{ip})_1^N = \begin{cases} \gamma_{ii} = 0 \\ \gamma_{ip} = \frac{\delta_{ip}}{1-\delta_{ii}}, \text{ if } i \neq p \end{cases}$$

and assume that Γ is nonnegative ($\gamma_{ip} \geq 0$). We introduce the notations

$$\begin{aligned} &\mathcal{G}(i,\mathbf{Y}) = \textstyle\sum_{p\neq i} \gamma_{ip}\,Y(p)\,, \\ &\mathbf{Y}_i \| Z = (Y(1),\ldots,Y(i-1),Z,Y(i+1),\ldots,Y(N)) \end{aligned}$$

The new iteration scheme applied to the equivalent system (7) is:

$$\begin{aligned} Y^{(k+1)}(i) &= \textstyle\sum_{l=0}^{r} \hat{A}_l(i,Y^{(k)}(i))^T\,Y^{(k)}(i)\,\hat{A}_l(i,Y^{(k)}(i)) \\ &\quad + T(i,Y^{(k)}(i)) + \mathcal{G}(i,1,\mathbf{Y}^{(k+1)}) + \mathcal{G}(i,2,\mathbf{Y}^{(k)})\,, \\ &\quad i = 1,\ldots,N. \end{aligned} \tag{9}$$

where

$$\mathcal{G}(i,1,\mathbf{Z}) = \sum_{j=1}^{i-1} \gamma_{ij}\,Z(j)\,, \text{ and } \mathcal{G}(i,2,\mathbf{Z}) = \sum_{j=i+1}^{N} \gamma_{ij}\,Z(j)\,.$$

The convergence properties of (9) are investigated. We will prove that the convergence of (9) takes place if the algorithm starts at any suitable initial point $\mathbf{Y}^{(0)}$. The new iteration (9) can be considered as an accelerated modification to iteration (5). The convergence result is given by the following theorem:

Theorem 2.3. *[11] We assume that Γ is a nonnegative matrix and $\frac{\lambda_{ii}}{1-\delta_{ii}}$ are positive numbers for all values of i. Letting there are symmetric matrices $\hat{\mathbf{Y}} = (\hat{Y}(1),\ldots,\hat{Y}(N))$ and $\mathbf{Y}^{(0)} = (Y^{(0)}(1),\ldots,Y^{(0)}(N))$ such that (a) $\mathcal{P}(i,\hat{\mathbf{Y}}) \geq \hat{Y}(i)$; (b) $\mathbf{Y}^{(0)} \geq \hat{\mathbf{Y}}$; (c)$\mathcal{P}(i,\mathbf{Y}^{(0)}) \leq Y(i)^{(0)}$ for $i = 1,\ldots,N$. Then for the matrix sequences $\{Y(1)^{(k)}\}_{k=1}^{\infty},\ldots,\{Y(N)^{(k)}\}_{k=1}^{\infty}$ defined by (9) the following properties hold:*

(i) *We have $\mathbf{Y}^{(k)} \geq \hat{\mathbf{Y}},\ \ \mathbf{Y}^{(k)} \geq \mathbf{Y}^{(k+1)}$ and*

$$\mathcal{P}(i,\mathbf{Y}^{(\mathbf{k})}) = Y^{(k+1)}(i) + \mathcal{G}(i,1,\mathbf{Y}^{(k)} - \mathbf{Y}^{(k+1)})$$

where $k = 0,1,2,\ldots$;

(ii) *The sequences $\{Y(1)^{(k)}\},\ldots,\{Y(N)^{(k)}\}$ converge to the solution $\tilde{\mathbf{Y}}$ of the equations $Y(i) = \mathcal{P}(i,\mathbf{Y})$ and $\tilde{\mathbf{Y}} \geq \hat{\mathbf{Y}}$.*

3. The LMI approach

There exists an increasing interest to consider a computational approach to stochastic algebraic Riccati equations via a semidefinite programming problem over linear matrix inequalities. Similar studies can be found in [12–14]. The main result from such type studies is the equivalence between the feasibility of the LMI and the solvability of the corresponding stochastic Riccati equation. Moreover, the maximal solution of a given stochastic algebraic Riccati equation can be obtained by solving a corresponding convex optimization problem (an LMI approach).

Further on, following the classical linear quadratic theory [13, 14] we know that the optimization problem is associated with (1) has the form (for example see [1, 7]):

$$\begin{aligned}
&\max \textstyle\sum_{i=1}^{N} \langle I, X(i)\rangle \\
&\text{subject to } i = 1, \ldots, N \\
&\begin{pmatrix} -X(i) + Q(i,\mathbf{X}) & S(i,\mathbf{X}) \\ S(i,\mathbf{X})^T & R(i,\mathbf{X}) \end{pmatrix} \geq 0 \qquad (10) \\
&R(i,\mathbf{X}) > 0, \\
&X(i) = X(i)^T .
\end{aligned}$$

However, we can apply the same approach to equivalent system (7). As a result we formulate a new optimization problem assigned to (7) and we will use it to find the maximal solution to (7).

The corresponding optimization problem, associated to the maximal solution to (7), is given by:

$$\begin{aligned}
&\max \textstyle\sum_{i=1}^{N} \langle I, Y(i)\rangle \\
&\text{subject to } i = 1, \ldots, N \\
&\begin{pmatrix} -Y(i) + \hat{C}^T(i)\hat{C}(i) + \sum_{j=1}^{N} \gamma_{ij} Y(j) \\ \sum_{l=0}^{r} \hat{A}_l(i)^T Y(i) \hat{A}_l(i) & \hat{S}(i, Y(i)) \\ \hat{S}(i, Y(i))^T & R(i) + \sum_{l=0}^{r} B_l(i)^T Y(i) B_l(i) \end{pmatrix} \geq 0 \qquad (11) \\
&R(i) + \textstyle\sum_{l=0}^{r} B_l(i)^T Y(i) B_l(i) > 0, \\
&Y(i) = Y(i)^T .
\end{aligned}$$

It is very important to analyze a case where the weighting matrices $R(i), Q(i), i = 1, \ldots, N$ are indefinite in the field of linear quadratic stochastic models. This case has a practical importance. There are studies where the cost matrices are allowed to be indefinite (see [3, 16] and reference there in). In this paper we will investigate this special case to considered general discrete-time Riccati equations (1). We will interpret iterations (6) and (9) in a case where matrices $R(i), i = 1, \ldots, N$ are indefinite and however, we will look for a maximal solution from $Dom\,\mathcal{P}$.

Based on the next example, we compare the LMI approach (through optimization problems (10) and (11)) for solving the maximal solution to set of nonlinear equations (1).

We take the $n \times n$ matrices $Q(1), Q(2)$ and $Q(3)$ as follows:

$$Q(1) = diag[0.0,\ 0.5,\ \dots\ 0.5],\quad Q(2) = diag[0.0,\ 1,\ \dots\ 1]$$
$$Q(3) = diag[0.0,\ 0.05,\ \dots\ 0.05].$$

and the same probability matrix as in Example 2.1 .

Example 3.1. *The coefficient matrices $A_0(i), A_1(i), B_0(i), B_1(i), L(i), i = 1,2,3$ for system (1) are given through formulas (using the* MATLAB *notations):*

$$A_0(1) = randn(n,n)/6;\ \ A_0(2) = randn(n,n)/6;\ \ A_0(3) = randn(n,n)/6;$$
$$A_1(1) = randn(n,n)/7;\ \ A_1(2) = randn(n,n)/7;\ \ A_1(3) = randn(n,n)/7;$$

$$B_0(1) = sprandn(n,2,0.3);\ \ B_0(2) = sprandn(n,2,0.3);\ \ B_0(3) = sprandn(n,2,0.3);$$
$$B_1(1) = randn(n,2)/8;\ \ B_1(2) = randn(n,2)/8;\ \ B_1(3) = randn(n,2)/8;$$

$$L(1) = randn(n,2)/8;\ \ L(2) = randn(n,2)/8;\ \ L(3) = randn(n,2)/8.$$

Test 3.1.1. $R(1) = diag[0.02,\ 0.04],\quad R(2) = diag[0.085,\ 0.2],\quad R(3) = diag[0.125,\ 0.1],$

Test 3.1.2. $R(1) = zeros(2,2),\quad R(2) = zeros(2,2),\quad R(3) = zeros(2,2),$

Test 3.1.3. $R(1) = diag[-0.002,\ 0.005],\quad R(2) = diag[-0.003,\ 0.010],$
$R(3) = diag[0.02,\ -0.0004],$

Test 3.1.4. $R(1) = diag[-0.00025,\ -0.00005],\quad R(2) = diag[-0.00035,\ -0.00010],$
$R(3) = diag[-0.0002,\ -0.00005].$

	Test 3.1.1				Test 3.1.2			
	LMI for (10)		LMI for (11)		LMI for (10)		LMI for (11)	
n	mIt	$avIt$	mIt	$avIt$	mIt	$avIt$	mIt	$avIt$
15	47	34.6	46	42.5	45	37.0	49	45.0
16	43	37.5	49	42.4	44	37.0	47	41.8
17	43	34.8	50	41.7	48	36.5	50	42.4
18	52	40.3	51	43.7	50	39.3	48	43.7
19	41	35.3	51	40.6	45	38.8	49	43.0
20	45	36.8	47	40.6	46	37.0	52	44.3
CPU time 10 runs (in seconds)								
20	1332.4		341.1		1431		338.67	

Table 2. Comparison between iterations for Example 3.1.

The MATLAB function mincx is applied with the relative accuracy equals to $1.e-10$ for solving the corresponding optimization problems. Our numerical experiments confirm the effectiveness of the LMI approach applied to the optimization problems (10) and (11). We have compared the results from these experiments in regard of number of iterations and time

	Test 3.1.3				Test 3.1.4			
	LMI for (10)		LMI for (11)		LMI for (10)		LMI for (11)	
n	*mIt*	*avIt*	*mIt*	*avIt*	*mIt*	*avIt*	*mIt*	*avIt*
15	45	36.8	50	44.4	46	36.7	51	43.9
16	47	37.5	50	43.6	48	37.8	52	43.9
17	49	38.7	57	46.4	48	38.6	55	46.3
18	61	42.2	59	46.4	53	41.4	61	48.2
19	46	35.4	50	43.6	46	35.6	53	43.0
20	46	38.6	50	43.3	44	39.2	49	43.0
CPU time 10 runs (in seconds)								
20	1401		355.66		1441.2		350.8	

Table 3. Comparison between iterations for Example 3.1.

of execution. The executed four tests of examples have demonstrated that LMI problem (10) performance needs more computational work than LMI problem (11) and thus, LMI method (11) is faster than LMI method (10) (see the CPU times displayed in tables 2 and 3).

4. The positive semidefinite case

We will investigate new iterations for computing the maximal solution to a set of Riccati equations (1) where the matrices $R(i,\mathbf{X}), i = 1,\ldots,N$ are positive semidefinite. It is well known the application of a special linear quadratic stochastic model in the finance [18] where the cost matrix R is zero and the corresponding matrix $R + B^T XB$ is singular. So, this special case where it is necessary to invert a singular matrix is important to the financial modelling process. Without loss of generality we assume that all matrices $R(i,\mathbf{X}), i = 1,\ldots,N$ in (1) are positive semidefinite. Thus, we will investigate set of equations (3)-(4) for existence a maximal solution. Investigations on similar type of generalized equations have been done by many authors (see [8, 9] and literature therein).

We introduce the following new iteration:

$$\begin{aligned} X(i)^{(k)} &= \textstyle\sum_{l=0}^{r} \left[\widehat{A}_l(i,\mathbf{X}^{(k-1)})\right]^T \left(\mathcal{E}_{i1}(\mathbf{X}^{(k)}) + p_{ii}X(i)^{(k-1)} + \mathcal{E}_{i2}(\mathbf{X}^{(k-1)})\right) \\ &\quad\times \left[\widehat{A}_l(i,\mathbf{X}^{(k-1)})\right] + T(i,\mathbf{X}^{(k-1)})\,, i = 1,2,\ldots,N,\ k = 1,2,3\ldots, \\ \text{where} &\quad \widehat{A}_l(i,\mathbf{Z}) = A_l(i) + B_l(i)F(i,\mathbf{Z}), \quad F(i,\mathbf{Z}) = -\left(R(i,\mathbf{Z})\right)^{\dagger} S(i,\mathbf{Z})^T, \\ &\quad T(i,\mathbf{Z}) = C^T(i)C(i) + F(i,\mathbf{Z})^T L(i)^T + L(i)\,F(i,\mathbf{Z}) \\ &\quad +F(i,\mathbf{Z})^T R(i)\,F(i,\mathbf{Z})\,. \end{aligned} \tag{12}$$

We will prove that the matrix sequence defined by (12) converges to the maximal solution $\tilde{\mathbf{X}}$ of (3)-(4) and $R(i,\tilde{\mathbf{X}}), i = 1,\ldots,N$ are positive semidefinite. Thus, the iteration (12) constructs a convergent matrix sequence.

Let us construct $\mathcal{P}(\mathbf{X}) = (\mathcal{P}(1,\mathbf{X}),\ldots,\mathcal{P}(N,\mathbf{X}))$ and define the set

$$dom\,\mathcal{P}^{\dagger} = \{\mathbf{X} \in \mathcal{H}^n \;:\; R(i,\mathbf{X}) \geq 0 \text{ and } Ker R(i,\mathbf{X}) \subseteq Ker S(i,\mathbf{X}),\ i = 1,\ldots,N\}\,.$$

Consider the rational operator $\mathcal{P} : dom\,\mathcal{P}^{\dagger} \to \mathcal{H}^n$ given by

$$\mathcal{P}(i,\mathbf{X}) = \sum_{l=0}^{r} A_l(i)^T \mathcal{E}_i(\mathbf{X}) A_l(i) + C^T(i)C(i) - S(i,\mathbf{X})\left(R(i,\mathbf{X})\right)^{\dagger} S(i,\mathbf{X})^T,$$
$$i = 1,\ldots,N$$

which has been investigated and some useful lemmas have been proved. We present some preliminary results from the matrix analysis.

Lemma 4.1. *[8, Lemma 4.2] Assume that Z is a $m \times n$ matrix and W is a $p \times n$ matrix. Then the following statements are equivalent:*

(i) $Ker\,Z \subseteq Ker\,W$;

(ii) $W = WZ^{\dagger}Z$;

(iii) $W^{\dagger} = Z^{\dagger}ZW^{\dagger}$.

Lemma 4.2. *[8, Lemma 4.3(i)] Let H be a hermitian matrix of size $n+m$ with $H = \begin{pmatrix} L & N \\ N^* & M \end{pmatrix}$ where L is $n \times n$ and M is $m \times m$. Then, H is positive semidefinite if and only if $M \geq 0$, $L - NM^{\dagger}N^* \geq 0$ and $Ker\,M \subseteq Ker\,N$.*

The next lemma generalizes lemma 3.1 derived by [9] in the following form:

Lemma 4.3. *If $\hat{\mathbf{X}} \in dom\,\mathcal{P}^{\dagger}$ and $Ker R(i,\hat{\mathbf{X}}) \subseteq Ker S(i,\hat{\mathbf{X}})$ for $i = 1,\ldots,N$, then $\mathbf{X} \in dom\,\mathcal{P}^{\dagger}$ for all $\mathbf{X} \geq \hat{\mathbf{X}}$.*

Proof. For $\mathbf{X} \geq \hat{\mathbf{X}}$ we have

$$R(i,\mathbf{X}) \geq R(i,\hat{\mathbf{X}}) \geq 0$$

and

$$Ker R(i,\mathbf{X}) \subseteq Ker R(i,\hat{\mathbf{X}}) \subseteq Ker S(i,\hat{\mathbf{X}}). \tag{13}$$

We apply lemma 4.2 for $H = \Pi(i,\mathbf{X}-\hat{\mathbf{X}}) \geq 0$ and we conclude $\sum_{l=0}^{r} B_l(i)^T \mathcal{E}_i(\mathbf{X}-\hat{\mathbf{X}}) B_l(i) \geq 0$ and

$$Ker \sum_{l=0}^{r} B_l(i)^T \mathcal{E}_i(\mathbf{X}-\hat{\mathbf{X}}) B_l(i) \subseteq Ker \sum_{l=0}^{r} A_l(i)^T \mathcal{E}_i(\mathbf{X}-\hat{\mathbf{X}}) B_l(i).$$

Moreover,

$$\begin{aligned} 0 &\leq R(i,\hat{\mathbf{X}}) \\ &= R(i) + \sum_{l=0}^{r} B_l(i)^T \mathcal{E}_i(\hat{\mathbf{X}} \pm \mathbf{X}) B_l(i) \\ &= R(i,\mathbf{X}) - \sum_{l=0}^{r} B_l(i)^T \mathcal{E}_i(\mathbf{X}-\hat{\mathbf{X}}) B_l(i). \end{aligned}$$

Thus

$$R(i,\mathbf{X}) \geq \sum_{l=0}^{r} B_l(i)^T \mathcal{E}_i(\mathbf{X}-\hat{\mathbf{X}}) B_l(i)$$

and

$$KerR(i,\mathbf{X}) \subseteq Ker \sum_{l=0}^{r} B_l(i)^T \mathcal{E}_i(\mathbf{X}-\hat{\mathbf{X}})B_l(i) \subseteq Ker \sum_{l=0}^{r} A_l(i)^T \mathcal{E}_i(\mathbf{X}-\hat{\mathbf{X}})B_l(i)\,. \tag{14}$$

Combining (13) and (14) we write down

$$KerR(i,\mathbf{X}) \subseteq Ker \left[S(i,\hat{\mathbf{X}}) + \sum_{l=0}^{r} A_l(i)^T \mathcal{E}_i(\mathbf{X}-\hat{\mathbf{X}})B_l(i) \right] = Ker\, S(i,\mathbf{X})\,.$$

□

We define

$$\widehat{W}_{\mathbf{X}}(i,\mathbf{H}) = [F(i,\mathbf{X}) - F(i,\mathbf{H})]^T\, R(i,\mathbf{H})\, [F(i,\mathbf{X}) - F(i,\mathbf{H})]$$

for $\mathbf{X} \in \mathcal{H}^n$ and $\mathbf{H} \in \mathcal{H}^n$. Obviously $\widehat{W}_{\mathbf{X}}(i,\mathbf{H}) \geq 0$ and $\widehat{W}_{\mathbf{X}}(i,\mathbf{X}) = 0$.

Lemma 4.4. *If $\mathbf{Y} \in \mathcal{H}^n$ and $\mathbf{Z} \in \mathcal{H}^n$ (or let $\mathbf{Y}$, $\mathbf{Z}$) be symmetric matrices with $KerR(i,\mathbf{Y}) \subseteq KerS(i,\mathbf{Y})$ and $KerR(i,\mathbf{Z}) \subseteq KerS(i,\mathbf{Z})$ for $i = 1,\ldots,N$. Then, the following identities hold:*

$$\begin{aligned} \mathcal{P}_{\mathbf{Z}}(i,\mathbf{Y}) &= \textstyle\sum_{l=0}^{r} \tilde{A}_l(i,\mathbf{Z})^T\, \mathcal{E}_i(\mathbf{Y})\tilde{A}_l(i,\mathbf{Z}) + T(i,\mathbf{Z}) - \widehat{W}_{\mathbf{Z}}(i,\mathbf{Y}) \\ \textit{where} \qquad & \tilde{A}_l(i,\mathbf{Z}) = A_l(i) + B_l(i)\, F(i,\mathbf{Z}) \end{aligned} \tag{15}$$

and

$$\mathcal{P}_{\mathbf{Z}}(i,\mathbf{Z}) - \mathcal{P}_{\mathbf{Z}}(i,\mathbf{Y}) = \sum_{l=0}^{r} \tilde{A}_l(i,\mathbf{Z})^T\, \mathcal{E}_i(\mathbf{Z}-\mathbf{Y})\tilde{A}_l(i,\mathbf{Z}) + \widehat{W}_{\mathbf{Z}}(i,\mathbf{Y})$$

for $i = 1,\ldots,N$.

Proof. Let us consider the difference

$$\begin{aligned} &\mathcal{P}(i,\mathbf{Y}) - T(i,\mathbf{Z}) \\ &= \sum_{l=0}^{r} A_l(i)^T \mathcal{E}_i(\mathbf{Y})A_l(i) - F(i,\mathbf{Y})^T\, R(i,\mathbf{Y})\, F(i,\mathbf{Y}) \\ &\quad -F(i,\mathbf{Z})^T\, L(i)^T - L(i)\, F(i,\mathbf{Z}) - F(i,\mathbf{Z})^T\, R(i)\, F(i,\mathbf{Z}) \\ &= \sum_{l=0}^{r} A_l(i)^T \mathcal{E}_i(\mathbf{Y})A_l(i) - F(i,\mathbf{Y})^T\, R(i,\mathbf{Y})\, F(i,\mathbf{Y}) \pm F(i,\mathbf{Z})^T\, R(i,\mathbf{Y})\, F(i,\mathbf{Y}) \\ &\quad -F(i,\mathbf{Z})^T\, L(i)^T - L(i)\, F(i,\mathbf{Z}) \pm F(i,\mathbf{Y})^T\, R(i,\mathbf{Y})\, F(i,\mathbf{Z}) \\ &\quad -F(i,\mathbf{Z})^T \left(R(i) \pm \sum_{l=0}^{r} B_l(i)^T \mathcal{E}_i(\mathbf{Y})B_l(i) \right) F(i,\mathbf{Z}). \end{aligned}$$

According to lemma 4.1 we obtain $F(i,\mathbf{Z})^T R(i,\mathbf{Z}) = -S(i,\mathbf{Z})$ and $F(i,\mathbf{Y})^T R(i,\mathbf{Y}) = -S(i,\mathbf{Y})$. We derive

$$\begin{aligned}&-F(i,\mathbf{Z})^T (L(i)^T + R(i,\mathbf{Y})\, F(i,\mathbf{Y})) \\ &= -F(i,\mathbf{Z})^T (L(i)^T - S(i,\mathbf{Y})^T) = F(i,\mathbf{Z})^T \sum_{l=0}^{r} B_l(i)^T \mathcal{E}_i(\mathbf{Y}) A_l(i),\end{aligned}$$

and

$$\begin{aligned}&-(L(i) + F(i,\mathbf{T})^T\, R(i,\mathbf{Y}))\, F(i,\mathbf{Z}) \\ &= -(L(i) - S(i,\mathbf{Y}))\, F(i,\mathbf{Z}) = \sum_{l=0}^{r} A_l(i)^T \mathcal{E}_i(\mathbf{Y}) B_l(i)\, F(i,\mathbf{Z}).\end{aligned}$$

Then

$$\mathcal{P}(i,\mathbf{Y}) - T(i,\mathbf{Z}) = \sum_{l=0}^{r} \tilde{A}_l(i,\mathbf{Z})^T\, \mathcal{E}_i(\mathbf{Y}) \tilde{A}_l(i,\mathbf{Z}) - \widehat{W}_{\mathbf{Z}}(i,\mathbf{Y})$$

and

$$\mathcal{P}(i,\mathbf{Y}) = \mathcal{P}_{\mathbf{Z}}(i,\mathbf{Y}) = \sum_{l=0}^{r} \tilde{A}_l(i,\mathbf{Z})^T\, \mathcal{E}_i(\mathbf{Y}) \tilde{A}_l(i,\mathbf{Z}) + T(i,\mathbf{Z}) - \widehat{W}_{\mathbf{Z}}(i,\mathbf{Y}),$$

i.e. the identity (15) holds for all values of i.

Further on, taking $\mathbf{Y} = \mathbf{Z}$ in (15) we obtain:

$$\mathcal{P}(i,\mathbf{Y}) = \sum_{l=0}^{r} \tilde{A}_l(i,\mathbf{Y})^T\, \mathcal{E}_i(\mathbf{Y}) \tilde{A}_l(i,\mathbf{Y}) - T(i,\mathbf{Y}).$$

Combining the last two equations it is received

$$\mathcal{P}_{\mathbf{Z}}(i,\mathbf{Z}) - \mathcal{P}_{\mathbf{Z}}(i,\mathbf{Y}) = \sum_{l=0}^{r} \tilde{A}_l(i,\mathbf{Z})^T\, \mathcal{E}_i(\mathbf{Z} - \mathbf{Y}) \tilde{A}_l(i,\mathbf{Z}) + \widehat{W}_{\mathbf{Z}}(i,\mathbf{Y}).$$

□

Now, we are ready to investigate recurrence equations (12) where $\mathbf{X}^{(0)}$ is a suitable matrix. We will prove some properties of the matrix sequence $\{X_i\}_{i=0}^{\infty}$ defined by the above recurrence equation. The limit of this matrix sequence is a solution to (3)-(4). We will derive the theorem:

Theorem 4.1. *Letting there are symmetric matrices* $\hat{\mathbf{X}} = (\hat{X}_1, \ldots, \hat{X}_N)$ *and* $\mathbf{X}^{(0)} = (X_1^{(0)}, \ldots, X_N^{(0)})$ *such that (for* $i = 1, \ldots, N$*):*

(a) $\hat{\mathbf{X}} \in dom\, \mathcal{P}^{\dagger}$ *with* $Ker R(i,\hat{\mathbf{X}}) \subseteq Ker S(i,\hat{\mathbf{X}})$;

(b) $\mathcal{P}(i,\, \hat{\mathbf{X}}) \geq \hat{X}(i)$;

(c) $\mathbf{X}^{(0)} \geq \hat{\mathbf{X}}$;

(d) $\mathcal{P}(i, \mathbf{X}^{(0)}) \leq X(i)^{(0)}$.

Then for the matrix sequences $\{X(1)^{(k)}\}_{k=1}^{\infty}, \ldots, \{X(N)^{(k)}\}_{k=1}^{\infty}$ *defined by (12) the following properties are satisfied:*

(i) *We have* $\mathbf{X}^{(k)} \geq \hat{\mathbf{X}}, \quad \mathbf{X}^{(k)} \geq \mathbf{X}^{(k+1)}$ *and*

$$\mathcal{P}(i, \mathbf{X}^{(k)}) = X(i)^{(k+1)} + \sum_{l=0}^{r} \tilde{A}_l(i, \mathbf{X}^{(k)})^T \mathcal{E}_{i1}(\mathbf{X}^{(k)} - \mathbf{X}^{(k+1)}) \tilde{A}_l(i, \mathbf{X}^{(k)}),$$

where $i = 1, 2, \ldots, N, \; k = 0, 1, 2, \ldots$;

(ii) *the sequences* $\{X(1)^{(k)}\}, \ldots, \{X(N)^{(k)}\}$ *converge to the maximal solution* $\tilde{\mathbf{X}}$ *of set of equations (3)-(4) and* $\tilde{\mathbf{X}} \geq \hat{\mathbf{X}}$.

Proof. Let $k = 0$. We will prove the inequality $\mathbf{X}^{(0)} \geq \mathbf{X}^{(1)}$. From iteration (12) for $k = 1$ with $\mathbf{X}^{(0)} \in dom\,\mathcal{P}^{\dagger}$ and for each i we get:

$$X(i)^{(1)} = \Sigma_{l=0}^{r} \left[\tilde{A}_l(i, \mathbf{X}^{(0)})\right]^T \left(\mathcal{E}_{i1}(\mathbf{X}^{(1)}) + p_{ii} X(i)^{(0)} + \mathcal{E}_{i2}(\mathbf{X}^{(0)})\right) \\ \times \left[\tilde{A}_l(i, \mathbf{X}^{(0)})\right] + T(i, \mathbf{X}^{(0)}).$$

We will derive an expression to $X(i)^{(0)} - X(i)^{(1)}$. We obtain

$$X(i)^{(0)} - X(i)^{(1)} = \Sigma_{l=0}^{r} \tilde{A}_l(i, \mathbf{X}^{(0)})^T \left(\mathcal{E}_{i1}(\mathbf{X}^{(0)} - \mathbf{X}^{(1)})\right) \tilde{A}_l(i, \mathbf{X}^{(0)}) + X(i)^{(0)} - \mathcal{P}(i, \mathbf{X}^{(0)}).$$

We conclude $X(i)^{(0)} - X(i)^{(1)} \geq 0$ for $i = 1, 2, \ldots, N$, under the assumption (d) of the theorem.

Beginning with $\mathbf{X}^{(0)}$ and using iteration (12) we construct two matrix sequences $\{F_{X_i}\}_{i=0}^{\infty}$, and $\{X_i\}_{i=1}^{\infty}$. We will prove by induction the following statements for $i = 1, \ldots, N$:

- $X(i)^{(p)} \geq \hat{X}(i)$ and thus $\mathbf{X}^{(p)} \in dom\,\mathcal{P}^{\dagger}$ and $Ker R(i, \mathbf{X}^{(p)}) \subseteq Ker S(i, \mathbf{X}^{(p)})$,
- $X(i)^{(p)} \geq X(i)^{(p+1)}$,
- $\mathcal{P}(i, \mathbf{X}^{(p)}) = X(i)^{(p+1)} + \Sigma_{l=0}^{r} \tilde{A}_l(i, \mathbf{X}^{(p)})^T \mathcal{E}_{i1}(\mathbf{X}^{(p)} - \mathbf{X}^{(p+1)}) \tilde{A}_l(i, \mathbf{X}^{(p)})$.

We will prove $X(i)^{(p)} \geq \hat{X}(i)$ for $i = 1, \ldots, N$. Using (15) with $\mathbf{Y} = \hat{\mathbf{X}}$ and $\mathbf{Z} = \mathbf{X}^{(p-1)}$ we from the difference

$$X(i)^{(p)} - \hat{X}(i) = \sum_{l=0}^{r} \tilde{A}_l(i, \mathbf{X}^{(p-1)})^T \left(\mathcal{E}_{i1}(\mathbf{X}^{(p)}) + p_{ii} X(i)^{(p-1)} + \mathcal{E}_{i2}(\mathbf{X}^{(p-1)})\right) \\ \times \tilde{A}_l(i, \mathbf{X}^{(p-1)}) + T(i, \mathbf{X}^{(p-1)}) - \hat{X}(i).$$

Based on identity (15) in the form

$$\mathcal{P}(i, \hat{\mathbf{X}}) = \mathcal{P}_{\mathbf{X}^{(p-1)}}(i, \hat{\mathbf{X}}) = \Sigma_{l=0}^{r} \tilde{A}_l(i, \mathbf{X}^{(p-1)})^T \mathcal{E}_i(\hat{\mathbf{X}}) \tilde{A}_l(i, \mathbf{X}^{(p-1)}) + T(i, \mathbf{X}^{(p-1)}) \\ - \widehat{W}_{\mathbf{X}^{(p-1)}}(i, \hat{\mathbf{X}}).$$

we derive

$$
\begin{aligned}
& X(i)^{(p)} - \hat{X}(i) - \mathcal{P}(i, \hat{\mathbf{X}}) \\
= & \sum_{l=0}^{r} \tilde{A}_l(i, \mathbf{X}^{(p-1)})^T \left(\mathcal{E}_{i1}(\mathbf{X}^{(p)} - \hat{\mathbf{X}}) + p_{ii}(X(i)^{(p-1)} - \hat{X}(i)) + \mathcal{E}_{i2}(\mathbf{X}^{(p-1)} - \hat{\mathbf{X}}) \right) \\
& \times \tilde{A}_l(i, \mathbf{X}^{(p-1)}) - \hat{X}(i) + \widehat{W}_{\mathbf{X}^{(p-1)}}(i, \hat{\mathbf{X}}) .
\end{aligned}
$$

We know that $\mathcal{P}(i, \hat{\mathbf{X}}) - \hat{X}(i) \geq 0$. Then $X(i)^{(p)} - \hat{X}(i) \geq 0$ for all $i = 1, \ldots, N$.

Lemma 4.3 confirms that $X(i)^{(p)} \in dom\, \mathcal{P}^{\dagger}$. We compute $F(i, \mathbf{X}^{(p)}) = -\left(R(i, \mathbf{X}^{(p)}) \right)^{\dagger} S(i, \mathbf{X}^{(p)})^T$. Next, we obtain the matrices $X(i)^{(p+1)}$ from (12) and we will prove $X(i)^{(p)} \geq X(i)^{(p+1)}$ for $i = 1, \ldots N$. After some matrix manipulations we derive

$$
\begin{aligned}
& X(i)^{(p)} - X(i)^{(p+1)} \\
= & \sum_{l=0}^{r} \tilde{A}_l(i, \mathbf{X}^{(p)})^T \, \mathcal{E}_{i1}(\mathbf{X}^{(p)} - \mathbf{X}^{(p+1)}) \, \tilde{A}_l(i, \mathbf{X}^{(p)}) \\
& + \sum_{l=0}^{r} \tilde{A}_l(i, \mathbf{X}^{(p-1)})^T \left(p_{ii}(X(i)^{(p-1)} - X(i)^{(p)}) + \mathcal{E}_{i2}(\mathbf{X}^{(p-1)} - \mathbf{X}^{(p)}) \right) \\
& \times \tilde{A}_l(i, \mathbf{X}^{(p-1)}) + \widehat{W}_{\mathbf{X}^{(p-1)}}(i, \mathbf{X}^{(p)}) .
\end{aligned}
$$

It is easy to see that $X(i)^{(p)} - X(i)^{(p+1)} \geq 0$ for $i = 1, 2, \ldots, N$ from the last equation.

Further on, we have to show that

$$
\mathcal{P}(i, \mathbf{X}^{(p)}) = X(i)^{(p+1)} + \sum_{l=0}^{r} \tilde{A}_l(i, \mathbf{X}^{(p)})^T \mathcal{E}_{i1}(\mathbf{X}^{(p)} - \mathbf{X}^{(p+1)}) \tilde{A}_l(i, \mathbf{X}^{(p)})
$$

for $i = 1, \ldots, N$.

We have

$$
\mathcal{P}(i, \mathbf{X}^{(p)}) = \sum_{l=0}^{r} \tilde{A}_l(i, \mathbf{X}^{(p)})^T \, \mathcal{E}_i(\mathbf{X}^{(p)}) \, \tilde{A}_l(i, \mathbf{X}^{(p)}) + T(i, \mathbf{X}^{(p)})
$$

and

$$
\begin{aligned}
X(i)^{(p+1)} = & \; T(i, \mathbf{X}^{(p)}) + \textstyle\sum_{l=0}^{r} \tilde{A}_l(i, \mathbf{X}^{(p)})^T \\
& \times \left(\mathcal{E}_{i1}(\mathbf{X}^{(p+1)}) + p_{ii} X(i)^{(p)} + \mathcal{E}_{i2}(\mathbf{X}^{(p)}) \right) \tilde{A}_l(i, \mathbf{X}^{(p)}) .
\end{aligned}
$$

Subtracting the last two equations we yield

$$
\mathcal{P}(i, \mathbf{X}^{(p)}) = X(i)^{(p+1)} + \textstyle\sum_{l=0}^{r} \tilde{A}_l(i, \mathbf{X}^{(p)})^T \mathcal{E}_{i1}(\mathbf{X}^{(p)} - \mathbf{X}^{(p+1)}) \tilde{A}_l(i, \mathbf{X}^{(p)}) ,
$$

for $i = 1, \ldots, N$.

Thus, it is received a nonincreasing matrix sequence $\{X_i\}_{i=1}^{\infty}$ of symmetric matrices bounded below by $\hat{\mathbf{X}}$ which converges to $\tilde{\mathbf{X}}$ and $\tilde{\mathbf{X}} \geq \hat{\mathbf{X}}$. Hence, $\tilde{\mathbf{X}} \in dom\,\mathcal{P}^{\dagger}$ by Lemma 4.3.

The theorem is proved. □

Thus, following the above theorem we could compute the maximal solution $\tilde{\mathbf{X}}$ of the set of equations (3)-(4). We should apply iteration (12). The next question is: How to apply the LMI method in this case?

Let us consider the modified optimization problem:

$$\begin{aligned} &\max \sum_{i=1}^{N} \langle I, X(i) \rangle \\ &\text{subject to } i = 1, \dots, N \\ &\begin{pmatrix} -X(i) + Q(i,\mathbf{X}) & S(i,\mathbf{X}) \\ & \\ S(i,\mathbf{X})^T & R(i,\mathbf{X}) \end{pmatrix} \geq 0 \\ & \\ &X(i) = X(i)^T . \end{aligned} \tag{16}$$

Theorem 4.2. *Assume that* $(\mathbf{A},\mathbf{B})$ *is stabilizable and there exists a solution to the inequalities* $\mathcal{P}(i,\mathbf{X}) - X(i) \geq 0$ *for* $i = 1,\dots,N$. *Then there exists a maximal solution* $\mathbf{X}^+$ *of (3)-(4) if and only if there exists a solution* $\tilde{\mathbf{X}}$ *for the above convex programming problem (16) with* $\mathbf{X}^+ \equiv \tilde{\mathbf{X}}$.

Proof. Note that the matrix $\mathbf{X} = (X(1),\dots,X(N))$ satisfies the restrictions of optimization problem (16) if and only if

$$\begin{cases} X(i) + Q(i,\mathbf{X}) - S(i,\mathbf{X})\,R(i,\mathbf{X})^{\dagger}\,S(i,\mathbf{X})^T = \mathcal{P}(i,\mathbf{X}) - X(i) \geq 0\,, \\ R(i,\mathbf{X}) \geq 0\,, \\ Ker\,R(i,\mathbf{X}) \subseteq Ker\,S(i,\mathbf{X}) \iff S(i,\mathbf{X})\left(I - R(i,\mathbf{X})\,R(i,\mathbf{X})^{\dagger}\right) = 0 \text{ (Lema 4.1 (ii))}, \\ \text{for } i = 1,\dots,N\,. \end{cases} \tag{17}$$

The last statement follows immediately by lemma 4.2.

Assume that $\mathbf{X}^+$ is the maximal solution of (3)-(4). Thus, $\mathbf{X}^+ \geq \mathbf{X}$ and $trX^+(1) + \dots + trX^+(N) \geq trX(1) + \dots + trX(N)$ for any solution $\mathbf{X}$ of (3)-(4), i.e. for any matrix $\mathbf{X}$ satisfies restrictions of (16) and then $\mathbf{X}^+$ is the solution of optimization problem (16).

Further on, suppose that $\tilde{\mathbf{X}}$ is a solution of optimization problem (16). The inequalities (17) are fulfilled for $\tilde{\mathbf{X}}$, i.e. $\tilde{\mathbf{X}} \in dom\,\mathcal{P}^{\dagger}$ and assumptions for theorem 4.1 hold for $\hat{\mathbf{X}} = \tilde{\mathbf{X}}$. Thus, there exists the maximal solution $\mathbf{X}^+$ with $\mathbf{X}^+ \geq \tilde{\mathbf{X}}$. Moreover, the optimality of $\tilde{\mathbf{X}}$ means that

$$tr\,(X^+(1) - \tilde{X}(1)) + \dots + tr\,(X^+(N) - \tilde{X}(N)) \leq 0$$

and then $X^+(j) - \tilde{X}(j) = 0$ for $j = 1,\dots,N$. Then $\mathbf{X}^+ \equiv \tilde{\mathbf{X}}$. □

Let us consider set of equations (7)-(8) under the assumption that $R(i) + \sum_{l=0}^{r} B_l(i)^T Y(i) B_l(i) \geq 0,\ i = 1, \ldots, N$. Thus, optimization problem (11) is transformed to the new optimization problem:

$$\max \sum_{i=1}^{N} \langle I, Y(i) \rangle$$

$$\text{subject to } i = 1, \ldots, N$$

$$\begin{pmatrix} -Y(i) + \hat{C}^T(i)\hat{C}(i) + \sum_{j=1}^{N} \gamma_{ij} Y(j) & \\ \sum_{l=0}^{r} \hat{A}_l(i)^T Y(i) \hat{A}_l(i) & \hat{S}(i, Y(i)) \\ \hat{S}(i, Y(i))^T & R(i) + \sum_{l=0}^{r} B_l(i)^T Y(i) B_l(i) \end{pmatrix} \geq 0 \tag{18}$$

$$Y(i) = Y(i)^T.$$

It is easy to verify that the solution of (18) is the maximal solution to (7)-(8) with the positive semidefinite assumption to matrices $R(i) + \sum_{l=0}^{r} B_l(i)^T Y(i) B_l(i) \geq 0,\ i = 1, \ldots, N$.

We investigate the numerical behavior of the LMI approach applied to the described optimization problems LMI: (16) and LMI(Y): (18) for finding the maximal solution to set of discrete-time generalized Riccati equations (3) - (4). In addition, we compare these LMI solvers with derived recurrence equations (12) for the maximal solution to the same set of equations. We will carry out some experiments for this purpose. In the experiments in this section we construct a family of examples $(N = 3, k = 3)$ with the weighting matrices

$$Q(1) = \text{diag}[0; 0.5; \ldots, 0.5], \quad Q(2) = \text{diag}[0; 1; 1; \ldots, 1],$$

$$Q(3) = \text{diag}[0; 0.05; 0.05; \ldots, 0.05],$$

$$R(1) = R(2) = R(3) = zeros(3,3),$$

and zero matrices $L(1), L(2), L(3)$ and the introduced transition probability matrix via Example 2.1.

Example 4.1. *We consider the case of $r = 1, n = 6, 7$, where the coefficient real matrices $A_0(i), A_1(i), A_2(i), B_0(i), B_1(i), B_2(i), L(i), i = 1, 2, 3$ are given as follows (using the* MATLAB *notations):*

$$A_0(1) = randn(n,n)/10; \ A_0(2) = randn(n,n)/5; \ A_0(3) = randn(n,n)/5;$$

$$A_1(1) = randn(n,n)/100; \ A_1(2) = randn(n,n)/50; \ A_1(3) = randn(n,n)/100;$$

$$B_0(1) = 100 * full(sprand(n,k,0.07)); \ B_0(2) = 100 * full(sprand(n,k,0.07));$$

$$B_0(3) = 100 * full(sprand(n,k,0.07));$$

$$B_1(1) = 100 * full(sprand(n,k,0.07)); \ B_1(2) = 100 * full(sprand(n,k,0.07));$$

$$B_1(3) = 100 * full(sprand(n,k,0.07)).$$

In our definitions the functions randn(p,k) and sprand(q,m,0.3) return a p-by-k matrix of pseudorandom scalar values and a q-by-m sparse matrix respectively (for more information see the MATLAB *description).*

Results from experiments are given in table 4. The parameters mIt and $avIt$ are the same as the previous tables. In addition, the CPU time in seconds is included. The optimization problems (16) and (18) need the equals iteration steps (the column $avIt$) for finding the maximal solution to set of equations (3) - (4). However, the executed examples have demonstrated that LMI problem (18) faster than LMI problem (16). Moreover, iterative method (12) is much faster than the LMI approaches and it achieves the same accuracy.

	(12)		LMI for (16)		LMI for (18)	
n	mIt	$avIt$	mIt	$avIt$	mIt	$avIt$
6	23	15.9	59	37.9	59	37.9
7	27	16.9	64	39.2	63	37.5
CPU time 20 runs (in seconds)						
20	0.41		72.12		14.48	

Table 4. Comparison between methods for the maximal solution in Example 4.1.

We introduce an additional example where the above optimization problems are compared.

Example 4.2. *The parameters of this system are presented as follows. The coefficient matrices are* $(r = 1, k = 3, n = 6)$*:*

$$A_0(1) = 0.001 * \begin{bmatrix} 58 & 20 & 66 & 60 & 45 & 13 \\ 7 & 33 & 45 & 3 & 33 & 45 \\ 21 & 19 & 36 & 20 & 11 & 42 \\ 58 & 34 & 26 & 38 & 28 & 20 \\ 7 & 51 & 53 & 31 & 59 & 59 \\ 40 & 16 & 56 & 17 & 27 & 29 \end{bmatrix}, \quad A_0(2) = 0.001 * \begin{bmatrix} 66 & 29 & 54 & 58 & 19 & 36 \\ 20 & 52 & 10 & 22 & 39 & 17 \\ 62 & 42 & 4 & 44 & 32 & 63 \\ 18 & 0 & 4 & 11 & 5 & 17 \\ 21 & 27 & 48 & 47 & 49 & 11 \\ 33 & 8 & 58 & 14 & 64 & 41 \end{bmatrix},$$

$$A_0(3) = 0.001 * \begin{bmatrix} 63 & 54 & 24 & 17 & 46 & 13 \\ 27 & 34 & 44 & 63 & 65 & 61 \\ 8 & 18 & 11 & 64 & 46 & 33 \\ 56 & 51 & 6 & 12 & 65 & 3 \\ 45 & 61 & 16 & 11 & 22 & 14 \\ 22 & 3 & 48 & 6 & 39 & 9 \end{bmatrix}, \quad A_1(1) = 0.001 * \begin{bmatrix} 5 & 7 & 1 & 5 & 8 & 3 \\ 1 & 6 & 7 & 5 & 8 & 6 \\ 1 & 3 & 7 & 10 & 9 & 5 \\ 2 & 6 & 7 & 3 & 9 & 7 \\ 6 & 7 & 9 & 8 & 3 & 0 \\ 2 & 7 & 8 & 4 & 7 & 1 \end{bmatrix},$$

$$A_1(2) = 0.001 * \begin{bmatrix} 9 & 8 & 5 & 5 & 8 & 9 \\ 9 & 9 & 10 & 3 & 3 & 1 \\ 8 & 6 & 9 & 0 & 9 & 2 \\ 6 & 7 & 3 & 8 & 10 & 7 \\ 10 & 7 & 9 & 7 & 4 & 8 \\ 6 & 7 & 6 & 9 & 6 & 2 \end{bmatrix}, \quad A_1(3) = 0.001 * \begin{bmatrix} 7 & 6 & 1 & 6 & 2 & 6 \\ 5 & 3 & 7 & 5 & 6 & 9 \\ 5 & 9 & 2 & 8 & 5 & 1 \\ 6 & 3 & 7 & 3 & 5 & 7 \\ 8 & 8 & 2 & 9 & 4 & 8 \\ 7 & 4 & 8 & 5 & 3 & 9 \end{bmatrix}.$$

Coefficient matrices $B_0(i), B_1(i),\ i = 1,2,3$ *are* 6×3 *zero matrices with nonzero elements:*

$$B_0(1)(5,3) = 10.07; \quad B_0(1)(3,1) = 2.56; \quad B_0(2)(1,2) = 6.428; \quad B_0(2)(4,2) = 5.48;$$

$$B_0(3)(6,3) = 13.498; \quad B_0(3)(5,3) = 1.285; \quad B_1(1)(2,3) = 6.525; \quad B_1(1)(4,3) = -5.2;$$
$$B_1(2)(2,1) = -22.99; \quad B_1(2)(4,1) = 3.25; \quad B_1(3)(6,1) = 6.8466; \quad B_1(3)(6,2) = 2.5\,.$$

This choice of the matrices $B_0(i), B_1(i),\ i = 1,2,3$ guaranteed that the matrices $R(i,\mathbf{X})$ are positive semidefinite, i.e. there are symmetric matrices $\mathbf{X}$ which belongs to dom $\mathrm{P}^{\dagger}$. The remain coefficient matrices are already in place.

We find the maximal solution to (3) - (4) for the constructed example with iterative method (12) and the LMI approach applied to optimization problems (16) and (18). The results are the following. Iteration (12) needs 15 iteration steps to achieve the maximal solution. The computed maximal solution $\mathbf{W}$ has the eigenvalues

$$\begin{aligned} \mathrm{Eig}\, W(1) &= (4.9558e-5;\ 0.50058;\ 0.50004;\ 0.50002;\ 0.50001;\ 0.5)\,, \\ \mathrm{Eig}\, W(2) &= (0.00019562;\ 1.0007;\ 1.0001;\ 1;\ 1;\ 1)\,, \\ \mathrm{Eig}\, W(3) &= (9.33e-5;\ 0.05041;\ 0.05005;\ 0.050018;\ 0.050011;\ 0.050003)\,. \end{aligned} \tag{19}$$

The LMI approach for optimization problem (16) does not give a satisfactory result. After 32 iteration step the calculations stop with the computed maximal solution $\mathbf{V}$. However, the norm of the difference between two solutions $\mathbf{W}$ and $\mathbf{V}$ is $\|W(1) - V(1)\| = 1.0122e-9$, $\|W(2) - V(2)\| = 2.8657e-6$, $\|W(3) - V(3)\| = 3.632e-6$.

The LMI approach for optimization problem (18) needs 28 iteration steps to compute the maximal solution to (1). This solution $\mathbf{Z}$ has the same eigenvalues as in (19). The norm of the difference between two solutions $\mathbf{W}$ and $\mathbf{Z}$ is $\|W(1) - Z(1)\| = 7.4105e-12$, $\|W(2) - Z(2)\| = 2.8982e-11$, $\|W(3) - Z(3)\| = 3.0796e-11$.

The results from this example show that the LMI approach applied to optimization problem (18) gives the more accurate results than the LMI method for (16). Moreover, the results obtained for problem (16) are not applicability. A researcher has to be careful when applied the LMI approach for solving a set of general discrete time equations in positive semidefinite case.

5. Conclusion

This chapter presents a survey on the methods for numerical computation the maximal solution of a wide class of coupled Riccati-type equations arising in the optimal control of discrete-time stochastic systems corrupted with state-dependent noise and with Markovian jumping. In addition, computational procedures to compute this solution for a set of discrete-time generalized Riccati equations (7)-(8) are derived. Moreover, the LMI solvers for this case are implemented and numerical simulations are executed. The results are compared and the usefulness of the proposed solvers are commented.

Author details

Ivan Ivanov
Sofia University "St.Kl.Ohridski", Bulgaria

6. References

[1] Costa, O.L.V. & Marques, R.P. (1999). Maximal and Stabilizing Hermitian Solutions for Discrete-Time Coupled Algebraic Ricacti Equations. *Mathematics of Control, Signals and Systems*, Vol. 12, 167-195, ISSN: 0932-4194.

[2] Costa, O.L.V., Fragoso, M.D., & Marques, R.P. (2005). Discrete-Time Markov Jump Linear Systems, Springer-Verlag, Berlin, ISBN: 978-1-85233-761-2.

[3] Costa, O.L.V. & de Paulo, W.L. (2007). Indefinite quadratic with linear costs optimal control of Markov jump with multiplicative noise systems, *Automatica*, Vol. 43, 587-597, ISSN: 0005-1098.

[4] Dragan, V. & Morozan, T. (2008). Discrete-time Riccati type Equations and the Tracking Problem, *ICIC Express. Letters*, Vol.2, 109Ű116, ISNN: 1881-803X.

[5] Dragan, V. & Morozan, T. (2010). A Class of Discrete Time Generalized Riccati Equations, *Journal of Difference Equations and Applications*, Vol.16, No.4, 291-320, ISSN: 1563-5120.

[6] Dragan, V., Morozan, T. & Stoica., A. M. (2010a). Iterative algorithm to compute the maximal and stabilising solutions of a general class of discrete-time Riccati-type equations, *International Journal of Control*, Vol.83, No.4, 837-847, ISSN: 1366-5820.

[7] Dragan, V., Morozan, T. & Stoica, A.M. (2010b). Mathematical Methods in Robust Control of Discrete-time Linear Stochastic Systems, Springer, ISBN: 978-1-4419-0629-8.

[8] Freiling, G. & Hochhaus, A. (2003). Properties of the Solutions of Rational Matrix Difference Equations, *Comput. Math. Appl.*, Vol.45, 1137-1154, ISSN: 0898-1221.

[9] Ivanov, I. (2007). Properties of Stein (Lyapunov) iterations for solving a general Riccati equation, *Nonlinear Analysis Series A: Theory, Methods & Applications*, Vol. 67, 1155-1166, ISSN: 0362-546X.

[10] Ivanov, I. (2011). An Improved Method for Solving a System of Discrete-Time Generalized Riccati Equations, Journal of Numerical Mathematics and Stochastics, Vol.3, No.1, 57-70, http://www.jnmas.org/jnmas3-7.pdf, ISNN: 2151-2302.

[11] Ivanov, I., & Netov, N. (2012), A new iteration to discrete-time coupled generalized Riccati equations, submitted to *Computational & Applied Mathematics*, ISNN: 0101-8205.

[12] Li, X., Zhou, X. & Rami, M. (2003). Indefinite stochastic linear quadratic control with Markovian jumps in infinite time horizon. *Journal of Global Optimization*, Vol. 27, 149-175, ISSN: 0925-5001.

[13] Rami, M. & Ghaoui, L. (1996). LMI Optimization for Nonstandard Riccati Equations Arising in Stochastic Control. *IEEE Transactions on Automatic Control*, Vol. 41, 1666-1671, ISSN: 0018-9286.

[14] Rami, M. & Zhou, X. (2000). Linear matrix inequalities, Riccati equations, and indefinite stochastic linear quadratic controls. *IEEE Transactions on Automatic Control*, Vol. 45, 1131-1143, ISSN: 0018-9286.

[15] Rami, M. A., Zhou, X.Y., & Moore, J.B. (2000). Well-posedness and attainability of indefinite stochastic linear quadratic control in infinite time horizon, *Systems & Control Letters*, Vol. 41, pp. 123-133, ISSN: 0167-6911.

[16] Song, X., Zhang, H. & Xie, L. (2009). Stochastic linear quadratic regulation for discrete-time linear systems with input delay, *Automatica*, Vol. 45, 2067-2073, ISSN: 0005-1098.

[17] do Val, J.B.R, Basar, T. (1999). Receding horizon control of jump linear systems and a macroeconomic policy problem, *Journal of Economic Dynamics and Control*, Vol. 23, 1099-1131, ISSN: 0165-1889.

[18] Yao, D.D., Zhang, S. & Zhou, X. (2006). Tracking a financial benchmark using a few assets. *Operations Research*, Vol. 54, 232-246, ISSN: 0030-364X.

Chapter 10

Discrete-Time Stochastic Epidemic Models and Their Statistical Inference

Raúl Fierro

Additional information is available at the end of the chapter

1. Introduction

There is a wide range of models, both stochastic and deterministic, for the spread of an epidemic. Usually, when the population is constituted of a large number of individuals, a deterministic model is useful as a first approximation, and random variations can be neglected. As an alternative, a stochastic model could be more appropriate for describing the epidemic, but it is less tractable and its mathematical analysis is usually possible only when the population size is very small. However, most populations are not large enough to neglect the effect of statistical fluctuations, nor are they small enough to avoid cumbersome mathematical calculations in the stochastic model. In these cases, it uses to be convenient to take into account both types of models and their relationship. The interplay between ordinary differential equations and Markovian counting processes has been widely investigated in the literature. Major references on this subject can be found in [11, 20–22]. Concerning the deterministic epidemic models, those using ordinary differential equations in their formulation have received special attention and a great number of epidemics is modeled by means of Markovian counting processes. For example, some epidemic models known as SIR, SI, SIS, and others derived from these ones, use differential equations and Markovian counting processes in their formulations. Furthermore, stochastic models based on Markovian counting processes and differential equations are mainly used to carry out the statistical analysis of the model parameters. The Mathematical Theory of Infectious Diseases by Bailey [2] represents a classical reference containing a presentation and analysis of these models. However a more recent book by Andersson and Britton [1] entitled Stochastic Epidemic Models and their Statistical Analysis is a more appropriate reference according to the point of view of this chapter. The spread of these epidemics is developed in a closed population, which is divided into three individual compartments, i.e. susceptible, infective and removed cases; different types of transitions can occur among these three groups of individuals. These models include the stochastic and deterministic versions of the Kermack and McKendrick model [19] and the SIS epidemic model, among others. Moreover, a number of variations of these models has been widely studied. Modeling of epidemics by continuous-time Markov chains has a long history; thus, it seems pertinent to cite the works by [4–6, 18, 24, 28]

Also, an epidemic can be modeled by means of discrete-time. This is the case of the classical Reed-Frost model, which is a Markovian discrete-time SIR epidemic model. However, this modeling has two differences with the corresponding one based on counting processes. First, its latent period is assumed constant and equal to the time unit. Secondly, there are no deterministic counterpart based on differential equations as it is the case of an epidemic modeled by means of a Markovian counting process. Another type of population modeling, which is applied to metapopulations, has been introduced by some authors such [9, 10, 27] and other references therein. These researchers derive an approximation that preserves the discrete time structure and reduces the complexity of the models. Probably, these results could be applied to epidemic models and asymptotic inference on the parameters of these models, could be carried out.

This chapter is a compendium of two works by the author, whose references are [12, 13]. A wide class of discrete-time stochastic epidemic models is introduced and analyzed from a statistical point of view. Just as some models based on ordinary differential equations involve a natural alternative through Markovian counting processes, this class includes a counterpart based on differential equations. Unlike those epidemic models where transitions occur at random times, our proposal involves the advantage of being suitable for epidemics that cannot be observed for a long period of time, as in some epidemics where observations are done at previously determined times. This is the main reason for preferentially considering these kind of stochastic models instead of those based on continuous time. It is expected the smaller the periods of time between transitions and the bigger the population, the more similar the stochastic and deterministic models would become. Indeed, one of the main aims of this paper is to prove such a similarity. As a second aim, we are highly interested in carry out statistical analysis on the parameters of the modeling. For this purpose, martingale estimators for the parameters involve in the modeling are derived and their asymptotic normality is proved.

Since the results stated here do not assume a distribution for the process modeling the epidemic, it is not possible to derive a likelihood ratio and hence maximum likelihood estimators cannot be obtained. Even, in many cases when the process representing the model is Markovian, the maximum likelihood estimators cannot be obtained in a closed form, which makes difficult to carry out statistical inference on the parameter of the model. As pointed out in [7], likelihood functions corresponding to epidemic data are often very complicated. In these cases, parameter estimation based on martingale estimators use to be an appropriate alternative to work out this difficulty. This method arises as a natural way of estimation when no distribution in the model is assumed or, when the maximum likelihood estimators cannot be obtained in a closed form.

This chapter is organized as follows. The general form of the model and two preliminary lemmas are introduced in Section 2. Section 3 contains brief definitions of some typical models included in the biomathematical literature. The deterministic counterpart of the general model along with its relationships with it is presented in Section 4. Indeed, the convergence of the stochastic model to the deterministic one and the asymptotic behavior of the corresponding fluctuations are proved. Moreover, in Section 4 a version of the SIS epidemic model is presented and numerical simulations are carried out. The parameter estimators are defined in Section 5, and their asymptotic normality is proved. The General Epidemic Model along with the statistical analysis on the parameters is stated in Section 6.

Results of Section 5 are applied here to some hypothesis tests. Section 7 is devoted to some numerical simulations. Finally, Section 8 contains some conclusions.

2. Modeling and preliminaries

2.1. Discrete-time modeling

The model which we introduce here is defined as follows: a community of individuals divided into three different compartments is considered, namely, susceptible, infective and removed individuals.

Suppose the size population is n, and for each $t \geq 0$, $S^n(t)$, $I^n(t)$ and $R^n(t)$ represent, respectively, the number of susceptible, infective and removed individuals at time t. Since it is assumed the population size is constant, then for each $t \geq 0$, should be $S^n(t) + I^n(t) + R^n(t) = n$. These processes are observed at discrete-time instants which are defined by the sequence $\{t_k^n\}_{k \in \mathbb{N}}$, where for each $k \in \mathbb{N}$, $t_k^n = k\Delta/n$, $(\Delta > 0)$, i.e. each time subinterval has length Δ/n. Let $(\Omega, \mathcal{F}, \mathbb{P})$ be a probability space. In the sequel, all stochastic processes and random variables are defined on this probability space and for a stochastic process Z, we denote $\Delta Z(t_k^n) = Z(t_k^n) - Z(t_{k-1}^n)$.

Let $M^\top$ denote the transpose of a matrix M and $X^n = (S^n, I^n, R^n)^\top$. Transitions of individuals among the three compartments are determined by m increasing stochastic processes $Z_1^n, \ldots, Z_m^n$, and the number of individuals in each compartment is obtained, for each $t \geq 0$, by means of

$$X^n(t) = X^n(0) + AZ^n(t), \tag{1}$$

where $Z^n(t) = (Z_1^n(t), \ldots, Z_m^n(t))^\top$ and A is a $3 \times m$-incidence matrix.

It is assumed $Z_1^n, \ldots, Z_m^n$ take values in the set of non-negative integer numbers, have right-continuous trajectories and start at zero, i.e. $Z_1^n(0) = \cdots = Z_m^n(0) = 0$. Let $\mathcal{F}_k^n$ be the σ-field $\sigma(Z^n(t_1^n), \ldots, Z^n(t_k^n))$ generated by $Z^n(t_1^n), \ldots, Z^n(t_k^n)$. The stochastic processes $Z_1^n, \ldots, Z_m^n$ increase according to m density dependent transition rates, which are defined by means of m non-negative functions $a_1, \ldots, a_m$, respectively. The domain of these functions is and open set of $\mathbb{R}^3$ containing the 3-simplex $E = \{(u, v, w)^\top \in [0,1]^3 : u + v + w = 1\}$ and it is assumed the following condition holds:

(C) For each $k \in \mathbb{N}$, $\Delta Z_1^n(t_k^n), \ldots, \Delta Z_m^n(t_k^n)$ are $\mathcal{F}_{k-1}^n$−conditionally independent and satisfy

$$\mathbb{E}(\Delta Z_i^n(t_k^n) | \mathcal{F}_{k-1}^n) = a_i(\chi^n(t_{k-1}^n)), \quad (i \in \{1, \ldots, m\}),$$

where $\chi^n(t) = (\sigma^n(t), \iota^n(t), \rho^n(t))^\top$, $\sigma^n(t) = S^n(t)/n$, $\iota^n(t) = I^n(t)/n$ and $\rho^n(t) = R^n(t)/n$.

A wide variety of stochastic models for epidemics satisfy condition (C). It is important to point out this condition does not determine the law or distribution of χ^n, i.e. there could be two or more processes satisfying this condition, though they have different transition probabilities. This fact enables this condition to be applied to a wide class of models, since in order to verify condition (C), the distribution of the process need not be known. Actually, a stochastic process satisfying condition (C) need not be Markovian. Nevertheless, some Markov chains, having density dependent transition rates, satisfy condition (C) and hence they may be included in our setting.

2.2. Two preliminary lemmas

In what follows, $[a]$ stands for the integer part of a real number a and for each $i = 1, \ldots, m$, we denote $L_i^n(t) = \sum_{k=1}^{[nt]} \xi_k^n(i)\Delta t^n$ and $L^n(t) = (L_1^n(t), \ldots, L_m^n(t))^\top$, $(t \geq 0)$, where $\xi_k^n(i) = \Delta Z_i^n(t_k^n) - a_i(\chi^n(t_{k-1}^n))$ and $\Delta t^n = 1/n$. From Condition (C), it is obtained that, by defining $\mathcal{G}_t^n = \mathcal{F}_{[nt]}^n$, $L^n = \{L^n(t); t \geq 0\}$ is an m-dimensional martingale with respect to $\{\mathcal{G}_t^n; t \geq 0\}$.

Through this chapter, for each $i = 1, \ldots, m$, $v_i^n(t)$ and $\langle L_i^n \rangle$ stand for the random variable

$$v_i^n(t) = \frac{1}{n} \sum_{k=1}^{[nt]} \mathbb{E}(\xi_k^n(i)^2 | \mathcal{F}_{k-1}^n)$$

and the predictable quadratic variation of L_i^n, respectively.

Lemma 2.1. *For each $t \geq 0$, $L_1^n(t), \ldots, L_m^n(t)$ are $\mathcal{G}_{t-}^n$-conditionally independent random variables and, the predictable quadratic variation matrix of L^n is given by*

$$\langle L^n \rangle(t) = \frac{1}{n} \begin{pmatrix} v_1^n(t) & \cdots & 0 \\ \vdots & \ddots & \vdots \\ 0 & \cdots & v_m^n(t) \end{pmatrix}, \quad (t \geq 0).$$

Proof. For $t \geq 0$, the $\mathcal{G}_{t-}^n$-conditional independence of $L_1^n(t), \ldots, L_m^n(t)$ follows from Assumption (C) and it is clear that for each $i = 1, \ldots, m$, the predictable quadratic variation of L_i^n is given by $\langle L_i^n \rangle(t) = \frac{1}{n^2} \sum_{k=1}^{[nt]} \mathbb{E}(\xi_k^n(i)^2 | \mathcal{F}_{k-1}^n)$. Hence $\langle L_i^n \rangle(t) = v_i^n(t)/n$, $(t \geq 0)$, which concludes the proof. □

In the sequel, for each $d \in \mathbb{N}$, $\|\cdot\|$ stands for the Euclidean vector norm in $\mathbb{R}^d$.

Lemma 2.2. *Let $T > 0$ and suppose for each $i = 1, \ldots, m$, $\{\frac{1}{n} v_i^n(T)\}_{n \in \mathbb{N}}$ converges in probability to zero, as n goes to ∞. Then, $\{\sup_{0 \leq t \leq T} \|L^n(t)\|\}_{n \in \mathbb{N}}$ converges in probability to zero.*

For each $T > 0$ and each $i = 1, \ldots, m$, $\{\frac{1}{n} v_i^n(T)\}_{n \in \mathbb{N}}$ converges in probability to zero, as n goes to ∞.

Proof. From Theorem 1 in [26], for any $\epsilon, \eta > 0$ we have

$$\mathbb{P}(\sup_{0 \leq t \leq T} \|L^n(t)\|^2 > \epsilon) \leq \frac{1}{\epsilon} \sum_{i=1}^{m} \mathbb{E}(\langle L_i^n \rangle(T) \wedge \eta) + \mathbb{P}(\sum_{i=1}^{m} \langle L_i^n \rangle(T) > \eta)$$

and hence, Lemma 2.1 implies

$$\mathbb{P}(\sup_{0 \leq t \leq T} \|L^n(t)\|^2 > \epsilon) \leq \frac{m\eta}{\epsilon} + \mathbb{P}(\sum_{i=1}^{m} \frac{1}{n} v_i^n(T) > \eta).$$

By assumption this lemma follows. □

3. Some epidemic models

Most of the typical models involved in the biomathematical literature have a version which belongs to the model class defined in this approach. Some of them are included below.

3.1. The general epidemic model

A deterministic version of this model based on differential equations, was introduced by [19], while the article by [18] was a pioneer in the stochastic version based on counting processes. This model has received the most attention in the literature, and its analysis can be found in [2], [1] and [23]. Two type of transitions are possible for any individual, from susceptible to infected and from infected to removed individuals. Thus under the perspective of this work, the modeling of this epidemic must satisfy (1) with

$$A = \begin{pmatrix} -1 & 0 \\ 1 & -1 \\ 0 & 1 \end{pmatrix} \quad \text{and} \quad Z^n = \begin{pmatrix} Z_1^n \\ Z_2^n \end{pmatrix}.$$

Transitions from susceptible to infected and from infected to removed cases are described by Z_1^n and Z_2^n, respectively, and the functions defining their transition rates are given by $a_1(u,v,w) = \beta uv$ and $a_2(u,v,w) = \gamma v$, where β and γ are two parameters denoting the infection and removal rates. This model is also known as SIR model.

3.2. The SIRS Model

This is a slight modification of the preceding model. Besides the transitions determined by Z_1^n and Z_2^n in the SIR model, a transition from removed to susceptible case is allowed and determined by an increasing stochastic process Z_3^n, where its transition rate is defined by $a_3(u,v,w) = \delta w$. In this case, some of the removed cases may become susceptible to be infected again. The incidence matrix defining this model is given by

$$A = \begin{pmatrix} -1 & 0 & 1 \\ 1 & -1 & 0 \\ 0 & 1 & -1 \end{pmatrix}.$$

3.3. The SIS Model

One of the simplest epidemic models is the SIS model, which uses to be suitable for infections resulting from bacteria such as gonorrhea, malaria, etc. In this case, only transitions from susceptible to infected individuals are allowed, as well as transitions from infected to susceptible individuals. According to the approach of this study, two transitions are allowed, and determined by Z_1^n and Z_2^n with transition rates defined by $a_1(u,v,w) = \beta uv$ and $a_2(u,v,w) = \gamma v$, respectively. In this case, the incidence matrix is

$$A = \begin{pmatrix} -1 & 1 \\ 1 & -1 \\ 0 & 0 \end{pmatrix}.$$

Notice the compartment corresponding to removed cases is considered having no individuals.

3.4. The modified SIR model

This is a modification of the general epidemic model and aims to AIDS modeling. As in the general epidemic model, two transitions Z_1^n and Z_2^n define the model with transition rates given by $a_1(u,v,w) = \beta uv/(u+v)$ and $a_2(u,v,w) = \gamma v$, respectively. As before, β and γ correspond to the model parameters. Some references concerning the deterministic version of this model are, for instance, [16, 17], while the stochastic version based on Markovian counting processes was introduced by [4]. The incidence matrix is defined as in the SIR model, i.e.

$$A = \begin{pmatrix} -1 & 0 \\ 1 & -1 \\ 0 & 1 \end{pmatrix}.$$

In general, as much in the SIR as the modified model, the transition rate due to infection is proportional to the susceptible density and the fraction of infected individuals with respect to individuals in circulation. Consequently, in the SIR model this fraction is $v/(u+v+w) = v$, while in an epidemic where to be removed is equivalent to be dead or out of circulation, this fraction is $v/(u+v)$. In modeling AIDS, removed cases are presumed to be so ill with AIDS that they no longer take part in transmission.

4. The deterministic counterpart

In this section, we examine the relationship between the model we are introducing here and an associated ordinary differential equation, which we call its deterministic counterpart.

Let $F(x) = Aa(x)$, where $a(x) = (a_1(x), \ldots, a_m(x))^\top$, $(x \in E)$, and consider the following ordinary differential equation:

$$\frac{\mathrm{d}\chi}{\mathrm{d}t}(t) = F(\chi(t)), \quad \chi(0) = \chi_0, \tag{2}$$

where $\chi_0 = (\sigma_0, \iota_0, \rho_0)^\top \in E$ is the initial condition.

In order to obtain existence and uniqueness of the solution to (2), it will be assumed the following usual Lipchitz condition holds:

(L) $\quad \|F(x) - F(y)\| \leq K\|x - y\|,$

for all $x, y \in E$, where K is a positive constant.

4.1. Comparison between the stochastic and deterministic models

The theorem below stated the consistency of the stochastic model with respect to the deterministic one and we will use it to study the asymptotic behavior of the estimators for the parameters of the model.

Theorem 4.1. *Let χ be the unique solution to (3) and assume conditions (C) and (L) are satisfied. Moreover, suppose the following two conditions hold:*

(4.1.1) *The sequence of initial conditions $\{\chi^n(0)\}_{n\in\mathbb{N}}$ converges in probability to $\chi_0 = (\sigma_0, \iota_0, \rho_0)^\top$, as n goes to ∞.*

(4.1.2) *For each $T > 0$ and each $i = 1, \ldots, m$, $\{\frac{1}{n} v_i^n(T)\}_{n\in\mathbb{N}}$ converges in probability to zero, as n goes to ∞.*

Then, $\{\chi^n\}_{n\in\mathbb{N}}$ converges in probability uniformly over compact subsets of $\mathbb{R}_+$ to χ, i.e. for each $T > 0$, $\{\sup_{0\le t\le T} \|\chi^n(t) - \chi(t)\|\}_{n\in\mathbb{N}}$ converges in probability to zero, as n goes to ∞.

Proof. Since, for each $i = 1, \ldots, m$, $\triangle Z_i^n(t_k^n) = a_i(\chi^n(t_{k-1}^n)) + \xi_k^n(i)$, we have

$$\begin{aligned} Z_i^n(t) &= n \sum_{k=1}^{[nt]} a_i(\chi^n(t_{k-1}^n))\triangle t^n + n \sum_{k=1}^{[nt]} \xi_k^n(i)\triangle t^n \\ &= n \int_0^t a_i(\chi^n(u))\, du - (nt - [nt])a_i(\chi^n(t)) + nL_i^n(t), \end{aligned}$$

and

$$\chi^n(t) = \chi^n(0) + \int_0^t F(\chi^n(u))\, du + AL^n(t) + \frac{(nt - [nt])}{n} F(\chi^n(t)).$$

Hence

$$\chi^n(t) - \chi(t) = \chi^n(0) - \chi(0) + \int_0^t [F(\chi^n(u)) - F(\chi(u))]\, du + AL^n(t) - \epsilon^n(t), \tag{3}$$

where $\epsilon^n(t) = \frac{(nt-[nt])}{n} F(\chi^n(t))$.

For any matrix B, let us denote $|||B||| = \sup_{\|x\|=1} \|Bx\|$. Fix $T > 0$ and let

$$g^n(t) = \sup_{0\le s\le t} \|\chi^n(s) - \chi(s)\|, \quad (t \in [0, T]).$$

From (3) and (L), for each $t \ge 0$, we have

$$g^n(t) \le \alpha_n + K \int_0^t g^n(u)\, du,$$

where $\alpha_n = g^n(0) + \sup_{0\le t\le T} \|AL^n(t)\| + O(1/n)$. Since $\{g^n(0)\}_{n\in\mathbb{N}}$ converges in probability to zero and $\sup_{0\le t\le T} \|AL^n(t)\| \le |||A||| \sup_{0\le t\le T} \|L^n(t)\|$, Lemma 2.1 and Gronwall's inequality imply $\{g^n(T)\}_{n\in\mathbb{N}}$ converges in probability to zero. This completes the proof. □

Remark 4.1. *By Chebishev's inequality, Condition (4.1.2) is satisfied whenever the following stronger condition holds: For each $T > 0$ and each $i = 1, \ldots, m$,*

$$\lim_{n\to\infty} \mathbb{E}(\langle L^n(i)\rangle(T)) = 0. \tag{4}$$

The following result aims to the problem of finding approximate confident bands for the solution to (2). Before stating it, for each $x \in E$, let $D(F)(x)$ denote the Jacobian matrix of F at $x = (x_1, x_2, x_3)^\top$, that is,

$$D(F)(x) = \begin{pmatrix} \frac{\partial F_1}{\partial x_1}(x) & \frac{\partial F_1}{\partial x_2}(x) & \frac{\partial F_1}{\partial x_3}(x) \\ \frac{\partial F_2}{\partial x_1}(x) & \frac{\partial F_2}{\partial x_2}(x) & \frac{\partial F_2}{\partial x_3}(x) \\ \frac{\partial F_3}{\partial x_1}(x) & \frac{\partial F_3}{\partial x_2}(x) & \frac{\partial F_3}{\partial x_3}(x) \end{pmatrix}.$$

In the sequel, $\{Y^n\}_{n\in\mathbb{N}}$ is the sequence defined as $Y^n = \sqrt{n}(\chi^n - \chi)$.

Theorem 4.2. *Let $\{Y^n\}_{n\in\mathbb{N}}$ be the sequence defined as $Y^n = \sqrt{n}(\chi^n - \chi)$ and suppose the following three conditions hold:*

(4.2.1) *For each $i, j = 1, 2, 3$, the partial derivative $\dfrac{\partial F_i}{\partial x_j}(x)$ exists and it is continuous at x in an open set containing E.*

(4.2.2) *For each $\epsilon > 0$ and each $i = 1, \ldots, m$, $\{\frac{1}{n}\sum_{k=1}^{n} \mathbb{E}(\xi_k^n(i)^2 \mathrm{I}_{\{|\xi_k^n(i)|>\epsilon\sqrt{n}\}}|\mathcal{F}_{k-1}^n)\}_{n\in\mathbb{N}}$ converges in probability to zero.*

(4.2.3) *$\{Y^n(0)\}_{n\in\mathbb{N}}$ converges in distribution to a three-variate random vector η.*

(4.2.4) *For each $t \geq 0$ and each $i = 1, \ldots, m$, $\{v_i^n(t)\}_{n\in\mathbb{N}}$ converges in probability to $v_i(t)$, where for each $i = 1, \ldots, m$, $v_i : [0, \infty[\to \mathbb{R}$ is an increasing continuous function such that $v_i(0) = 0$.*

Then, $\{Y^n\}_{n\in\mathbb{N}}$ converges in law to the solution Y satisfying the following stochastic differential equation:

$$\mathrm{d}Y(t) = \mathrm{D}(F)(\chi(t))Y(t)\mathrm{d}t + \mathrm{d}M(t), \quad Y(0) = \eta, \tag{5}$$

where M is a continuous martingale with predictable quadratic variation given by the matrix

$$\langle M\rangle(t) = A \cdot \mathrm{Diag}(v_1(t), \ldots, v_m(t)) \cdot A^\top$$

and $\mathrm{Diag}(v_1(t), \ldots, v_m(t))$ stands for the diagonal matrix with entries $v_1(t), \ldots, v_m(t)$ at its diagonal.

Proof. Let $M^n = \sqrt{n}AL^n$. By making use of Corollary 12, Chapter II in [29], (4.2.2) and (4.2.4) imply that $\{\sqrt{n}L^n\}_{n\in\mathbb{N}}$ converges in law to a continuous martingale Q with predictable quadratic variation $\langle Q\rangle$ given by $\langle Q\rangle(t) = \mathrm{Diag}(v_1(t), \ldots, v_m(t))$. Consequently, $\{M^n\}_{n\in\mathbb{N}}$ converges in law to a continuous martingale $M = AQ$ with predictable quadratic variation $\langle M\rangle$ given, for each $t \geq 0$, by $\langle M\rangle(t) = A \cdot \mathrm{Diag}(v_1(t), \ldots, v_m(t)) \cdot A^\mathrm{T}$.

From (3), we have

$$Y^n(t) = Y^n(0) + \int_0^t \mathrm{D}(F)(\theta^n(s))Y^n(s)\,\mathrm{d}s + M^n(t) - U^n(t), \tag{6}$$

where $\theta^n(s)$ is between $\chi^n(s)$ and $\chi(s)$, and $U^n(t) = \sqrt{n}\epsilon^n(t) = \frac{(nt-[nt])}{\sqrt{n}}F(\chi^n(t))$.

Put $C_1 = \sup_{x\in E} \|\mathrm{D}(F)(x))\|$ and let $C_2 > 0$ such that $\sup_{t\geq 0} \|U^n(t)\| \leq C_2$. We have

$$\sup_{0\leq u\leq t} \|Y^n(u)\| \leq \|Y^n(0)\| + C_1 \int_0^t \sup_{0\leq u\leq s} \|Y^n(u)\|\,\mathrm{d}s + \sup_{0\leq u\leq t} \|M^n(u)\| + C_2.$$

Consequently, from a standard application of the Gronwall inequality, we obtain

$$\sup_{0\leq u\leq t} \|Y^n(u)\| \leq (\|Y^n(0)\| + \sup_{0\leq u\leq t} \|M^n(u)\| + C_2)\,\mathrm{e}^{C_1 t}. \tag{7}$$

Since $\{Y^n(0)\}_{n\in\mathbb{N}}$ and $\{\sup_{0\leq t\leq 1} \|M^n(t)\|\}_{n\in\mathbb{N}}$ are sequences converging in distribution, Theorem 6.2 in [8] implies

$$\lim_{a\to\infty} \sup_{n\in\mathbb{N}} \mathbb{P}(\|Y^n(0)\| > a) = 0, \tag{8}$$

and, from (18), for each $\epsilon > 0$ and any $t \geq 0$,

$$\lim_{\delta \to 0} \sup_{n \in \mathbb{N}} \mathbb{P}(\sup_{0 \leq u \leq t} \|Y^n(u)\| > \epsilon/\delta) = 0, \tag{9}$$

In order to prove the convergence in law of $\{Y^n\}_{n \in \mathbb{N}}$, fix $T > 0$ and let us define the modulus of continuity $\omega_T^D : \mathrm{D}([0,T], \mathbb{R}^3) \times]0, \infty[\to \mathbb{R}$ as

$$\omega_T^D(x, \delta) = \inf_{\{t_i\}} \max_{0 < i \leq r} \sup_{t_{i-1} \leq s,t < t_i} \|x(s) - x(t)\|,$$

where the infimum extends over the finite sets $\{t_i\}$ of points satisfying

$$\begin{cases} 0 = t_0 < t_1 < \cdots < t_r = T, \\ t_i - t_{i-1} > \delta, \quad i = 1, \ldots, r. \end{cases}$$

Here, $\mathrm{D}([0,T], \mathbb{R}^3)$ stands for the Skorohod space of all functions from $[0,T]$ into $\mathbb{R}^3$, which are right continuous and left-hand limited.

From (6) we have

$$\omega_T^D(Y^n, \delta) \leq \delta C_1 \sup_{0 \leq t \leq T} \|Y^n(t)\| + \omega_T^D(M^n, \delta) + 2C_2/\sqrt{n}. \tag{10}$$

Since $\{M^n\}_{n \in \mathbb{N}}$ converges in law to M, it follows from Theorem 15.2 by [8] that for each $\epsilon > 0$, $\lim_{\delta \to 0} \sup_{n \in \mathbb{N}} \mathbb{P}(\omega_T^D(M^n, \delta) > \epsilon) = 0$. Hence, from (9) and (10), for each $\epsilon > 0$, we have

$$\lim_{\delta \to 0} \sup_{n \in \mathbb{N}} \mathbb{P}(\omega_T^D(Y^n, \delta) > \epsilon) = 0. \tag{11}$$

Conditions (8) and (11) imply the sequence $\{Y^n\}_{n \in \mathbb{N}}$ satisfies the hypotheses of Theorem 15.2 in [8] and hence, the sequence of probabilities measures $\{\mathbb{P}(Y^n \in \cdot)\}_{n \in \mathbb{N}}$ is tight. This fact, along Theorem 6.1 in [8], imply that for the convergence in law, $\{Y^n\}_{n \in \mathbb{N}}$ is relatively compact. Let Y be a process and $\{Y^{n_k}\}_{k \in \mathbb{N}}$ a subsequence of $\{Y^n\}_{n \in \mathbb{N}}$ such that $\{Y^{n_k}\}_{k \in \mathbb{N}}$ converges in law to Y. Since $\{\sup_{t \geq 0} \|U^n(t)\|\}_{n \in \mathbb{N}}$ converges to zero, it follows from (6), (4.2.1) and Theorem 4.1 that

$$Y(t) = Y(0) + \int_0^t \mathrm{D}(F)(\chi(s)) Y(s) \, \mathrm{d}s + M(t).$$

Moreover, since $\{Y^n(0)\}_{n \in \mathbb{N}}$ converges in distribution to η and $Y(0)$ equals η in distribution, we have Y is a solution to (5). Finally, uniqueness of solutions to (5) implies $\{Y^n\}_{n \in \mathbb{N}}$ converges in distribution to Y, which concludes the proof. □

Remark 4.2. *By Itô's rule, the unique solution to (5) is given by*

$$Y(t) = \Psi(t) \left[\eta + \int_0^t \Psi(s)^{-1} \, \mathrm{d}M(s) \right], \quad 0 \leq t \leq 1,$$

where Ψ is the unique solution to the matrix differential equation

$$\Psi'(t) = \mathrm{D}(F)(\chi(t)) \Psi(t), \quad \Psi(0) = \textit{identity matrix}.$$

The stochastic process $\{Y(t); t \geq 0\}$ allows us to give confidence bands for the deterministic model defined by the solution χ to (2). In Section 6 such band is constructed for the SIS epidemic model, where some simulations are carried out.

4.2. Numerical simulations for the SIS epidemic model

In this section, our attention is focused on the SIS epidemic model. As explained in Subsection 3.3, it is assumed that only susceptible and infective individuals are in a closed homogeneously mixing population. Let $\sigma(t)$ and $\iota(t)$ be the densities of susceptible and infective individuals, respectively. The deterministic model is defined by the following system of ordinary differential equations:

$$\begin{aligned} \tfrac{\mathrm{d}\sigma}{\mathrm{d}t}(t) &= -\beta\sigma(t)\iota(t) + \gamma\iota(t) \\ \tfrac{\mathrm{d}\iota}{\mathrm{d}t}(t) &= \beta\sigma(t)\iota(t) - \gamma\iota(t). \end{aligned}$$

Since for each $t \geq 0$, $\sigma(t) + \iota(t) = 1$, this model is completely determined by the ordinary differential equation:

$$\frac{\mathrm{d}\iota}{\mathrm{d}t}(t) = \beta(1-\iota(t))\iota(t) - \gamma\iota(t),$$

which, given $\iota(0) = \iota_0 \in]0,1[$, has the unique solution

$$\iota(t) = \begin{cases} \frac{\iota_0}{\iota_0\beta t+1} & \text{if } \beta = \gamma \\ \frac{\iota_0(\beta-\gamma)\,\mathrm{e}^{(\beta-\gamma)t}}{\beta-\gamma+\iota_0\beta(\mathrm{e}^{(\beta-\gamma)t}-1)} & \text{if } \beta \neq \gamma. \end{cases} \tag{12}$$

The relative removal-rate, see for instance [2] and [3], is defined as $\tau = \gamma/\beta$, where γ and β represent the removal and infection rates, respectively. We note that $\iota(t) \to 0$ as $t \to \infty$ if $\tau \geq 1$, while $\iota(t) \to 1-\tau > 0$ as $t \to \infty$ if $\tau < 1$. For this reason, $n(1-\tau)$ is called the endemic level of the process.

Let $\sigma^n(t)$ and $\iota^n(t)$ be the densities of susceptible and infective individuals in the stochastic version of the SIS epidemic model. According to our setting, σ^n and ι^n are defined as

$$\begin{aligned} \sigma^n(t) &= \sigma^n(0) + \tfrac{1}{n}(Z_2^n(t) - Z_1^n(t)) \\ \iota^n(t) &= \iota^n(0) + \tfrac{1}{n}(Z_1^n(t) - Z_2^n(t)), \end{aligned}$$

where $\mathbb{E}(\Delta Z_1^n(t_k^n)|\mathcal{F}_{k-1}^n) = \beta\sigma(t_{k-1}^n)\iota(t_{k-1}^n)$ and $\mathbb{E}(\Delta Z_2^n(t_k^n)) = \gamma\iota(t_{k-1}^n)$.

Let us assume $n\iota^n(0) = [n\iota_0]$. From Theorem 4.1, $\{\iota^n\}_{n\in\mathbb{N}}$ converges uniformly in probability to ι, over compact subsets of $\mathbb{R}_+$, whenever $\iota(0) = \iota_0$. It is worth noting $1-\tau$ is an asymptotically stable equilibrium state for the deterministic model and not for the stochastic model. However, it is expected $\iota^n(t)$ is close to this value due to Theorem 4.1, for a large enough n. In Figure 1, the deterministic and stochastic models, for $\beta = 3$, $\gamma = 1$, $\iota_0 = (1-\tau)/2$ and $n = 1{,}000$, are compared. Let $y^n = \sqrt{n}(\iota^n - \iota)$. It follows from Theorem 4.2 that $\{y^n\}_{n\in\mathbb{N}}$ converges in law to y, the solution to the following Langevin equation:

$$\mathrm{d}y(t) = (\beta - \gamma - 2\beta\iota(t))y(t)\,\mathrm{d}t + \sqrt{\beta(1-\iota(t))\iota(t) + \gamma\iota(t)}\,\mathrm{d}W(t), \quad \iota(0) = 0,$$

where W is a one dimensional standard Brownian motion. From Remark 4.2, the solution to this equation is

$$y(t) = \int_0^t b(u)\,\mathrm{e}^{\int_u^t a(s)\,\mathrm{d}s}\,\mathrm{d}W(u), \tag{13}$$

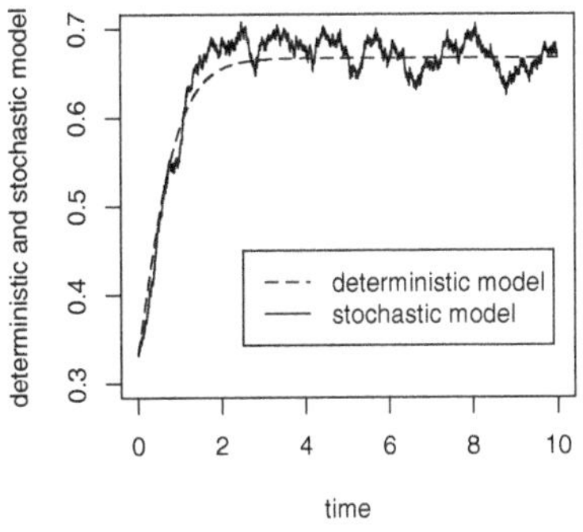

Figure 1. Comparing the deterministic and stochastic models.

where $a(s) = \beta - \gamma - 2\beta y(s)$ and $b(u) = \sqrt{\beta(1 - y(u))y(u) + \gamma y(u)}$. Hence, $y(t)$ has a normal distribution with mean zero and variance

$$\mathrm{Var}(y(t)) = \int_0^t b(u)^2 \,\mathrm{e}^{2\int_u^t a(s)\,\mathrm{d}s}\, du.$$

Suppose $\tau < 1$ and the process $n\iota^n$ starts close to the endemic level, i.e. $\iota_0 = 1 - \tau$. Consequently, $\iota(t) = 1 - \tau$ for all $t \geq 0$, and (13) becomes

$$y(t) = \sqrt{2\gamma(1 - \gamma/\beta)} \int_0^t \mathrm{e}^{-(\beta-\gamma)(t-u)}\, \mathrm{d}W(u).$$

In this case, a simple expression for the variance of $y(t)$ can be obtained and a confidence interval for $\iota(t)$ derived. Indeed, for each $t \geq 0$, $y(t)$ has normal distribution with mean zero and variance

$$\mathrm{Var}(y(t)) = \frac{\gamma}{\beta}(1 - \mathrm{e}^{-2(\beta-\gamma)t}).$$

Since for each $t > 0$ $\mathrm{Var}(y(t)) \leq \tau < 1$, $\iota^n(t)$ differs from $\iota(t)$ at most $w_{\alpha/2}\sqrt{\tau/n}$ with an approximate probability equals $1 - \alpha$.

In Figure 2, a simulation of ι^n is carried out for $\beta = 3$, $\gamma = 1$ and $n = 10$, and we note in this case random fluctuations are important to be neglected.

For $\alpha \in]0, 1[$, $[u^-_{\alpha,n}(t), u^+_{\alpha,n}(t)]$ is a confidence interval for $\iota(t)$, with an approximate confidence level $1 - \alpha$, where

$$u^{\pm}_{\alpha,n}(t) = \iota^n(t) \pm w_{\alpha/2}\sqrt{\mathrm{Var}(y(t))/n}, \tag{14}$$

$1 - \Phi(w_{\alpha/2}) = \alpha/2$ and Φ is the cumulative distribution function of a standard normal random variable. The random bounds given by (14) allow to construct (nonuniform) confidence bands for the solution ι given by (12). In turn, by defining $u^{\pm}_{\alpha}(t) = \iota(t) \pm w_{\alpha/2}\sqrt{\mathrm{Var}(y(t))/n}$, we have $\iota^n(t) \in [u^-_{\alpha}(t), u^+_{\alpha}(t)]$ with an approximate probability $1 - \alpha$ for large values of n.

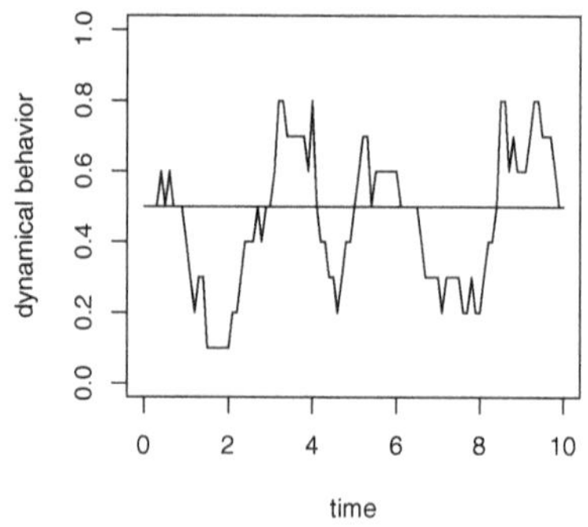

Figure 2. Simulation for $\beta = 3, \gamma = 1$ and $n = 10$.

In Figure 3, another simulation of ι^n is carried out for $\beta = 3$ and $\gamma = 1$. However, in order to appreciate the convergence of ι^n to the equilibrium $1 - \tau$ for the deterministic model, a population of size $n = 5,000$ is now considered. The bounds u_α^- and u_α^+ are pictured with dash lines for $\alpha = .05$, and hence, in this case $w_{\alpha/2} = 1.96$.

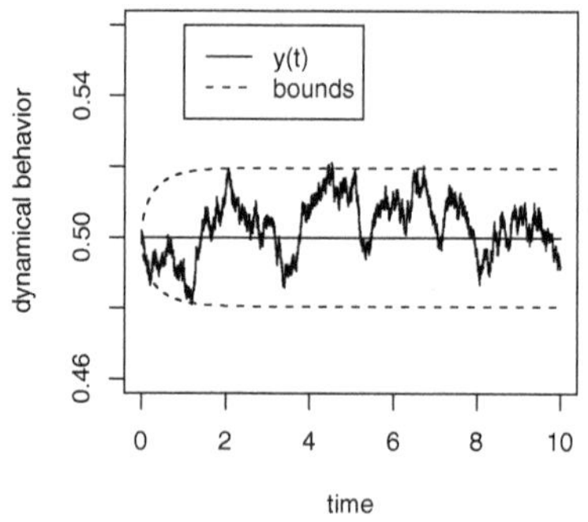

Figure 3. Simulation for $\beta = 3, \gamma = 1$ and $n = 5,000$.

Figure 3 confirms $|\iota^n(t) - \iota(t)| \leq C$ with an approximate probability bigger than $1 - \alpha = .95$, where $C = w_{\alpha/2}\sqrt{\tau/n} = .0226$.

5. Estimators and their asymptotic behavior

The main results of this article are stated in this section; however their proofs are deferred to the last section.

5.1. Preliminaries

Let us suppose for each $i = 1, \ldots, m$, a_i splits as $a_i = \beta_i b_i$, where β_i is a parameter of the model and b_i is a non-negative continuous function defined on an open set containing E.

Since for each $i = 1, \ldots, m$, L_i^n is a martingale and

$$Z_i^n(t)/n = \beta_i \sum_{k=1}^{[nt]} b_i(\chi^n(t_{k-1}^n))\Delta t^n + L_i^n(t), \quad (t > 0),$$

by observing the epidemic through the time interval $[0, T]$, $(T > 0)$, we have

$$\widehat{\beta_i^n}(T) = \frac{Z_i^n(T)}{n\sum_{k=1}^{[nT]} b_i(\chi^n(t_{k-1}^n))\Delta t^n} \tag{15}$$

is a martingale estimator of β_i.

For each $T > 0$, let $\widehat{\beta^n}(T) = (\widehat{\beta_1^n}(T), \ldots, \widehat{\beta_m^n}(T))^\top$. Proposition 5.1 below states that $\{\widehat{\beta^n}(T)\}_{n\in\mathbb{N}}$ is a consistent sequence of estimators for $\beta = (\beta_1, \ldots, \beta_m)^\top$.

Proposition 5.1. *Let assume the hypotheses of Theorem 4.1 are satisfied. Then, for each $T > 0$ the sequence $\{\widehat{\beta^n}(T)\}_{n\in\mathbb{N}}$ converges in probability to β.*

Proof. Note that for each $i = 1, \ldots, m$,

$$\widehat{\beta_i^n}(T) = \beta_i + \frac{L_i^n(T)}{\sum_{k=1}^{[nT]} b_i(\chi^n(t_{k-1}^n))\Delta t^n}.$$

From Theorem 4.1, $\{\sum_{k=1}^{[nT]} b_i(\chi^n(t_{k-1}^n))\Delta t^n\}_{n\in\mathbb{N}}$ converges in probability to $\int_0^T b_i(\chi(t))\,dt$, and from (4.1.1) along with Lemma 2.2 imply $\{L_i^n(T)\}_{n\in\mathbb{N}}$ converges in probability to zero. Therefore, the proof is complete. □

5.2. Asymptotic normality of estimators

In this subsection we state two asymptotic normality results for the martingale estimators for β.

Theorem 5.1. *Let χ be the unique solution to (3) and assume conditions (C) and (L) are satisfied. Moreover, suppose the following three conditions hold:*

(5.1.1) $\{\chi^n(0)\}_{n\in\mathbb{N}}$ *converges in probability to* $\chi(0) = (\sigma_0, \iota_0, \rho_0)^\top$, *as n goes to* ∞.

(5.1.2) *For each* $\epsilon > 0$ *and each* $i = 1, \ldots, m$, $\{\frac{1}{n}\sum_{k=1}^{n} \mathbb{E}(\xi_k^n(i)^2 \mathrm{I}_{\{|\xi_k^n(i)|>\epsilon\sqrt{n}\}} | \mathcal{F}_{k-1}^n)\}_{n\in\mathbb{N}}$ *converges in probability to zero, as n goes to* ∞.

(5.1.3) *For each* $i = 1, \ldots, m$, *there exists a continuous increasing function* $v_i : [0, \infty[\to \mathbb{R}$ *such that* $v_i(0) = 0$ *and for each* $t > 0$, $\{v_i^n(t)\}_{n\in\mathbb{N}}$ *converges in probability to* $v_i(t)$, *as n goes to* ∞.

Then, for each $T > 0$, $\{\sqrt{n}(\widehat{\beta^n}(T) - \beta)\}_{n\in\mathbb{N}}$ *converges in distribution, as* n *goes to* ∞, *to a normal random vector* $N(0,\Sigma)$ *having mean zero and variance-covariance matrix* $\Sigma = \{\sigma_{ij}(T)\}_{1\le i,j\le m}$ *satisfying*

$$\sigma_{ij}(T) = \begin{cases} v_i(T)/[\int_0^T b_i(\chi(t))\,dt]^2 & \text{if } i = j \\ 0 & \text{if } i \neq j. \end{cases}$$

Proof. Let $Q^n = (Q_1^n, \ldots, Q_m^n)^\top$, where $Q_i^n = \sqrt{n}L_i^n$, $(i = 1, \ldots, m)$. From Lemma 2.1, for each $t \geq 0$, $Q_1^n(t), \ldots, Q_m^n(t)$ are $\mathcal{G}_{t-}^n$-conditionally independent random variables and for each $i = 1, \ldots, m$, the predictable quadratic variation of the martingale Q_i^n is given by $\langle Q_i^n \rangle = v_i^n$. Condition (5.1.3) indicates for each $i = 1, \ldots, m$, $\{\langle Q_i^n \rangle(t)\}_{n\in\mathbb{N}}$ converges in probability to $v_i(t)$. This fact and Condition (5.1.2) enable to conclude that the hypotheses of Corollary 12 in Chapter II in (Rebolledo, 1979) hold. Consequently, $\{Q^n(T)\}_{n\in\mathbb{N}}$ converges in distribution to a normal random vector $Q(T) = (Q_1(T), \ldots, Q_m(T))^\top$ with mean zero and satisfying

$$\mathbb{E}(Q_i(T)Q_j(T)) = \begin{cases} v_i(T) & \text{if } i = j \\ 0 & \text{if } i \neq j. \end{cases}$$

We have $\sqrt{n}(\widehat{\beta^n}(T) - \beta) = D^n Q^n(T)$, where $D^n = Diag(d_1^n, \ldots, d_m^n)$ is the diagonal random matrix with entries $d_1^n, \ldots, d_m^n$ in its diagonal given by

$$d_i^n = \frac{1}{\sum_{k=1}^{[nT]} b_i(\chi^n(t_{k-1}^n))\Delta t^n}.$$

Condition (5.1.3) implies Condition (4.1.2) and consequently, the hypotheses of Theorem 4.1 are satisfied. Thus, for each $i = 1, \ldots, m$, $\{\sum_{k=1}^{[nT]} b_i(\chi^n(t_{k-1}^n))\Delta t^n\}_{n\in\mathbb{N}}$ converges in probability to $\int_0^T b_i(\chi(t))\,dt$ and Slutzky's theorem (see Theorem 5.1.6 in [25], for instance) enables us to conclude $\{D^n Q^n(T)\}_{n\in\mathbb{N}}$ converges in distribution to a normal random vector with mean zero and variance-covariance matrix $\Sigma = \{\sigma_{ij}\}_{1\le i,j\le m}$, which satisfies

$$\sigma_{ij} = \begin{cases} v_i(T)/[\int_0^T b_i(\chi(t))\,dt]^2 & \text{if } i = j \\ 0 & \text{if } i \neq j. \end{cases}$$

This concludes the proof of Theorem 5.1. □

Remark 5.1. *Note that Condition (5.1.2) holds whenever for each* $\epsilon > 0$ *and each* $i = 1, \ldots, m$,

$$\{\max_{k\le n} \mathbb{E}(\xi_k^n(i)^2 \mathrm{I}_{\{|\xi_k^n(i)| > \epsilon\sqrt{n}\}})\}_{n\in\mathbb{N}}$$

converges in probability to zero, as n *goes to* ∞. *In particular, this condition is satisfied when the double sequence* $\{\xi_k^n(i)^2; 0 \leq k \leq n, n \in \mathbb{N}\}$ *is uniformly integrable.*

6. The general epidemic model

In order to carry out asymptotical inference for a great number of epidemic models, results stated in Section 5 can be applied. In this subsection, statistical inference for the General Epidemic Model is developed.

6.1. The model

As mentioned in Subsection 3.1, this model contains two parameters β and γ denoting the infection and removal rate, respectively, and it is defined by two increasing integer valued processes, which we denote by A^n and B^n, so that

$$\begin{aligned} S^n(t) &= S^n(0) - A^n(t), \\ I^n(t) &= I^n(0) + A^n(t) - B^n(t), \\ R^n(t) &= R^n(0) + B^n(t). \end{aligned}$$

From (15) and the definition of this model given in Subsection 3.1, the martingale estimators of β and γ in $[0,T]$, are respectively given by

$$\widehat{\beta^n}(T) = \frac{A^n(T)}{\sum_{k=1}^{[nT]} \sigma^n(t_{k-1}^n)\iota^n(t_{k-1}^n)} \quad \text{and} \quad \widehat{\gamma^n}(T) = \frac{B^n(T)}{\sum_{k=1}^{[nT]} \iota^n(t_{k-1}^n)}. \tag{16}$$

In order to verify $\widehat{\beta^n}(T)$ and $\widehat{\gamma^n}(T)$ are martingale estimators, Condition (C) has to be hold. By taking into account some heuristic considerations, which are related to the infection spreading, the distribution of the process can be determined. This fact is sufficient, although not necessary as mentioned previously, to obtain Condition (C). Let β denote the average of effective contacts per time unit between an infected person and any other individual in the population. This constant is known as the contact rate, cf. [15]. Hence, β/n is the average number of effective contacts per time unit per capita of an infected, and it is natural to assume the probability of a susceptible individual to become infective in a time interval $]t_{k-1}^n, t_k^n]$ is $(\beta/n)\iota^n(t_{k-1}^n)$. On the other hand, since the total number of adequate contacts in t_{k-1}^n that may produce an infection in t_k^n equals the susceptible number $S^n(t_{k-1}^n)$, it is assumed that $\Delta A^n(t_k^n)$ and $\Delta B^n(t_k^n)$ conditionally on $\mathcal{F}_{k-1}^n$ have independent Binomial distribution with parameters $(S^n(t_{k-1}^n), (\beta/n)\iota^n(t_{k-1}^n))$ and $(I^n(t_{k-1}^n), \gamma\Delta t_n)$, respectively.

Note that it satisfies

$$\mathbb{E}(\Delta A^n(t_k^n)|\mathcal{F}_{k-1}^n) = a_1(\chi^n(t_{k-1}^n)) \quad \text{and} \quad \mathbb{E}(\Delta B^n(t_k^n)|\mathcal{F}_{k-1}^n) = a_2(\chi^n(t_{k-1}^n)),$$

where $a_1(u,v,w) = \beta uv$ and $a_2(u,v,w) = \gamma v$. Hence, conditions (C) and (L) hold and, $\widehat{\beta^n}(T)$ and $\widehat{\gamma^n}(T)$ are martingales estimators. Also, in the next subsection, we see that this approach satisfies the hypotheses of Theorem 5.1 and hence, asymptotic normality of $\widehat{\beta^n}(T)$ and $\widehat{\gamma^n}(T)$ is obtained.

Regarding the initial state of the epidemic, two natural assumptions can be made and both of them satisfy Condition (5.1.1) in Theorem 5.1. The first consists in assuming $\{\chi^n(0)\}_{n\in\mathbb{N}}$ is a deterministic sequence converging to $\chi(0)$ in E. This assumption is quite reasonable whenever a good knowledge about the initial numbers of susceptible, infected and removed individuals is involved. For instance, this assumption can be done when the population is small and the proportion of individuals belonging to each compartment can be observed and calculated at time zero. The second possible assumption consists in assuming $(S^n(0), I^n(0), R^n(0))^\top$ has multinomial distribution with parameters n, p_1, p_2 and $1 - p_1 - p_2$, $(0 < p_1 + p_2 < 1)$, i.e. for each $(s,i) \in \{0,\ldots,n\} \times \{0,\ldots,n\}$ such that $s + i \leq n$,

$$\mathbb{P}(S^n(0) = s, I^n(0) = i) = \frac{n!}{s!i!(n-s-i)!} p_1^s p_2^i (1 - p_1 - p_2)^{n-s-i}.$$

This second assumption will be held from now on, and although the parameters of this distribution can be estimated by taking a sample at time zero, in the sequel it will be assumed these parameters are known.

Note that the solution $\chi = (\sigma, \iota, \rho)^\top$ to (3) satisfies:

$$\begin{aligned} \tfrac{d\sigma}{dt}(t) &= \quad -\beta\sigma(t)\iota(t) \\ \tfrac{d\iota}{dt}(t) &= \beta\sigma(t)\iota(t) - \gamma\iota(t) \\ \tfrac{d\rho}{dt}(t) &= \qquad \gamma\iota(t), \end{aligned} \tag{17}$$

with initial condition $(\sigma(0), \iota(0), \rho(0)) = (p_1, p_2, 1 - p_1 - p_2)$.

6.2. Asymptotic normality

As a consequence of Theorem 5.1, the following proposition is stated, which shows the parameter estimators for the SIR epidemic model are asymptotically normal. In this subsection all notations and facts given on the preceding subsection will be maintained.

Proposition 6.1. *Let* $\lambda = (\beta, \gamma)^\top$, $T > 0$ *and* $\widehat{\lambda^n} = (\widehat{\beta^n}, \widehat{\gamma^n})^\top$. *Then,* $\{\sqrt{n}(\widehat{\lambda^n}(T) - \lambda)\}_{n\in\mathbb{N}}$ *converges in distribution to a normal random bivariate vector with mean zero and variance-covariance matrix*

$$\Sigma(T) = \begin{pmatrix} \beta^2/(\sigma(0) - \sigma(T)) & 0 \\ 0 & \gamma^2/(\rho(T) - \rho(0)) \end{pmatrix}.$$

Proof. From the Kolmogorov Law of Large Numbers, $\{\chi^n(0)\}_{n\in\mathbb{N}}$ converges almost sure to $\chi_0 = (p_1, p_2, 1 - p_1 - p_2)^\top$ and hence, Condition (5.1.1) holds.

Let $\xi_k^n(1) = \Delta A^n(t_k^n) - \beta\sigma^n(t_{k-1}^n)\iota^n(t_{k-1}^n)$ and $\xi_k^n(2) = \Delta B^n(t_k^n) - \gamma\iota^n(t_{k-1}^n)$. In order to verify Condition (5.1.2), it suffices to prove for each $i = 1, \dots, m$, the double sequence $\{\xi_k^n(i)^2; 0 \le k \le n, n \in \mathbb{N}\}$ is uniformly integrable. For this purpose, we note that, for a random variable X with Binomial distribution, the following inequality holds:

$$\mathbb{E}(|X - \mathbb{E}(X)|^3) \le 8\,\mathbb{E}(X)^3 + 6\,\mathbb{E}(X)^2 + \mathbb{E}(X).$$

Consequently, according to our approach, for each $i = 1, 2$, we have

$$\sup\{\mathbb{E}(|\xi_k^n(i)|^3) : n, k \in \mathbb{N}\} \le 8\zeta^3 + 6\zeta^2 + \zeta,$$

where

$$\zeta = \begin{cases} \beta \text{ if } i = 1 \\ \gamma \text{ if } i = 2. \end{cases}$$

Hence, $\{\xi_k^n(i)^2\}_{n,k\in\mathbb{N}}$ is uniformly integrable and Condition (5.1.2) holds.

Let

$$v_A^n(t) = \frac{1}{n}\sum_{k=1}^{[nt]} \mathbb{E}([\Delta A^n(t_k^n) - \beta S^n(t_{k-1}^n)\iota^n(t_{k-1}^n)\Delta t_n]^2 | \mathcal{F}_{k-1}^n)$$

and

$$v_B^n(t) = \frac{1}{n}\sum_{k=1}^{[nt]} \mathbb{E}([\Delta B^n(t_k^n) - \gamma I^n(t_{k-1}^n)\Delta t_n]^2 | \mathcal{F}_{k-1}^n).$$

We have,

$$v_A^n(t) = \frac{1}{n}\sum_{k=1}^{[nt]} \beta\sigma^n(t_{k-1}^n)\iota^n(t_{k-1}^n)(1-\beta\iota^n(t_{k-1}^n)\Delta t_n),$$

and

$$v_B^n(t) = \frac{1}{n}\sum_{k=1}^{[nt]} \gamma\iota^n(t_{k-1}^n)(1-\gamma\Delta t_n).$$

Since $0 \le v_A^n(t) \le \beta t$ and $0 \le v_B^n(t) \le \gamma t$, the sequences $\{v_A^n(t)/n\}_{n\in\mathbb{N}}$ and $\{v_B^n(t)/n\}_{n\in\mathbb{N}}$ converge to zero and hence, Condition (4.1.2) holds. Thus, assumptions of Theorem 4.1 are satisfied and consequently, $\{\chi^n\}_{n\in\mathbb{N}}$ converges to χ uniformly in probability over compact subsets of $\mathbb{R}_+$. This fact along with

$$v_A^n(t) = \int_0^t \beta\sigma^n(u)\iota^n(u)\,du + O(1/n) \quad \text{and} \quad v_B^n(t) = \int_0^t \gamma\iota^n(u)\,du + O(1/n),$$

allow to conclude for each $t \ge 0$, $\{v_A^n(t)\}_{n\in\mathbb{N}}$ and $\{v_B^n(t)\}_{n\in\mathbb{N}}$ converge in probability to

$$v_A(t) = \int_0^t \beta\sigma(u)\iota(u)\,du \quad \text{and} \quad v_B(t) = \int_0^t \gamma\iota(u)\,du,$$

respectively.

Thus, as a consequence of Theorem 5.1, we have that $\{\sqrt{n}(\widehat{\gamma^n}(T) - \gamma)\}_{n\in\mathbb{N}}$ converges in distribution to a normal random bivariate vector having mean zero and variance-covariance matrix

$$\Sigma(T) = \begin{pmatrix} \beta/\int_0^T \sigma(u)\iota(u)\,du & 0 \\ 0 & \gamma/\int_0^T \iota(u)\,du \end{pmatrix}.$$

From (6) we have $\sigma(0) - \sigma(T) = \beta\int_0^T \sigma(u)\iota(u)\,du$ and $\rho(T) - \rho(0) = \gamma\int_0^T \iota(u)\,du$. Hence,

$$\Sigma(T) = \begin{pmatrix} \beta^2/(\sigma(0)-\sigma(T)) & 0 \\ 0 & \gamma^2/(\rho(T)-\rho(0)) \end{pmatrix},$$

and this concludes the proof. □

Since Proposition 6.1 involves a limit distribution depending on the solution of (17), this proposition may be of limited utility. However, Proposition 4.1 and Slutsky's theorem have, as a consequence, the following corollary, which is an alternative to this possible difficulty.

Corollary 6.1. *Let $T > 0$ and*

$$Q_n(T) = \sqrt{n}\begin{pmatrix} (\sigma(0)-\sigma^n(T))^{1/2}(\widehat{\beta^n}(T)-\beta) \\ (\rho^n(T)-\rho(0))^{1/2}(\widehat{\gamma^n}(T)-\gamma) \end{pmatrix}.$$

Then, $\{Q_n(T)\}_{n\in\mathbb{N}}$ converges in distribution to a normal random bivariate vector with mean zero and covariance matrix

$$I = \begin{pmatrix} \beta^2 & 0 \\ 0 & \gamma^2 \end{pmatrix}.$$

6.3. Hypothesis test for the infection rate

In this section, we are interested in proving whether the parameter β of this SIR epidemic model belongs to a subset of the parametric set. This analysis, which is known as hypothesis test, will intend to define whether the infection rate is bigger than a fixed value $\beta_0 > 0$, i.e. the null hypothesis is stated as $H_0 : \beta = \beta_0$ against the one-side alternative $H_1 : \beta > \beta_0$. Since under H_0, $\{\widehat{\beta^n}(T)\}_{n\in\mathbb{N}}$ converges in probability to β_0, H_0 shall be rejected in favor of H_1 when $\widehat{\beta^n}(T)$ is too large, i.e. when $\widehat{\beta^n}(T) > u_n$, being u_n determined by $\mathbb{P}(\widehat{\beta^n}(T) > u_n|H_0) \leq \alpha$. Here, α is the preassigned level of significance which controls the probability of falsely rejecting H_0 when H_0 is true.

By means of Corollary 6.1, it is possible to obtain an approximate value u_n when n is large. For this purpose, $\mathbb{P}(\widehat{\beta^n}(T) > u_n|H_0) \leq \alpha$ is replaced by the following weaker requirement:

$$\mathbb{P}(\widehat{\beta^n}(T) > u_n|H_0) \to \alpha, \quad \text{as} \quad n \to \infty.$$

From the Proposition 6.1, under H_0, $\{\sqrt{n}(\sigma(0) - \sigma(T))^{1/2}(\widehat{\beta^n}(T) - \beta_0)/\beta_0\}_{n\in\mathbb{N}}$ converges in distribution to a random variable having mean zero and variance one. Therefore, with an asymptotic level of significance α,

$$\sqrt{n}(\sigma(0) - \sigma(T))^{1/2}(u_n - \beta_0)/\beta_0 \to t_\alpha \text{ as } n \to \infty,$$

where $1 - \Phi(t_\alpha) = \alpha$. Here, Φ is the cumulative distribution function of a standard normal variable. Hence, $u_n = \beta_0 + \beta_0 t_\alpha / \sqrt{n(\sigma(0) - \sigma(T))} + o(1/n)$. In particular, we can choose $u_n = \beta_0 + \beta_0 t_\alpha / \sqrt{n(\sigma(0) - \sigma(T))}$ and therefore, with an asymptotic level of significance α, a critical region for the test is

$$R_1(n) = \{\sqrt{n}(\sigma(0) - \sigma(T))^{1/2}(\widehat{\beta^n}(T) - \beta_0)/\beta_0 \geq t_\alpha\}.$$

Under H_0, $\sigma(T)$ is known but it can not be explicitly obtained due to the fact that (6) does not admit a closed-form solution. However, we can take advantage of the fact that $\{\sigma^n(T)\}_{n\in\mathbb{N}}$ converges in probability to $\sigma(T)$, and by Slutzky's theorem, it can be obtained that

$$R_2(n) = \{\sqrt{n}(\sigma(0) - \sigma^n(T))^{1/2}(\widehat{\beta^n}(T) - \beta_0)/\beta_0 \geq t_\alpha\}$$

is a critical region for the test with an asymptotic level of significance α.

Let $\Delta > 0$ and let us consider the alternative hypothesis

$$H_1 : \beta = \beta_0 + \Delta/\sqrt{n}.$$

In this case, under H_1 the power of the test $\pi_n(\beta) = \mathbb{P}(R(n)|H_1)$, where $R(n) = R_1(n)$ or $R(n) = R_2(n)$, converges to $\Phi(\frac{(\sigma(0)-\sigma(T))^{1/2}\Delta - t_\alpha\beta_0}{\beta})$.

6.4. Relative removal-rate and basic reproduction number

As pointed out in Subsection 4.2, a fundamental concept resulting from the mathematical theory of the general deterministic epidemic model is the relative removal-rate, see for instance [2] and [3]. This number is defined as $\tau = \gamma/\beta$, where γ and β represent the removal and infection rates, respectively, and it plays a crucial part in determining the probable occurrence of an epidemic outbreak. Actually, due to the fact that in the deterministic version of this model, we have (see (17))

$$\frac{d\iota}{dt}(t) = \beta\sigma(t)\iota(t) - \gamma\iota(t),$$

by assuming $\iota(0) > 0$, one has the derivative of ι at zero is bigger than zero, if and only if, $\tau < \sigma(0)$. Consequently, unless the initial density of susceptible individuals is bigger than the relative removal-rate, no epidemics can start. It is worth pointing out that this parameter is connected with the basic reproduction number, which is defined as the number of cases generated by one infective over the period of infectivity when that infective was introduced into a large population of susceptible individuals. See for instance, [16], or, [14]. Actually, it becomes $R_0 = \sigma(0)/\tau$. Notice $\sigma(0) \approx 1$ when a large population of susceptible individuals is considered. Hence, an epidemic can start with a positive probability, if $R_0 > 1$, and it will die out quickly if $R_0 \leq 1$. The knowledge of the basic reproduction number allows the establishment of vaccination policies when necessary, in order to reduce the number of susceptible individuals in a population to such a level that R_0 is brought below the unity threshold. Moreover, it is useful to carry out the following hypothesis test:

$$\mathrm{H}_0 : R_0 = 1 \quad \text{against} \quad \mathrm{H}_1 : R_0 > 1.$$

A natural critical region for the test is $\{\widehat{R_0^n}(T) > u_n\}$, where $\widehat{R_0^n}(T) = \sigma(0)\widehat{\beta^n}(T)/\widehat{\gamma^n}(T)$ and u_n is a constant which should be determined by $\mathbb{P}(\widehat{R_0^n}(T) > u_n|\mathrm{H}_0) \leq \alpha$, being α a preassigned level of significance controlling the probability of falsely rejecting H_0 when H_0 is true. The following proposition allows us to carry out the mentioned hypothesis test.

Proposition 6.2. *Let Δ be a non-negative real number and $\widehat{U}_n(T) = \sqrt{n}(\widehat{R_0^n}(T) - 1)$. Then, under local alternatives having the form*

$$\mathrm{H}_\Delta : R_0 = 1 + \Delta/\sqrt{n},$$

$\widehat{U_n}(T)$ has as asymptotically normal distribution with mean Δ and variance $v_0(T)^2$, where

$$v_0(T) = \left(\frac{1}{\sigma(0) - \sigma(T)} + \frac{1}{\rho(T) - \rho(0)}\right)^{1/2}.$$

Proof. Under H_Δ we have $\widehat{U_n}(T) = \sqrt{n}(\widehat{R_0^n}(T) - R_0) + \Delta$. Let $f(x,y) = x/y$, $(x, y \in \mathbb{R}, y \neq 0)$. By Taylor's theorem, we have

$$\begin{aligned}\sqrt{n}(\widehat{R_0^n}(T) - R_0) &= \sqrt{n}(f(\sigma(0)\widehat{\beta^n}(T), \widehat{\gamma^n}(T)) - f(\sigma(0)\beta, \gamma)) \\ &= \frac{\sigma(0)}{\gamma}\sqrt{n}(\widehat{\beta^n}(T) - \beta) - \frac{\sigma(0)\beta}{\gamma^2}\sqrt{n}(\widehat{\gamma^n}(T) - \gamma) + E_n,\end{aligned}$$

where

$$E_n = -\frac{\sigma(0)\sqrt{n}}{\gamma_n^2}(\widehat{\beta^n}(T) - \beta)(\widehat{\gamma^n}(T) - \gamma) + \frac{\sigma(0)^2\delta_n\sqrt{n}}{\gamma_n^3}(\widehat{\gamma^n}(T) - \gamma)^2$$

and (δ_n, γ_n) is a random point between $(\sigma(0)\widehat{\beta^n}(T), \widehat{\gamma^n}(T))$ and $(\sigma(0)\beta, \gamma)$.

From Propositions 5.1 and 6.1, $\{E_n\}_{n\in\mathbb{N}}$ converges in probability to zero. Hence, it follows from Proposition 6.1 that $\widehat{U_n}(T)$ is asymptotically normal with mean Δ and variance $v_0^2(T)$. Therefore, this proof is complete. □

The constant u_n can be chosen as $u_n = 1 + t_\alpha v_0(T)/\sqrt{n} + o(1/\sqrt{n})$, where t_α satisfies $1 - \Phi(t_\alpha) = \alpha$. In particular, we can choose $u_n = 1 + t_\alpha v_0(T)/\sqrt{n}$, and therefore, with an asymptotic level of significance α, a critical region for the test is $\{\widehat{U}_n(T)/v_0(T) > t_\alpha\}$. Note also that, under H_Δ, the power of the test $\pi_n(\Delta) = \mathbb{P}(\widehat{U}_n(T)/v_0(T) > t_\alpha|H_\Delta)$ converges to $\pi(\Delta) = \Phi(\Delta/v_0(T) - t_\alpha)$.

Remark 6.1. *An application of the preceding proposition is to calculate the approximate power of the test, with a level of significance α, relative to*

$$H_0 : R_0 = 1 \quad \textit{against} \quad H_1 : R_0 = r,$$

where $r > 1$.

We interpret Δ in H_Δ as $\sqrt{n}(r-1)$ and approximate the power of the test by means of $\pi = \mathbb{P}(\mathcal{N} > t_\alpha|H_\Delta)$, where $\mathcal{N}$ is a normal random variable with mean $\frac{\sqrt{n}(r-1)}{rv_0(T)}$ and variance 1. Consequently, the power of the test can be approximated by

$$\pi = \Phi\left(\frac{\sqrt{n}(r-1)}{rv_0(T)} - t_\alpha\right).$$

7. Numerical simulations for the general epidemic model

In order to carry out asymptotical inference for a great number of epidemic models, results stated in Section 5 can be applied. In this subsection, statistical inference for the General Epidemic Model is developed. In this section we maintain notations used in Section 6.

7.1. Validating the population size

In this subsection, some numerical simulations are carried out in order to validate the appropriate population size under which the results contained in this work are applicable.

Let

$$X_n(T) = \begin{pmatrix} \frac{\sqrt{\sigma(0)-\sigma(T)}}{\beta} & 0 \\ 0 & \frac{\sqrt{\rho(T)-\rho(0)}}{\gamma} \end{pmatrix} \begin{pmatrix} \widehat{\beta^n}(T) \\ \widehat{\gamma^n}(T) \end{pmatrix}.$$

According to Proposition 6.1, the square Euclidean norm of $X_n(T)$, namely $\|X_n(T)\|^2$, has asymptotically $\chi^2(2)$ distribution. The positive part of the real straight line is partitioned in m subintervals, which are determined by $0 = t_0 < t_1 < \cdots < t_{m-1} < \infty$. Let F denote the $\chi^2(2)$ distribution function. For $\beta = 2$, $\gamma = 1$ and $T = 10$, $\|X_n(T)\|^2$ is simulated 1,000 times with $m = 10$, where $t_1, \ldots, t_9$ have been chosen in such a way that $F(t_i) = i/10$, i.e., $t_1 = 0.21$, $t_2 = 0.45$, $t_3 = 0.71$, $t_4 = 1.02$, $t_5 = 1.39$, $t_6 = 1.83$, $t_7 = 2.41$, $t_8 = 3.22$ and $t_9 = 4.61$. In order to $F(t_{10}) = 1$, t_{10} is defined as ∞. For different population sizes ($n = 20, 50, 100, 300, 500, 1000$), percentages obtained into the corresponding subintervals are presented in Table 1. From Proposition 6.1, better approximations correspond to percentages close to 10%. These closeness are measured by means of the standard deviation (SD) of the observations corresponding to the different population sizes, which are indicated at the last column of Table 1 below.

n	$[t_0,t_1[$	$[t_1,t_2[$	$[t_2,t_3[$	$[t_3,t_4[$	$[t_4,t_5[$	$[t_5,t_6[$	$[t_6,t_7[$	$[t_7,t_8[$	$[t_8,t_9[$	$[t_9,t_{10}[$	SD
30	10.7	8.9	8.3	9.0	7.4	7.1	9.1	8.4	9.1	22.0	43.3
50	10.8	7.6	10.0	9.3	7.8	10.0	10.6	9.1	8.3	16.5	25.4
100	11.0	9.6	10.6	9.3	10.3	10.2	8.3	8.1	10.8	11.8	11.8
300	10.0	10.0	9.3	8.6	8.9	10.3	10.0	9.8	10.8	12.3	10.4
500	9.3	10.8	9.2	9.2	9.9	11.0	9.5	10.2	11.3	9.6	7.9
1000	9.4	9.3	10.0	10.3	10.8	10.1	9.8	9.0	9.9	11.4	7.2

Table 1. Percentages of observations of $\|X_n(T)\|^2$ for the indicated population size, $\beta = 2$, $\gamma = 1$ and $T = 10$.

By carrying out five simulations of a random variable $\chi^2(2)$, $1,000$ times each, the percentages of values that resulted in the corresponding subintervals for each of the five simulations, with the corresponding SD at the last column, are showed in Table 2 below. From Tables 1 and

N	$[t_0,t_1[$	$[t_1,t_2[$	$[t_2,t_3[$	$[t_3,t_4[$	$[t_4,t_5[$	$[t_5,t_6[$	$[t_6,t_7[$	$[t_7,t_8[$	$[t_8,t_9[$	$[t_9,t_{10}[$	SD
1	10.3	10.3	11.4	9.3	9.7	7.8	8.8	11.2	11.5	9.6	12.0
2	10.4	9.6	9.8	10.4	10.3	11.4	10.5	8.7	10.2	8.4	9.1
3	10.5	11.3	10.3	8.8	9.7	11.0	11.0	9.4	9.4	8.6	9.6
4	9.7	12.3	8.8	9.5	9.7	9.8	8.9	11.7	9.8	9.8	11.2
5	10.6	10.4	10.3	10.1	9.5	8.1	9.9	10.3	9.7	11.1	8.1
										Average SD	10.0

Table 2. Percentages of $\chi^2(2)$ observations into each subintervals for 5 series of $1,000$ trials each.

2, we can conclude that, for a population size over 300, the distribution of $\|X_n(T)\|^2$ is quite approximate to the $\chi^2(2)$ distribution. In Table 3, estimates of β, γ and R_0 have been simulated for different population size. Table 3 corroborates the convergence in probability demands

n	$\widehat{\beta^n}(T)$	$\widehat{\gamma^n}(T)$	$\widehat{R_0^n}(T)$
100	2.87	0.82	1.76
200	1.73	1.17	0.74
500	1.82	0.99	0.93
1,000	1.91	0.98	0.99
2,000	1.95	1.01	0.97
5,000	2.01	0.99	1.01
10,000	2.00	1.00	1.01

Table 3. Estimates for $\widehat{\beta^n}(T)$, $\widehat{\gamma^n}(T)$ and $\widehat{R_0^n}(T)$ for the indicated population size, $\beta = 2$, $\gamma = 1$ and $T = 10$.

bigger population size in order to obtain a suitable approximation of the parameters. We appreciate, appropriate estimates for β, γ and R_0 are obtained for population sizes over $n = 5,000$.

For $\beta = 2$, $\gamma = 1$ and $T = 10$, $X_n(T)$ is simulated $1,000$ times again. Let $X_{n,1}(T)$ and $X_{n,2}(T)$ the first and second row of $X_n(T)$, i.e. $X_n(T) = (X_{n,1}(T), X_{n,2}(T))^\top$. In Figure 4 below, histograms of $X_{n,1}(T)$ and $X_{n,2}(T)$ are showed for $n = 300$. It is observed that their corresponding frequencies of simulated values are quite close to the standard normal density. Let $\widehat{U}_n(T)$ and $v_0(T)$ be as in Proposition 6.2 and denote $X_{n,3}(T) = \widehat{U}_n(T)/(R_0 v_0(T))$. By observing in Figure 5 the histogram of $X_{n,3}(10)$, we notice the graphical frequency of its simulated values is also quite close to the standard normal density.

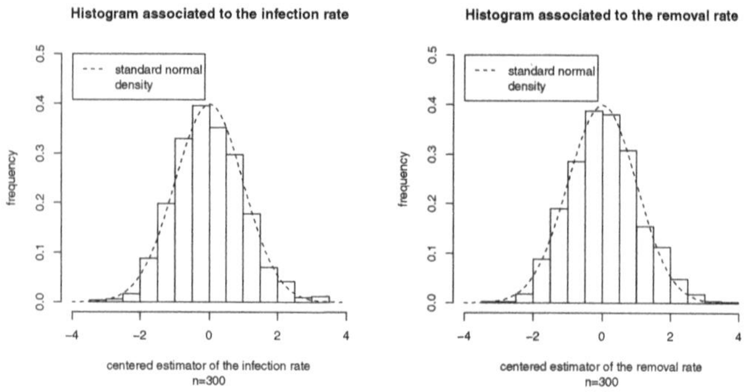

Figure 4. Histograms of $X_{n,1}(10)$ and $X_{n,2}(10)$ for a population size equals 300.

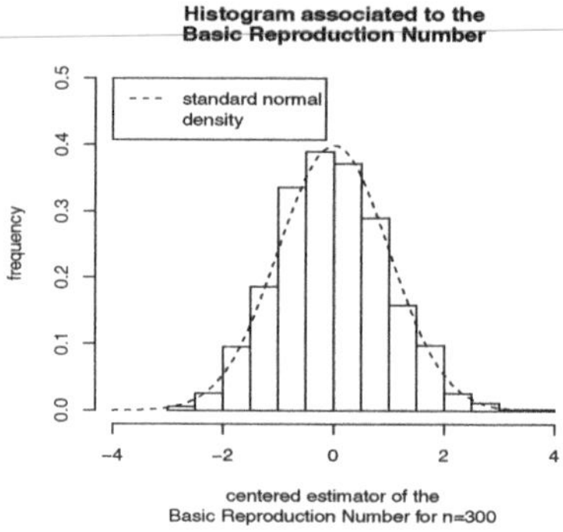

Figure 5. Histograms of $X_{n,3}(10)$ for a population size equals 300.

7.2. The power function of the test for the basic reproduction number

According to Proposition 6.2, the power function $\pi_n(\Delta)$ converges to the asymptotic power function (apf) $\pi(\Delta) = \Phi(\Delta/v_0(T) - t_\alpha)$. The level of significance of the test is assumed to be $\alpha = 0.05$ and consequently, $t_\alpha = 1.644853$. Notice $\pi_n(\Delta)$ can be estimated by $\widehat{\pi}_n(\Delta)$, the frequency H_0 is rejected (rf). Hence, $\widehat{\pi}_n(\Delta)$ is the frequency of times that $\widehat{U}_n(T)/v_0(T) > 1.644853$, under $H_\Delta : R_0 = 1 + \Delta/\sqrt{n}$. By the Law of Large Number, a strong consistent estimator of $\pi_n(\Delta)$ is given by

$$\widehat{\pi}_n(\Delta) = \frac{1}{M}\sum_{k=1}^{M} I_{\{\widehat{U}_{n,k}(T)/v_0(T) > 1.644853\}},$$

where $\widehat{U}_{n,1}(T), \ldots, \widehat{U}_{n,M}(T)$ are independent and identically distributed random variables, which have the same distribution as $\widehat{U}_n(T)$ under H_Δ. Even, it can prove that the convergence of $\widehat{\pi}_n$ to π_n, as M goes to ∞, is uniform on compact subsets of $\mathbb{R}_+$, I.e., for each $T > 0$,

$$\lim_{M\to\infty} \sup_{0\le\Delta\le T} |\widehat{\pi}_n(\Delta) - \pi_n(\Delta)| = 0, \quad a.s.$$

In Table 4 below, $M = 1,000$ simulations of $\widehat{\pi}_n(\Delta)$ are carried out for a population size equals 300 and for 150 values of Δ between 0 and 14.9, with step sizes of equal length $\Delta = 0.1$. Since

Δ	.0	.1	.2	.3	.4	.5	.6	.7	.8	.9
0	0.000	0.001	0.001	0.004	0.002	0.005	0.002	0.007	0.008	0.005
1	0.004	0.014	0.004	0.021	0.010	0.015	0.015	0.016	0.018	0.020
2	0.023	0.040	0.032	0.036	0.042	0.047	0.052	0.059	0.063	0.060
3	0.078	0.075	0.104	0.085	0.120	0.105	0.136	0.140	0.143	0.150
4	0.151	0.183	0.190	0.202	0.201	0.210	0.247	0.267	0.289	0.282
5	0.296	0.329	0.343	0.357	0.342	0.376	0.415	0.431	0.427	0.446
6	0.467	0.463	0.509	0.536	0.524	0.558	0.579	0.570	0.604	0.595
7	0.632	0.628	0.655	0.663	0.703	0.706	0.723	0.719	0.730	0.775
8	0.762	0.787	0.765	0.807	0.814	0.798	0.805	0.842	0.845	0.846
9	0.881	0.854	0.874	0.880	0.890	0.900	0.913	0.901	0.913	0.904
10	0.911	0.927	0.934	0.937	0.941	0.941	0.944	0.942	0.958	0.957
11	0.972	0.961	0.969	0.967	0.988	0.973	0.978	0.986	0.988	0.981
12	0.982	0.987	0.982	0.987	0.988	0.989	0.981	0.995	0.990	0.996
13	0.991	0.992	0.993	0.998	0.996	0.998	0.998	0.997	0.994	0.997
14	0.997	1.000	1.000	0.999	0.999	0.999	0.999	1.000	0.998	0.998

Table 4. Frequency of times $\widehat{U}_n(T)/v_0(T) > 1.644853$ for different values of Δ.

for $\Delta = 0$ the power of the test should be close to 0.05, the random function $\widehat{\pi}_n$ cannot be considered a good estimator for π_n, at least for $n = 300$. However, the asymptotic power function π seems to give a better approximation for π_n. The values of $\pi(\Delta)$ are given in Table 5 below for 150 values of Δ between 0 and 14.9, with step sizes of equal length $\Delta = 0.1$. In

Δ	.0	.1	.2	.3	.4	.5	.6	.7	.8	.9
0	0.050	0.053	0.056	0.059	0.063	0.067	0.071	0.075	0.079	0.083
1	0.088	0.093	0.098	0.103	0.108	0.114	0.119	0.125	0.132	0.138
2	0.144	0.151	0.158	0.165	0.173	0.180	0.188	0.196	0.205	0.213
3	0.222	0.230	0.239	0.249	0.258	0.268	0.277	0.287	0.297	0.307
4	0.318	0.328	0.339	0.350	0.361	0.372	0.383	0.394	0.405	0.417
5	0.428	0.440	0.451	0.463	0.475	0.486	0.498	0.510	0.521	0.533
6	0.545	0.556	0.568	0.579	0.591	0.602	0.613	0.625	0.636	0.647
7	0.657	0.668	0.679	0.689	0.699	0.709	0.719	0.729	0.739	0.748
8	0.757	0.766	0.775	0.784	0.792	0.801	0.809	0.817	0.824	0.832
9	0.839	0.846	0.853	0.860	0.866	0.872	0.878	0.884	0.890	0.895
10	0.900	0.905	0.910	0.915	0.919	0.923	0.928	0.931	0.935	0.939
11	0.942	0.946	0.949	0.952	0.955	0.957	0.960	0.962	0.965	0.967
12	0.969	0.971	0.973	0.974	0.976	0.978	0.979	0.981	0.982	0.983
13	0.984	0.985	0.986	0.987	0.988	0.989	0.990	0.991	0.991	0.992
14	0.992	0.993	0.994	0.994	0.994	0.995	0.995	0.996	0.996	0.996

Table 5. Values of $\pi(\Delta)$ for different values of Δ.

Figure 6 below, a graphical comparison between π and $\widehat{\pi}_n$ is given for a population size equals $n = 300$.

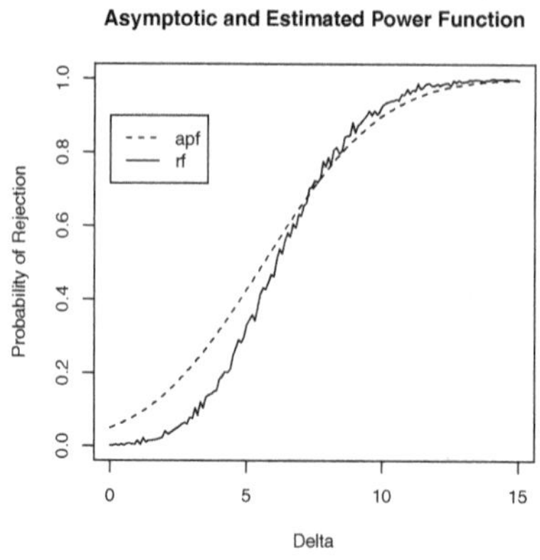

Figure 6. Grafics of $\pi(\Delta)$ and $\widehat{\pi}_n(\Delta)$ for a population size equals $n = 300$.

8. Conclusions

A wide class of discrete-time stochastic epidemic models was introduced and analyzed in this work, the convergence of the stochastic model to the deterministic one as the population increases was proved. Moreover, it was proved the convergence in distribution of a process, depending on the fluctuations between the stochastic and deterministic models, to the solution to a stochastic differential equation. The statistical analysis was focused in proving asymptotic normality of natural martingale estimators, for the parameters of the model, and these results were applied to hypothesis tests for the General Epidemic Model. As a consequence, a hypothesis test for the reproduction number was stated and numerical simulations validated the size populations under which the results are useful.

The stochastic models considered here represent an alternative modeling to those using counting processes and having transitions occurring at random times. However, both are asymptotically consistent with their deterministic counterparts. The advantage of considering stochastic models at discrete times is that it is not necessary to observe the epidemic over a long period of time; however, in order to attain the asymptotical consistency mentioned, our model requires frequent observation when dealing with a large population. Another important conclusion is obtained from Subsection 4.2, where numerical simulations was carried out for the SIS epidemic model. Indeed, the simulation showed large fluctuations of the SIS stochastic epidemic model, for $n = 10$, regarding the deterministic one. However, by simulating the process for $n = 1,000$ it could be appreciated the trajectories of the deterministic and stochastic models were quite closed. Moreover, Theorem 2 provides confidence bounds which give an insight of the fluctuations of the stochastic model regarding the deterministic one. The simulation in Subsection 4.2 for the SIS epidemic model shows the coherence between these bounds and the simulated process (Figure 3). As a conclusion, the deterministic and stochastic model are quite different for small populations, while for large populations both models perform similarly. However, it is worth noting, although the stochastic and deterministic trajectories are similar for large size populations, an attractor state for the deterministic model is not necessarily an attractor for the stochastic model. As shown in Subsection 4.2, this is the case for the SIS epidemic model.

In order to develop statistical inference on the parameters of the model, martingale estimators are proposed. This fact allows to obtain closed form for the estimators. Even, such estimators can be obtained when the distribution of the process governing the epidemic is not completely known.

All the models belong to the class that we are presenting in this chapter assume a stochastic latent period, and consequently, the Reed-Frost model are not included here. As a matter of fact, none of these models could be considered an extension or a particular case of the Reed-Frost model. Moreover, the modeling presented in this chapter need not be Markovian, even though some Markovian epidemic models are included in this setting.

Acknowledgements

This work was partially supported by Instituto de Matemática of the Pontificia Universidad Católica de Valparaíso and by FONDECYT under project 1120879.

Author details

Raúl Fierro
Instituto de Matemática, Pontificia Universidad Católica de Valparaíso, Centro de Investigación y Modelamiento de Fenómenos Aleatorios, Universidad de Valparaíso, Chile

9. References

[1] Andersson, H. & Britton, T. [2000]. *Stochastic Epidemic Models and Their Statistical Analysis. Lecture Notes in Statistics 151*, Springer, New York.
[2] Bailey, N. [1975]. *The Mathematical Theory of Infectious Diseases*, Griffin & Co., New York.
[3] Ball, F. [1983]. The threshold behavior of epidemic models, *Journal of Applied Probability* Vol. 20(No. 2): 227–241.
[4] Ball, F. & O'Neill, P. [1993]. A modification of the general stochastic epidemic motivated by aids modelling, *Journal of Applied Probability* Vol. 25: 39–62.
[5] Barbour, A. [1972]. The principle of the diffusion of arbitrary constants, *Journal of Applied Probability* Vol. 9(No. 3): 519–541.
[6] Barbour, A. [1974]. On a functional central limit theorem for markov population processes, *Advances in Applied Probability* Vol. 6(No. 1): 21–39.
[7] Becker, N. [1993]. Parametric inference for epidemic models, *Mathematical Biosciences* Vol. 117(No. 1-2): 239–251.
[8] Billingsley, P. [1968]. *Convergence of Probability Measures*, Wiley & Sons, New York.
[9] Buckley, F. & Pollet, P. [2009]. Analytical methods for a stochastic mainland-island metapopulation model, *in* R. Anderssen, R. Braddock & L. Newham (eds), *Modelling and Simulation Society of Australia and New Zealand and International Association for Mathematics and Computers in Simulation*, Camberra, Australia, pp. 1767–1773.
[10] Buckley, F. & Pollet, P. [2010]. Limit theorems for discrete-time metapopulation models, *Probability Surveys* Vol. 7: 53–83.
[11] Ethier, S. & Kurtz, N. [1986]. *The Mathematical Theory of Infectious Diseases*, Wiley & Sons, New York.
[12] Fierro, R. [2010]. A class of stochastic epidemic models and its deterministic counterpart, *Journal of the Korean Statistical Society* Vol. 39(No. 4): 397–407.

[13] Fierro, R. [2012]. Asymptotic distribution of martingale estimators for a class of epidemic models, *Statistical Methods and Applications* (to appear).
[14] Heesterbeek, J. & Dietz, K. [1996]. The concept of r_0 in epidemic theory, *Statistica Neerlandica* Vol. 50(No. 1): 89–110.
[15] Hetcote, H. & Van Ark, J. [1987]. Epidemiological models for heterogeneous populations: proportionate mixing, parameter estimation and immunization programs, *Mathematical Biosciences* Vol. 84(No. 1): 85–118.
[16] Jacquez, J. & O'Neill, P. [1991]. Reproduction numbers and thresholds in stochastic epidemic models i. homogeneous populations, *Mathematical Biosciences* Vol. 107(No. 2): 161–186.
[17] Jacquez, J., Simon, C., Koopman, J., Sattenspiel, L. & Perry, T. [1988]. Modelling and analysing hiv transmission: The effect of contact patterns, *Mathematical Biosciences* Vol. 92(No. 2): 119–199.
[18] Kendall, D. [1956]. Deterministic and stochastic epidemic in closed populations, *Proceedings of Third Berkeley Symposium in Mathematical Statistics and Probability* Vol. 4: 149–165.
[19] Kermack, W. & McKendrick, A. [1927]. Contribution to the mathematical theory of epidemics, *Proceedings of the Royal Society of London A* Vol. 115: 700–721.
[20] Kurtz, T. [1970]. Solutions of ordinary differential equations as limits of pure jump markov processes, *Journal of Applied Probability* Vol. 7(No. 1): 49–58.
[21] Kurtz, T. [1971]. Limit theorems for sequences of jump markov processes approximating ordinary differential processes, *Journal of Applied Probability* Vol. 8(No. 2): 344–356.
[22] Kurtz, T. [1981]. *Approximation of Population Processes*, SIAM, Philadelphia.
[23] Lefèvre, C. [1990]. Stochastic epidemic models for s-i-r infectious diseases: a brief survey of the recent general theory, *in* J. Gabriel, C. Lefèvre & P. Picard (eds), *Stochastic Processes in Epidemic Theory. Lecture Notes in Biomathematics 86*, Springer-Verlag, New York, pp. 1–12.
[24] Lefèvre, C. & Picard, P. [1993]. An epidemic model with fatal risk, *Mathematical Biosciences* Vol. 117(No. 1-2): 127–145.
[25] Lehmann, E. [1999]. *Elements of Large-Sample Theory*, Springer-Verlag, New York.
[26] Lenglart, E. [1977]. Relation de domination entre deux processus, *Annales de l'Institut Henri Poincaré (B) Probabilités et Statistiques* Vol. 13(No. 2): 171–179.
[27] McVinish, R. & Pollett, P. [2010]. Limits of large metapopulations with patch dependent extintion probabilities, *Advances in Applied Probability* Vol. 42(No. 4): 1172–1186.
[28] Picard, P. & Lefèvre, C. [1993]. Distribution of the final state and severity of epidemics with fatal risk, *Stochastic Processes and their Applications* Vol. 48(No. 2): 277–294.
[29] Rebolledo, R. [1979]. La méthode des martingales appliquée á l'etude de la convergence en loi de processus, *Mémoires de la Société Mathématique de France* Vol. 62(No. 4): 1–125.

Chapter 11

Stochastic Control and Improvement of Statistical Decisions in Revenue Optimization Systems

Nicholas A. Nechval and Maris Purgailis

Additional information is available at the end of the chapter

1. Introduction

A large number of problems in production planning and scheduling, location, transportation, finance, and engineering design require that decisions be made in the presence of uncertainty. From the very beginning of the application of optimization to these problems, it was recognized that analysts of natural and technological systems are almost always confronted with uncertainty. Uncertainty, for instance, governs the prices of fuels, the availability of electricity, and the demand for chemicals. A key difficulty in optimization under uncertainty is in dealing with an uncertainty space that is huge and frequently leads to very large-scale optimization models. Decision-making under uncertainty is often further complicated by the presence of integer decision variables to model logical and other discrete decisions in a multi-period or multi-stage setting.

Approaches to optimization under uncertainty have followed a variety of modeling philosophies, including expectation minimization, minimization of deviations from goals, minimization of maximum costs, and optimization over soft constraints. The main approaches to optimization under uncertainty are stochastic programming (recourse models, robust stochastic programming, and probabilistic models), fuzzy programming (flexible and possibilistic programming), and stochastic dynamic programming.

This paper is devoted to improvement of statistical decisions in revenue management systems. Revenue optimization – or revenue management as it is also called – is a relatively new field currently receiving much attention of researchers and practitioners. It focuses on how a firm should set and update pricing and product availability decisions across its various selling channels in order to maximize its profitability. The most familiar example probably comes from the airline industry, where tickets for the same flight may be sold at many different fares throughout the booking horizon depending on product restrictions as well as the remaining time until departure and the number of unsold seats. Since the tickets for a flight have to be sold before the plane takes off, the product is perishable and cannot be stored for future use. The use of the above strategies has transformed the transportation and hospitality industries, and has become increasingly important in retail, telecommunications, entertainment, financial services, health care and manufacturing. In parallel, pricing and revenue optimization has become a rapidly expanding practice in consulting services, and a growing area of software and IT development, where the revenue optimization systems are tightly integrated in the existing Supply Chain Management solutions.

Most stochastic models, which are used in revenue optimization systems, are developed under the assumptions that the parameter values of the models are known with certainty. When these models are applied to solve real-world problems, the parameters are estimated and then treated as if they were the true values. The risk associated with using estimates rather than the true parameters is called estimation risk and is often ignored. When data are limited and/or unreliable, estimation risk may be significant, and failure to incorporate it into the model design may lead to serious errors. Its explicit consideration is important since decision rules which are optimal in the absence of uncertainty need not even be approximately optimal in the presence of such uncertainty.

In this paper, we consider the cases where it is known that the underlying probability distributions belong to a parameterized family of distributions. However, unlike in the Bayesian approach, we do not assume any prior knowledge on the parameter values. The primary purpose of the paper is to introduce the idea of embedding of sample statistics (say, sufficient statistics or maximum likelihood estimators) in a performance index of revenue optimization problem. In this case, we find the optimal stochastic control policy directly. We demonstrate the fact that the traditional approach, which separates the estimation and the optimization tasks in revenue optimization systems (i.e., when we use the estimates as if they were the true parameters) can often lead to pure results. It will be noted that the optimal statistical decision rules depend on data availability.

For constructing the improved statistical decisions, a new technique of invariant embedding of sample statistics in a performance index is proposed (Nechval et al., 1999; 2004; 2008; 2010a; 2010b; 2010c; 2010d; 2010e; 2011a; 2011b). This technique represents a simple and computationally attractive statistical method based on the constructive use of the invariance principle in mathematical statistics. Unlike the Bayesian approach, an invariant embedding technique is independent of the choice of priors, i.e., subjectivity of investigator is eliminated from the problem. The technique allows one to eliminate unknown parameters from the problem and to find the improved invariant statistical decision rules, which has smaller risk than any of the well-known traditional statistical decision rules.

In order to obtain improved statistical decisions for revenue management under parametric uncertainty, it can be considered the three prediction situations: "new-sample" prediction, "within-sample" prediction, and "new-within-sample" prediction. For the new-sample prediction situation, the data from a past sample are used to make predictions on a future unit or sample of units from the same process or population. For the within-sample prediction situation, the problem is to predict future events in a sample or process based on early data from that sample or process. For the new-within-sample prediction situation, the problem is to predict future events in a sample or process based on early data from that sample or process as well as on a past data sample from the same process or population. Some mathematical preliminaries for the within-sample prediction situation are given below.

2. Mathematical preliminaries for the within-sample prediction situation

Theorem 1. Let $X_1 \le \dots \le X_k$ be the first k ordered observations (order statistics) in a sample of size m from a continuous distribution with some probability density function $f_\theta(x)$ and distribution function $F_\theta(x)$, where θ is a parameter (in general, vector). Then the joint probability density function of $X_1 \le \dots \le X_k$ and the *l*th order statistics X_l $(1 \le k < l \le m)$ is given by

$$f_\theta(x_1, \dots, x_k, x_l) = f_\theta(x_1, \dots, x_k) g_\theta(x_l \mid x_k), \tag{1}$$

where

$$f_\theta(x_1, \dots, x_k) = \frac{m!}{(m-k)!} \prod_{i=1}^{k} f_\theta(x_i)[1 - F_\theta(x_k)]^{m-k}, \tag{2}$$

$$g_\theta(x_l \mid x_k) = \frac{(m-k)!}{(l-k-1)!(m-l)!} \left[\frac{F_\theta(x_l) - F_\theta(x_k)}{1 - F_\theta(x_k)}\right]^{l-k-1} \left[1 - \frac{F_\theta(x_l) - F_\theta(x_k)}{1 - F_\theta(x_k)}\right]^{m-l} \frac{f_\theta(x_l)}{1 - F_\theta(x_k)}$$

$$= \frac{(m-k)!}{(l-k-1)!(m-l)!} \sum_{j=0}^{l-k-1} \binom{l-k-1}{j} (-1)^j \left[\frac{1 - F_\theta(x_l)}{1 - F_\theta(x_k)}\right]^{m-l+j} \frac{f_\theta(x_l)}{1 - F_\theta(x_k)}$$

$$= \frac{(m-k)!}{(l-k-1)!(m-l)!} \sum_{j=0}^{m-l} \binom{m-l}{j} (-1)^j \left[\frac{F_\theta(x_l) - F_\theta(x_k)}{1 - F_\theta(x_k)}\right]^{l-k-1+j} \frac{f_\theta(x_l)}{1 - F_\theta(x_k)} \tag{3}$$

represents the conditional probability density function of X_l given $X_k = x_k$.

Proof. The joint density of $X_1 \le \dots \le X_k$ and X_l is given by

$$f_\theta(x_1, \dots, x_k, x_l) = \frac{m!}{(l-k-1)!(m-l)!} \prod_{i=1}^{k} f_\theta(x_i)[F_\theta(x_l) - F_\theta(x_k)]^{l-k-1} f_\theta(x_l)[1 - F_\theta(x_l)]^{m-l}$$

$$= f_\theta(x_1, \dots, x_k) g_\theta(x_l \mid x_k). \tag{4}$$

It follows from (4) that

$$f_\theta(x_l \mid x_1, \dots, x_k) = \frac{f_\theta(x_1, \dots, x_k, x_l)}{f_\theta(x_1, \dots, x_k)} = g_\theta(x_l \mid x_k), \tag{5}$$

i.e., the conditional distribution of X_l, given $X_i = x_i$ for all $i = 1,\ldots, k$, is the same as the conditional distribution of X_l, given only $X_k = x_k$. This ends the proof. □

Corollary 1.1. The conditional probability distribution function of X_l given $X_k=x_k$ is

$$P_\theta\{X_l \le x_l \mid X_k = x_k\} = 1 - \frac{(m-k)!}{(l-k-1)!(m-l)!}$$

$$\times \sum_{j=0}^{l-k-1} \binom{l-k-1}{j} \frac{(-1)^j}{m-l+1+j} \left[\frac{1-F_\theta(x_l)}{1-F_\theta(x_k)}\right]^{m-l+1+j}$$

$$= \frac{(m-k)!}{(l-k-1)!(m-l)!} \sum_{j=0}^{m-l} \binom{m-l}{j} \frac{(-1)^j}{l-k+j} \left[\frac{F_\theta(x_l)-F_\theta(x_k)}{1-F_\theta(x_k)}\right]^{l-k+j}. \quad (6)$$

Corollary 1.2. If $l = k + 1$,

$$g_\theta(x_{k+1} \mid x_k) = (m-k)\left[\frac{1-F_\theta(x_{k+1})}{1-F_\theta(x_k)}\right]^{m-k-1} \frac{f_\theta(x_{k+1})}{1-F_\theta(x_k)}$$

$$= (m-k)\sum_{j=0}^{m-k-1} \binom{m-k-1}{j} (-1)^j \left[\frac{F_\theta(x_{k+1})-F_\theta(x_k)}{1-F_\theta(x_k)}\right]^j \frac{f_\theta(x_{k+1})}{1-F_\theta(x_k)}, \quad 1 \le k \le m-1, \quad (7)$$

and

$$P_\theta\{X_{k+1} \le x_{k+1} \mid X_k = x_k\} = 1 - \left[\frac{1-F_\theta(x_{k+1})}{1-F_\theta(x_k)}\right]^{m-k}$$

$$= (m-k)\sum_{j=0}^{m-k-1} \binom{m-k-1}{j} \frac{(-1)^j}{1+j} \left[\frac{F_\theta(x_{k+1})-F_\theta(x_k)}{1-F_\theta(x_k)}\right]^{1+j}, \quad 1 \le k \le m-1. \quad (8)$$

Corollary 1.3. If $l = m$,

$$g_\theta(x_m \mid x_k) = (m-k)\sum_{j=0}^{m-k-1} \binom{m-k-1}{j} (-1)^j$$

$$\times \left[\frac{1-F_\theta(x_m)}{1-F_\theta(x_k)}\right]^j \frac{f_\theta(x_m)}{1-F_\theta(x_k)} = (m-k)\left[\frac{F_\theta(x_m)-F_\theta(x_k)}{1-F_\theta(x_k)}\right]^{m-k-1} \frac{f_\theta(x_m)}{1-F_\theta(x_k)}, \quad 1 \le k \le m-1, \quad (9)$$

and

$$P_\theta\{X_m \le x_m \mid X_k = x_k\} = 1 - (m-k)\sum_{j=0}^{m-k-1} \binom{m-k-1}{j} \frac{(-1)^j}{1+j} \left[\frac{1-F_\theta(x_m)}{1-F_\theta(x_k)}\right]^{1+j}$$

$$= \left[\frac{F_\theta(x_m) - F_\theta(x_k)}{1 - F_\theta(x_k)} \right]^{m-k}, \quad 1 \le k \le m-1. \tag{10}$$

2.1. Exponential distribution

In order to use the results of Theorem 1, we consider, for illustration, the exponential distribution with the probability density function

$$f_\theta(x) = \frac{1}{\theta}\exp\left(-\frac{x}{\theta}\right), \quad x \ge 0, \ \theta > 0, \tag{11}$$

and the probability distribution function

$$F_\theta(x) = 1 - \exp\left(-\frac{x}{\theta}\right), \quad x \ge 0, \ \theta > 0. \tag{12}$$

Theorem 2. Let $X_1 \le \ldots \le X_k$ be the first k ordered observations (order statistics) in a sample of size m from the exponential distribution (11). Then the conditional probability density function of the lth order statistics X_l $(1 \le k < l \le m)$ given $X_k = x_k$ is

$$g_\theta(x_l \mid x_k) = \frac{1}{B(l-k, (m-l+1))} \sum_{j=0}^{l-k-1} \binom{l-k-1}{j} (-1)^j \frac{1}{\theta} \exp\left(-\frac{(m-l+1+j)(x_l - x_k)}{\theta}\right)$$

$$= \frac{1}{B(l-k, (m-l+1))} \sum_{j=0}^{m-l} \binom{m-l}{j} (-1)^j \frac{1}{\theta} \left[1 - \exp\left(-\frac{x_l - x_k}{\theta}\right)\right]^{l-k-1+j} \exp\left(\frac{x_l - x_k}{\theta}\right),$$

$$x_l \ge x_k, \tag{13}$$

and the conditional probability distribution function of the lth order statistics X_l given $X_k = x_k$ is

$$P_\theta\{X_l \le x_l \mid X_k = x_k\} = 1 - \frac{1}{B(l-k, (m-l+1))} \sum_{j=0}^{l-k-1} \binom{l-k-1}{j}$$

$$\times \frac{(-1)^j}{m-l+1+j} \exp\left(-\frac{(m-l+1+j)(x_l - x_k)}{\theta}\right)$$

$$= \frac{1}{B(l-k, (m-l+1))} \sum_{j=0}^{m-l} \binom{m-l}{j} \frac{(-1)^j}{l-k+j} \left[1 - \exp\left(-\frac{x_l - x_k}{\theta}\right)\right]^{l-k+j}. \tag{14}$$

Proof. It follows from (3) and (6), respectively. □

Corollary 2.1. If $l = k + 1$,

$$g_\theta(x_{k+1} \mid x_k) = (m-k)\frac{1}{\theta}\exp\left(-\frac{(m-k)(x_{k+1}-x_k)}{\theta}\right)$$

$$= (m-k)\sum_{j=0}^{m-k-1}\binom{m-k-1}{j}(-1)^j\frac{1}{\theta}\left[1-\exp\left(-\frac{x_{k+1}-x_k}{\theta}\right)\right]^j \exp\left(\frac{x_{k+1}-x_k}{\theta}\right),$$

$$x_{k+1} \geq x_k, \quad 1 \leq k \leq m-1, \tag{15}$$

and

$$P_\theta\{X_{k+1} \leq x_{k+1} \mid X_k = x_k\} = 1-\exp\left(-\frac{(m-k)(x_{k+1}-x_k)}{\theta}\right)$$

$$= (m-k)\sum_{j=0}^{m-k-1}\binom{m-k-1}{j}\frac{(-1)^j}{1+j}\left[1-\exp\left(-\frac{x_{k+1}-x_k}{\theta}\right)\right]^{1+j}, \quad 1 \leq k \leq m-1. \tag{16}$$

Corollary 2.2. If $l = m$,

$$g_\theta(x_m \mid x_k) = (m-k)$$

$$\times \sum_{j=0}^{m-k-1}\binom{m-k-1}{j}(-1)^j\frac{1}{\theta}\exp\left(-\frac{(1+j)(x_m-x_k)}{\theta}\right)$$

$$= (m-k)\frac{1}{\theta}\left[1-\exp\left(-\frac{x_m-x_k}{\theta}\right)\right]^{m-k-1} \exp\left(-\frac{x_m-x_k}{\theta}\right), \quad x_m \geq x_k, \quad 1 \leq k \leq m-1, \tag{17}$$

and

$$P_\theta\{X_m \leq x_m \mid X_k = x_k\} = 1-(m-k)\sum_{j=0}^{m-k-1}\binom{m-k-1}{j}\frac{(-1)^j}{1+j}\exp\left(-\frac{(1+j)(x_m-x_k)}{\theta}\right)$$

$$= \left[1-\exp\left(-\frac{x_m-x_k}{\theta}\right)\right]^{m-k}, \quad 1 \leq k \leq m-1. \tag{18}$$

Theorem 3. Let $X_1 \leq \ldots \leq X_k$ be the first k ordered observations (order statistics) in a sample of size m from the exponential distribution (11), where the parameter θ is unknown. Then the predictive probability density function of the lth order statistics X_l $(1 \leq k < l \leq m)$ is given by

$$g_{s_k}(x_l \mid x_k) = \frac{k}{B(l-k,(m-l+1)}\sum_{j=0}^{l-k-1}\binom{l-k-1}{j}(-1)^j$$

$$\times\left[1+(m-l+1+j)\frac{x_l-x_k}{s_k}\right]^{-(k+1)}\frac{1}{s_k}, \quad x_l \geq x_k, \tag{19}$$

where

$$S_k = \sum_{i=1}^{k} X_i + (m-k)X_k \tag{20}$$

is the sufficient statistic for θ, and the predictive probability distribution function of the *l*th order statistics X_l is given by

$$P_{s_k}\{X_l \le x_l \mid X_k = x_k\} = 1 - \frac{1}{B(l-k,(m-l+1)}\sum_{j=0}^{l-k-1}\binom{l-k-1}{j}\frac{(-1)^j}{m-l+1+j}$$

$$\times\left[1+(m-l+1+j)\frac{x_l - x_k}{s_k}\right]^{-k}. \tag{21}$$

Proof. Using the technique of invariant embedding (Nechval et al., 1999; 2004; 2008; 2010a; 2010b; 2010c; 2010d; 2010e; 2011a; 2011b), we reduce (13) to

$$g_\theta(x_l \mid x_k) = \frac{1}{B(l-k,(m-l+1)}\sum_{j=0}^{l-k-1}\binom{l-k-1}{j}(-1)^j$$

$$\times v\exp\left(-\frac{(m-l+1+j)(x_l - x_k)}{s_k}v\right)\frac{1}{s_k} = g_{s_k}(x_l \mid x_k, v), \tag{22}$$

where

$$V = S_k / \theta \tag{23}$$

is the pivotal quantity, the probability density function of which is given by

$$f(v) = \frac{1}{\Gamma(k)} v^{k-1}\exp(-v), \quad v \ge 0. \tag{24}$$

Then

$$g_{s_k}(x_l \mid x_k) = E\{g_{s_k}(x_l \mid x_k, v)\} = \int_0^\infty g_{s_k}(x_l \mid x_k, v) f(v)dv = g_{s_k}(x_l \mid x_k). \tag{25}$$

This ends the proof. □

Corollary 3.1. If l = k + 1,

$$g_{s_k}(x_{k+1} \mid x_k) = k(m-k)\left[1+(m-k)\frac{x_{k+1} - x_k}{s_k}\right]^{-(k+1)}\frac{1}{s_k},$$

$$x_{k+1} \ge x_k, \quad 1 \le k \le m-1, \tag{26}$$

and

$$P_{s_k}\{X_{k+1} \le x_{k+1} \mid X_k = x_k\} = 1 - \left[1 + (m-k)\frac{x_{k+1} - x_k}{s_k}\right]^{-k}, \quad 1 \le k \le m-1. \tag{27}$$

Corollary 3.2. If l = m,

$$g_{s_k}(x_m \mid x_k) = k(m-k)\sum_{j=0}^{m-k-1}\binom{m-k-1}{j}(-1)^j\left[1 + (1+j)\frac{x_m - x_k}{s_k}\right]^{-(k+1)}\frac{1}{s_k},$$

$$x_m \ge x_k, \quad 1 \le k \le m-1, \tag{28}$$

and

$$P_{s_k}\{X_m \le x_m \mid X_k = x_k\} = 1 - (m-k)\sum_{j=0}^{m-k-1}\binom{m-k-1}{j}\frac{(-1)^j}{1+j}\left[1 + (1+j)\frac{x_m - x_k}{s_k}\right]^{-k}$$

$$= \sum_{j=0}^{m-k}\binom{m-k}{j}(-1)^j\left[1 + j\frac{x_m - x_k}{s_k}\right]^{-k}, \quad 1 \le k \le m-1. \tag{29}$$

2.2. Cumulative customer demand

The primary purpose of this paper is to introduce the idea of cumulative customer demand in inventory control problems to deal with the order statistics from the underlying distribution. It allows one to use the above results to improve statistical decisions for inventory control problems under parametric uncertainty.

Assumptions. The customer demand at the *i*th period represents a random variable Y_i, $i \in \{1, \ldots, m\}$. It is assumed (for the cumulative customer demand) that the random variables

$$X_1 = Y_1, \ldots, X_k = \sum_{i=}^{k} Y_i, \ldots, X_l = \sum_{i=1}^{l} Y_i, \ldots, X_m = \sum_{i=1}^{m} Y_i \tag{30}$$

represent the order statistics ($X_1 \le \ldots \le X_m$) from the exponential distribution (11).

Inferences. For the above case, we have the following inferences.

Conditional probability density function of Y_{k+1}, $k \in \{1, \ldots, m-1\}$, is given by

$$g_\theta(y_{k+1} \mid k) = \frac{m-k}{\theta}\exp\left(-\frac{(m-k)y_{k+1}}{\theta}\right), \quad y_{k+1} \ge 0; \tag{31}$$

Conditional probability distribution function of Y_{k+1}, $k \in \{1, \ldots, m-1\}$, is given by

$$G_\theta\{y_{k+1} \mid k\} = 1 - \exp\left(-\frac{(m-k)y_{k+1}}{\theta}\right). \quad (32)$$

Conditional probability density function of $Z_m = \sum_{i=k+1}^{m} Y_i$ is given by

$$g_\theta(z_m \mid k) = (m-k)\frac{1}{\theta}\left[1 - \exp\left(-\frac{z_m}{\theta}\right)\right]^{m-k-1} \exp\left(-\frac{z_m}{\theta}\right),\ z_m \geq 0,\ 1 \leq k \leq m-1; \quad (33)$$

Conditional probability distribution function of $Z_m = \sum_{i=k+1}^{m} Y_i$ is given by

$$G_\theta(z_m \mid k) = \left[1 - \exp\left(-\frac{z_m}{\theta}\right)\right]^{m-k},\ 1 \leq k \leq m-1. \quad (34)$$

Predictive probability density function of Y_{k+1}, $k \in \{1, \ldots, m-1\}$, is given by

$$g_{s_k}(y_{k+1} \mid k) = k(m-k)\left[1 + (m-k)\frac{y_{k+1}}{s_k}\right]^{-(k+1)} \frac{1}{s_k},\ y_{k+1} \geq 0; \quad (35)$$

Predictive probability distribution function of Y_{k+1}, $k \in \{1, \ldots, m-1\}$, is given by

$$G_{s_k}(y_{k+1} \mid k) = 1 - \left[1 + (m-k)\frac{y_{k+1}}{s_k}\right]^{-k}. \quad (36)$$

Predictive probability density function of Z_m is given by

$$g_{s_k}(z_m \mid k) = k(m-k)\sum_{j=0}^{m-k-1}\binom{m-k-1}{j}(-1)^j\left[1 + (1+j)\frac{z_m}{s_k}\right]^{-(k+1)} \frac{1}{s_k},$$

$$z_m \geq 0,\ 1 \leq k \leq m-1; \quad (37)$$

Predictive probability distribution function of Z_m is given by

$$G_{s_k}(z_m \mid k) = \sum_{j=0}^{m-k}\binom{m-k}{j}(-1)^j\left[1 + j\frac{z_m}{s_k}\right]^{-k},\ 1 \leq k \leq m-1. \quad (38)$$

3. Stochastic inventory control problem

Most of the inventory management literature assumes that demand distributions are specified explicitly. However, in many practical situations, the true demand distributions are not known, and the only information available may be a time-series of historic demand data. When the demand distribution is unknown, one may either use a parametric approach (where it is assumed that the demand distribution belongs to a parametric family of distributions) or a non-parametric approach (where no assumption regarding the parametric form of the unknown demand distribution is made).

Under the parametric approach, one may choose to estimate the unknown parameters or choose a prior distribution for the unknown parameters and apply the Bayesian approach to incorporating the demand data available. Scarf (1959) and Karlin (1960) consider a Bayesian framework for the unknown demand distribution. Specifically, assuming that the demand distribution belongs to the family of exponential distributions, the demand process is characterized by the prior distribution on the unknown parameter. Further extension of this approach is presented in (Azoury, 1985). Application of the Bayesian approach to the censored demand case is given in (Ding et al., 2002; Lariviere & Porteus, 1999). Parameter estimation is first considered in (Conrad, 1976) and recent developments are reported in (Agrawal & Smith, 1996; Nahmias, 1994). Liyanage & Shanthikumar (2005) propose the concept of operational statistics and apply it to a single period newsvendor inventory control problem.

This section deals with inventory items that are in stock during a single time period. At the end of the period, leftover units, if any, are disposed of, as in fashion items. Two models are considered. The difference between the two models is whether or not a setup cost is incurred for placing an order. The symbols used in the development of the models include:

c = setup cost per order,
c_1= holding cost per held unit during the period,
c_2= penalty cost per shortage unit during the period,
$g_\theta(y_{k+1}|k)$ = conditional probability density function of customer demand, Y_{k+1}, during the (k+1)th period,
θ = parameter (in general, vector),
u = order quantity,
q = inventory on hand before an order is placed.

3.1. No-setup model (Newsvendor model)

This model is known in the literature as the *newsvendor* model (the original classical name is the *newsboy* model). It deals with stocking and selling newspapers and periodicals. The assumptions of the model are:

1. Demand occurs instantaneously at the start of the period immediately after the order is received.
2. No setup cost is incurred.

The model determines the optimal value of u that minimizes the sum of the expected holding and shortage costs. Given optimal u (= u^*), the inventory policy calls for ordering $u^* - q$ if $q < u^*$; otherwise, no order is placed.

If $Y_{k+1} \le u$, the quantity $u - Y_{k+1}$ is held during the (k+1)th period. Otherwise, a shortage amount $Y_{k+1} - u$ will result if $Y_{k+1} > u$. Thus, the cost per the (k+1)th period is

$$C(u)=\begin{cases} c_1\dfrac{u-Y_{k+1}}{\theta} & \text{if } Y_{k+1}\le u, \\ c_2\dfrac{Y_{k+1}-u}{\theta} & \text{if } Y_{k+1}>u. \end{cases} \tag{39}$$

The expected cost for the (k+1)th period, $E_\theta\{C(u)\}$, is expressed as

$$E_\theta\{C(u)\}=\frac{1}{\theta}\left(c_1\int_0^u (u-y_{k+1})g_\theta(y_{k+1}\,|\,k)dy_{k+1}+c_2\int_u^\infty (y_{k+1}-u)g_\theta(y_{k+1}\,|\,k)dy_{k+1}\right). \tag{40}$$

The function $E_\theta\{C(u)\}$ can be shown to be convex in u, thus having a unique minimum. Taking the first derivative of $E_\theta\{C(u)\}$ with respect to u and equating it to zero, we get

$$\frac{1}{\theta}\left(c_1\int_0^u g_\theta(y_{k+1}\,|\,k)dy_{k+1}-c_2\int_u^\infty g_\theta(y_{k+1}\,|\,k)dy_{k+1}\right)=0 \tag{41}$$

or

$$c_1P_\theta\{Y_{k+1}\le u\}-c_2(1-P_\theta\{Y_{k+1}\le u\})=0 \tag{42}$$

or

$$P_\theta\{Y_{k+1}\le u\}=\frac{c_2}{c_1+c_2}. \tag{43}$$

It follows from (31), (32), (40), and (43) that

$$u^*=\frac{\theta}{m-k}\ln\left(1+\frac{c_2}{c_1}\right) \tag{44}$$

and

$$E_\theta\{C(u^*)\}=\frac{1}{\theta}\left(c_2E_\theta\{Y_{k+1}\}-(c_1+c_2)\int_0^{u^*} y_{k+1}g_\theta(y_{k+1}\,|\,k)dy_{k+1}\right)$$

$$=\frac{c_1}{m-k}\ln\left(1+\frac{c_2}{c_1}\right). \tag{45}$$

Parametric uncertainty. Consider the case when the parameter θ is unknown. To find the best invariant decision rule u^{BI}, we use the invariant embedding technique (Nechval et al., 1999; 2004; 2008; 2010a; 2010b; 2010c; 2010d; 2010e; 2011a; 2011b) to transform (39) to the form, which is depended only on the pivotal quantities V, V_1, and the ancillary factor η. In statistics, a pivotal quantity or pivot is a function of observations and unobservable parameters whose probability distribution does not depend on unknown parameters. Note that a pivotal quantity need not be a statistic—the function and its value can depend on parameters of the model, but its distribution must not. If it is a statistic, then it is known as an ancillary statistic.

Transformation of C(u) based on the pivotal quantities V, V_1 is given by

$$C^{(1)}(\eta) = \begin{cases} c_1(\eta V - V_1) & \text{if} \quad V_1 \le \eta V, \\ c_2(V_1 - \eta V) & \text{if} \quad V_1 > \eta V, \end{cases} \tag{46}$$

where

$$\eta = \frac{u}{S_k}, \tag{47}$$

$$V_1 = \frac{Y_{k+1}}{\theta} \sim g(v_1 \mid k) = (m-k)\exp[-(m-k)v_1], \quad v_1 \ge 0. \tag{48}$$

Then $E\{C^{(1)}(\eta)\}$ is expressed as

$$E\{C^{(1)}(\eta)\} = \int_0^\infty \left(c_1 \int_0^{\eta v} (\eta v - v_1) g(v_1 \mid k) dv_1 + c_2 \int_{\eta v}^\infty (v_1 - \eta v) g(v_1 \mid k) dv_1 \right) f(v) dv. \tag{49}$$

The function $E\{C^{(1)}(\eta)\}$ can be shown to be convex in η, thus having a unique minimum. Taking the first derivative of $E\{C^{(1)}(\eta)\}$ with respect to η and equating it to zero, we get

$$\int_0^\infty v \left(c_1 \int_0^{\eta v} g(v_1 \mid k) dv_1 - c_2 \int_{\eta v}^\infty g(v_1 \mid k) dv_1 \right) f(v) dv = 0 \tag{50}$$

or

$$\frac{\int_0^\infty v P(V_1 \le \eta v) f(v) dv}{\int_0^\infty v f(v) dv} = \frac{c_2}{c_1 + c_2}. \tag{51}$$

It follows from (47), (49), and (51) that the optimum value of η is given by

$$\eta^* = \frac{1}{m-k}\left[\left(1 + \frac{c_2}{c_1}\right)^{1/(k+1)} - 1\right], \tag{52}$$

the best invariant decision rule is

$$u^{BI} = \eta^* S_k = \frac{S_k}{m-k}\left[\left(1 + \frac{c_2}{c_1}\right)^{1/(k+1)} - 1\right], \tag{53}$$

and the expected cost, if we use u^{BI}, is given by

$$E_\theta\{C(u^{BI})\} = \frac{c_1(k+1)}{m-k}\left[\left(1+\frac{c_2}{c_1}\right)^{1/(k+1)} - 1\right] = E\{C^{(1)}(\eta^*)\}. \tag{54}$$

It will be noted that, on the other hand, the invariant embedding technique (Nechval et al., 1999; 2004; 2008; 2010a; 2010b; 2010c; 2010d; 2010e; 2011a; 2011b) allows one to transform equation (40) as follows:

$$E_\theta\{C(u)\} = \frac{1}{\theta}\left(c_1\int_0^u (u-y_{k+1})g_\theta(y_{k+1}\,|\,k)dy_{k+1} + c_2\int_u^\infty (y_{k+1}-u)g_\theta(y_{k+1}\,|\,k)dy_{k+1}\right)$$

$$= \frac{1}{s_k}\left(c_1\int_0^u (u-y_{k+1})v^2(m-k)\exp\left(-\frac{v(m-k)y_{k+1}}{s_k}\right)\frac{1}{s_k}dy_{k+1}\right.$$

$$\left. + c_2\int_u^\infty (y_{k+1}-u)v^2(m-k)\exp\left(-\frac{v(m-k)y_{k+1}}{s_k}\right)\frac{1}{s_k}dy_{k+1}\right). \tag{55}$$

Then it follows from (55) that

$$E\{E_\theta\{(C(u)\}\} = \int_0^\infty E_\theta\{C(u)\}f(v)dv = E_{s_k}\{C^{(1)}(u)\}, \tag{56}$$

where

$$E_{s_k}\{C^{(1)}(u)\} = \frac{k}{s_k}\left(c_1\int_0^u (u-y_{k+1})g^\bullet_{s_k}(y_{k+1}\,|\,k)dy_{k+1} + c_2\int_u^\infty (y_{k+1}-u)g^\bullet_{s_k}(y_{k+1}\,|\,k)dy_{k+1}\right) \tag{57}$$

represents the expected prediction cost for the (k+1)th period. It follows from (57) that the cost per the (k+1)th period is reduced to

$$C^{(2)}(u) = \begin{cases} c_1\dfrac{u-Y_{k+1}}{s_k/k} & \text{if } Y_{k+1}\le u, \\ c_2\dfrac{Y_{k+1}-u}{s_k/k} & \text{if } Y_{k+1}> u, \end{cases} \tag{58}$$

and the predictive probability density function of Y_{k+1} (compatible with (40)) is given by

$$g^\bullet_{s_k}(y_{k+1}\,|\,k) = (k+1)(m-k)\left[1+(m-k)\frac{y_{k+1}}{s_k}\right]^{-(k+2)}\frac{1}{s_k},\quad y_{k+1}\ge 0. \tag{59}$$

Minimizing the expected prediction cost for the (k+1)th period,

$$E_{s_k}\{C^{(2)}(u)\} = \frac{k}{s_k}\left(c_1\int_0^u (u - y_{k+1}) g^{\bullet}_{s_k}(y_{k+1} \mid k) dy_{k+1} + c_2 \int_u^{\infty} (y_{k+1} - u) g^{\bullet}_{s_k}(y_{k+1} \mid k) dy_{k+1} \right), \quad (60)$$

with respect to u, we obtain u^{BI} immediately, and

$$E_{s_k}\{C^{(2)}(u^{BI})\} = \frac{c_1(k+1)}{m-k}\left[\left(1+\frac{c_2}{c_1}\right)^{1/(k+1)} - 1\right]. \quad (61)$$

It should be remarked that the cost per the (k+1)th period, $C^{(2)}(u)$, can also be transformed to

$$C^{(3)}(\eta) = \begin{cases} c_1 k\left(\frac{u}{s_k} - \frac{Y_{k+1}}{s_k}\right) & \text{if } \frac{Y_{k+1}}{s_k} \le \frac{u}{s_k} \\ c_2 k\left(\frac{Y_{k+1}}{s_k} - \frac{u}{s_k}\right) & \text{if } \frac{Y_{k+1}}{s_k} > \frac{u}{s_k} \end{cases} = \begin{cases} c_1 k(\eta - W) & \text{if } W \le \eta \\ c_2 k(W - \eta) & \text{if } W > \eta, \end{cases} \quad (62)$$

where the probability density function of the ancillary statistic W (compatible with (40)) is given by

$$g^{\circ}_{s_k}(w \mid k) = (k+1)(m-k)\left[1 + (m-k)w\right]^{-(k+2)}, \quad w \ge 0. \quad (63)$$

Then the best invariant decision rule $u^{BI} = \eta^* S_k$, where η^* minimizes

$$E\{C^{(3)}(\eta)\} = k\left(c_1 \int_0^{\eta} (\eta - w) g^{\circ}(w \mid k) dw + c_2 \int_{\eta}^{\infty} (w - \eta) g^{\circ}(w \mid k) dw \right). \quad (64)$$

Comparison of statistical decision rules. For comparison, consider the maximum likelihood decision rule that may be obtained from (44),

$$u^{ML} = \frac{\hat{\theta}}{m-k}\ln\left(1+\frac{c_2}{c_1}\right) = \eta_j^{ML} S_k, \quad (65)$$

where $\hat{\theta} = S_k / k$ is the maximum likelihood estimator of θ,

$$\eta^{ML} = \frac{1}{m-k}\ln\left(1+\frac{c_2}{c_1}\right)^{1/k}. \quad (66)$$

Since u^{BI} and u^{ML} belong to the same class,

$$C = \{u : u = \eta S_k\}, \quad (67)$$

it follows from the above that u^{ML} is inadmissible in relation to u^{BI}. If, say, k=1 and $c_2 / c_1 = 100$, we have that

$$\text{Rel.eff.}\{u^{ML}, u^{BI}, q\} = E_\theta\{C(u^{BI})\} / E_\theta\{C(u^{ML})\} = 0.838. \tag{68}$$

Thus, in this case, the use of u^{BI} leads to a reduction in the expected cost of about 16.2 % as compared with u^{ML}. The absolute expected cost will be proportional to θ and may be considerable.

3.2. Setup model (s-S policy)

The present model differs from the one in Section 3.1 in that a setup cost c is incurred. Using the same notation, the total expected cost per the (k+1)th period is

$$E_\theta\{\bar{C}(u)\} = c + E_\theta\{C(u)\}$$

$$= c + \frac{1}{\theta}\left(c_1 \int_0^u (u - y_{k+1}) g_\theta(y_{k+1} \mid k) dy_{k+1} + c_2 \int_u^\infty (y_{k+1} - u) g_\theta(y_{k+1} \mid k) dy_{k+1} \right). \tag{69}$$

As shown in Section 3.1, the optimum value u^* must satisfy (43). Because c is constant, the minimum value of $E_\theta\{\bar{C}(u)\}$ must also occur at u^*. In Fig.1,

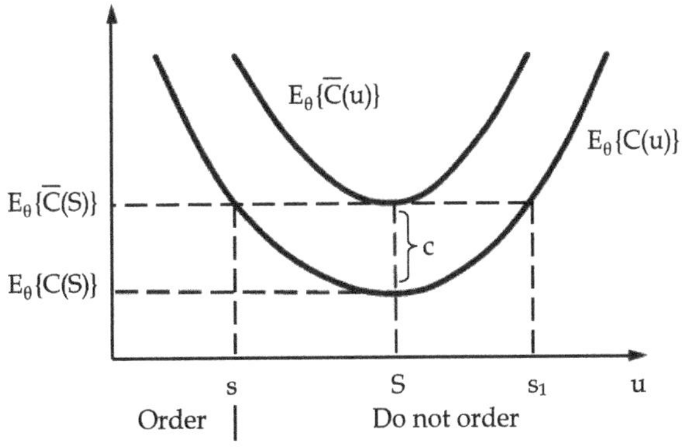

$S = u^*$, and the value of s (< S) is determined from the equation

Figure 1. (s–S) optimal ordering policy in a single-period model with setup cost

$$E_\theta\{C(s)\} = E_\theta\{\bar{C}(S)\} = c + E_\theta\{C(S)\}, \quad s < S. \tag{70}$$

The equation yields another value s_1 (> S), which is discarded.

Assume that q is the amount on hand before an order is placed. How much should be ordered? This question is answered under three conditions: 1) $q < s$; 2) $s \le q \le S$; 3) $q > S$.

Case 1 (q < s). Because q is already on hand, its equivalent cost is given by $E_\theta\{C(q)\}$. If any additional amount u – q (u > q) is ordered, the corresponding cost given u is $E_\theta\{\bar{C}(u)\}$, which includes the setup cost c. From Fig. 1, we have

$$\min_{u>q} E_\theta\{\bar{C}(u)\} = E_\theta\{\bar{C}(S)\} < E_\theta\{C(q)\}. \tag{71}$$

Thus, the optimal inventory policy in this case is to order S – q units.

Case 2 (s ≤ q ≤ S). From Fig. 1, we have

$$E_\theta\{C(q)\} \le \min_{u>q} E_\theta\{\bar{C}(u)\} = E_\theta\{\bar{C}(S)\}. \tag{72}$$

Thus, it is not advantageous to order in this case and $u^* = q$.

Case 3 (q > S). From Fig. 1, we have for u > q,

$$E_\theta\{C(q)\} < E_\theta\{\bar{C}(u)\}. \tag{73}$$

This condition indicates that, as in case (2), is not advantageous to place an order – that is, $u^* = q$.

The optimal inventory policy, frequently referred to as the *s – S policy*, is summarized as

$$\begin{aligned} &\text{If } x < S, \text{ order } S - x, \\ &\text{If } x \ge s, \text{ do not order.} \end{aligned} \tag{74}$$

The optimality of the s – S policy is guaranteed because the associated cost function is convex.

Parametric uncertainty. In the case when the parameter θ is unknown, the total expected prediction cost for the (*k*+1)th period,

$$E_{s_k}\{\bar{C}^{(1)}(u)\} = c + E_{s_k}\{C^{(1)}(u)\}$$

$$= c + \frac{k}{s_k}\left(c_1 \int_0^u (u - y_{k+1}) g^{\bullet}_{s_k}(y_{k+1} \mid k) dy_{k+1} + c_2 \int_u^\infty (y_{k+1} - u) g^{\bullet}_{s_k}(y_{k+1} \mid k) dy_{k+1} \right), \tag{75}$$

is considered in the same manner as above.

4. Airline revenue management problem

The process of revenue management has become extremely important within the airline industry. It consists of setting fares, setting overbooking limits, and controlling seat inventory to increase revenues. It has allowed the airlines to survive deregulation by allowing them to respond to competitors' deep discount fares on a rational basis.

An airline, typically, offers tickets for many origin–destination itineraries in various fare classes. These fare classes not only include business and economy class, which are settled in separate parts of the plane, but also include fare classes for which the difference in fares is explained by different conditions regarding for example cancellation options or overnight stay arrangements. Therefore the seats on a flight are products, which can be offered to different customer segments for different prices. Since the tickets for a flight have to be sold before the plane takes off, the product is perishable and cannot be stored for future use. The same is true for most other service industries, such as hotels, hospitals and schools.

4.1. Airline seat inventory control

At the heart of airline revenue management lies the airline seat inventory control problem. It is common practice for airlines to sell a pool of identical seats at different prices according to different booking classes to improve revenues in a very competitive market. In other words, airlines sell the same seat at different prices according to different types of travelers (first class, business and economy) and other conditions. The question then arises whether to offer seats at a relatively low price at a given time with a given number of seats remaining or to wait for the possible arrival of a higher paying customer. Assigning seats in the same compartment to different fare classes of passengers in order to improve revenues is a major problem of airline seat inventory control. This problem has been considered in numerous papers. For details, the reader is referred to a review of yield management, as well as perishable asset revenue management, by Weatherford et al. (1993), and a review of relevant mathematical models by Belobaba (1987).

The problem of finding an optimal airline seat inventory control policy for multi-leg flight with multiple fare classes, which allows one to maximize the expected profit of this flight, is one of the most difficult problems of air transport logistics. On the one hand, one must have reasonable assurance that the requirements of customers for reservations will be met under most circumstances. On the other hand, one is confronted with the limitation of the capacity of the cabin, as well as with a host of other less important constraints. The problem is normally solved by the application of judgment based on past experience. The question arises whether or not it is possible to construct a simple mathematical theory of the above problem, which will allow one better to use the available data based upon airline statistics. Two models (dynamic model and static one) of airline data may be considered. In the dynamic model, the problem is formulated as a sequential decision process. In this case, an optimal dynamic reservation policy is used at each stage prior to departure time for multi-leg flights with several classes of passenger service. The essence of determining the optimal dynamic reservation policy is maximization of the expected gain of the flight, which is carried out at each stage prior to departure time using the available data. The term (dynamic reservation policy) is used in this paper to mean a decision rule, based on the available data, for determining whether to accept a given reservation request made at a particular time for some future date. An optimal static reservation policy is based on the static model. The models proposed here contain a simple and natural treatment of the airline reservation process and may be appropriate in practice.

4.2. Expected marginal seat revenue model (EMSR)

Different approaches were developed for solving the airline seat inventory control problem. The most important and widely used model – EMSR – was originally proposed by Littlewood (1972). To explain the basic ideas of the Littlewood model, we will consider the seat allocation problem on a nonstop flight with two airfares. We denote by U the number of seats in the aircraft, by u_L the number of seats reserved for passengers with a lower fare and by p the probability that a passenger who would pay the higher fare cannot find a seat because it was sold to a passenger paying a lower fare. A rise in variable u_L means a rise in probability p. A rise in the number of seats for passengers with the lower fare, i.e. a rise in variable p, decreases the number of seats for passengers with the higher fare, i.e., variable U – u_L decreases. A reduction in variable U – u_L increases the probability that a passenger paying the higher fare cannot find a vacant seat on their desired flight because it has been sold to a passenger with a lower fare. Probability p is given by

$$p = \int_{U-u_L}^{\infty} f_\theta(x)dx, \tag{76}$$

where the $f_\theta(x)$ is the underlying probability density function for the total number of requests X of passengers who would pay the higher fare, θ is the parameter (in general, vector). Littlewood (1972) proposed the following way to determine the reservation level u_L to which reservations are accepted for passengers with the lower fare. We denote by c_1 the revenue from passengers with the lower fare and by c_2 the revenue from passengers with the higher fare. Let u_L seats be reserved for low fare passengers. The revenue per lower fare seat is c_1. Expected revenue from a potential high fare passenger is c_2p. Passengers with lower fares should be accepted until

$$c_1 \geq c_2 p, \tag{77}$$

i.e.,

$$p \leq \frac{c_1}{c_2}. \tag{78}$$

Reservation level u_L is determined by solving the following equality:

$$\Pr\{X > h\} = \gamma, \tag{79}$$

where

$$h = U - u_L, \tag{80}$$

$$\gamma = \frac{c_1}{c_2}. \tag{81}$$

Richter (1982) gave a marginal analysis, which proved that (77) gives an optimal allocation (assuming certain continuity conditions). Optimal policies for more than two classes have been presented independently by Curry (1990), Wollmer (1992), Brumelle & McGill (1993), and Nechval et al. (2006).

Parametric uncertainty. In order to solve (79) under parametric uncertainty, it can be used, for example, the following results.

Theorem 4. Let $X_1 \le \ldots \le X_r$ be the first r ordered past observations from a previous sample of size n from the two-parameter Weibull distribution

$$f(x\,|\,\beta,\delta)=\frac{\delta}{\beta}\left(\frac{x}{\beta}\right)^{\delta-1}\exp\left[-\left(\frac{x}{\beta}\right)^{\delta}\right]\quad (x>0), \tag{82}$$

where both distribution parameters (β – scale, δ - shape) are positive. Then a lower one-sided conditional γ prediction limit h on the lth order statistic X_l^f in a set of m future ordered observations also from the distribution (82) is given by

$$h=z_h^{1/\hat{\delta}}\hat{\beta}, \tag{83}$$

where z_h satisfies the equation

$$\Pr\{X_l^f>h\,|\,z^{(r)}\}=\Pr\{X_l^f>z_h^{1/\hat{\delta}}\hat{\beta}\,|\,z^{(r)}\}$$

$$=\frac{\left[\int_0^{\infty} v_2^{r-2}\prod_{i=1}^{r} z_i^{v_2}\sum_{k=0}^{l-1}\binom{m}{k}\sum_{j=0}^{k}\binom{k}{j}(-1)^j\left((m-k+j)z_h^{v_2}+\sum_{i=1}^{r}z_i^{v_2}+(n-r)z_r^{v_2}\right)^{-r}dv_2\right]}{\int_0^{\infty} v_2^{r-2}\prod_{i=1}^{r} z_i^{v_2}\left(\sum_{i=1}^{r}z_i^{v_2}+(n-r)z_r^{v_2}\right)^{-r}dv_2}=\gamma, \tag{84}$$

$$z^{(r)}=(z_1,z_2,\ldots,z_r), \tag{85}$$

$$Z_i=\left(\frac{X_i}{\hat{\beta}}\right)^{\hat{\delta}},\quad i=1,\ldots,r, \tag{86}$$

are ancillary statistics, any r−2 of which form a functionally independent set (for notational convenience we include all of $z_1, \ldots, z_r$ in (85); z_{r-1} and z_r can be expressed as function of $z_1, \ldots, z_{r-2}$ only), $\hat{\beta}$ and $\hat{\delta}$ are the maximum likelihood estimators of β and δ based on the first r ordered past observations ($X_1 \le \ldots \le X_r$) from a sample of size n from the two-parameter Weibull distribution, which can be found from solution of

$$\hat{\beta}=\left(\left[\sum_{i=1}^{r}x_i^{\hat{\delta}}+(n-r)x_r^{\hat{\delta}}\right]\Big/ r\right)^{1/\hat{\delta}}, \tag{87}$$

and

$$\widehat{\delta}=\left[\left(\sum_{i=1}^{r}x_i^{\widehat{\delta}}\ln x_i+(n-r)x_r^{\widehat{\delta}}\ln x_r\right)\left(\sum_{i=1}^{r}x_i^{\widehat{\delta}}+(n-r)x_r^{\widehat{\delta}}\right)^{-1}-\frac{1}{r}\sum_{i=1}^{r}\ln x_i\right]^{-1}, \tag{88}$$

(Observe that an upper one-sided conditional α prediction limit h on the lth order statistic X_l^f from a set of m future ordered observations $X_1^f\le\ldots\le X_m^f$ may be obtained from a lower one-sided conditional γ prediction limit by replacing γ by $1-\gamma$.)

Proof. The joint density of $X_1 \le \ldots \le X_r$ is given by

$$f(x_1,\ldots,x_r\,|\,\beta,\delta)=\frac{n!}{(n-r)!}\prod_{i=1}^{r}\frac{\delta}{\beta}\left(\frac{x_i}{\beta}\right)^{\delta-1}\exp\left(-\left(\frac{x_i}{\beta}\right)^{\delta}\right)\exp\left(-(n-r)\left(\frac{x_r}{\beta}\right)^{\delta}\right). \tag{89}$$

Let $\widehat{\beta}$, $\widehat{\delta}$ be the maximum likelihood estimates of β, δ, respectively, based on $X_1 \le \ldots \le X_r$ from a complete sample of size n, and let

$$V_1=\left(\frac{\widehat{\beta}}{\beta}\right)^{\delta},\quad V_2=\frac{\delta}{\widehat{\delta}},\quad \text{and}\quad Z_i=\left(\frac{X_i}{\widehat{\beta}}\right)^{\widehat{\delta}},\quad i=1,\ldots,r, \tag{90}$$

Parameters β and δ in (90) are scale and shape parameters, respectively, and it is well known that if $\widehat{\beta}$ and $\widehat{\delta}$ are estimates of β and δ, possessing certain invariance properties, then V_1 and V_2 are the pivotal quantities whose distributions depend only on n. Most, if not all, proposed estimates of β and δ possess the necessary properties; these include the maximum likelihood estimates and various linear estimates.

Using (90) and the invariant embedding technique (Nechval et al., 1999; 2004; 2008; 2010a; 2010b; 2010c; 2010d; 2010e; 2011a; 2011b), we then find in a straightforward manner, that the joint density of V_1, V_2, conditional on fixed $\mathbf{z}=(z_1,z_2,\ldots,z_r)$, is

$$f(v_1,v_2\,|\,z^{(r)})$$

$$=\vartheta^{\bullet}(z^{(r)})v_2^{r-2}\prod_{i=1}^{r}z_i^{v_2}v_1^{r-1}\exp\left(-v_1\left[\sum_{i=1}^{r}z_i^{v_2}+(n-r)z_r^{v_2}\right]\right),\quad v_1\in(0,\infty),\ v_2\in(0,\infty), \tag{91}$$

where

$$\vartheta^{\bullet}(z^{(r)})=\left[\int_0^{\infty}\Gamma(r)v_2^{r-2}\prod_{i=1}^{r}z_i^{v_2}\left(\sum_{i=1}^{r}z_i^{v_2}+(n-r)z_r^{v_2}\right)^{-r}dv_2\right]^{-1}, \tag{92}$$

is the normalizing constant. Writing

$$\Pr\{X_l^f>h\,|\,\beta,\delta\}=\sum_{k=0}^{l-1}\binom{m}{k}[F(h\,|\,\beta,\delta)]^k[1-F(h\,|\,\beta,\delta)]^{m-k}$$

$$=\sum_{k=0}^{l-1}\binom{m}{k}\left[1-\exp\left(-\left(\frac{h}{\beta}\right)^{\delta}\right)\right]^{k}\left[\exp\left(-\left(\frac{h}{\beta}\right)^{\delta}\right)\right]^{m-k}$$

$$=\sum_{k=0}^{l-1}\binom{m}{k}\sum_{j=0}^{k}\binom{k}{j}(-1)^{j}\exp[-v_1(m-k+j)z_h^{v_2}]$$

$$=\Pr\{X_l^f>h\,|\,v_1,v_2\}, \tag{93}$$

where

$$z_h=\left(\frac{h}{\hat{\beta}}\right)^{\hat{\delta}}, \tag{94}$$

we have from (91) and (93) that

$$\Pr\{X_l^f>h\,|\,z^{(r)}\}=\int_0^{\infty}\int_0^{\infty}\Pr\{X_l^f>h\,|\,v_1,v_2\}f(v_1,v_2\,|\,z^{(r)})dv_1dv_2. \tag{95}$$

Now v_1 can be integrated out of (95) in a straightforward way to give

$$\Pr\{X_l^f>h\,|\,z^{(r)}\}$$

$$=\frac{\left[\int_0^{\infty}v_2^{r-2}\prod_{i=1}^{r}z_i^{v_2}\sum_{k=0}^{l-1}\binom{m}{k}\sum_{j=0}^{k}\binom{k}{j}(-1)^{j}\left((m-k+j)\left(\frac{h}{\hat{\beta}}\right)^{\hat{\delta}v_2}+\sum_{i=1}^{r}z_i^{v_2}+(n-r)z_r^{v_2}\right)^{-r}dv_2\right]}{\int_0^{\infty}v_2^{r-2}\prod_{i=1}^{r}z_i^{v_2}\left(\sum_{i=1}^{r}z_i^{v_2}+(n-r)z_r^{v_2}\right)^{-r}dv_2}. \tag{96}$$

This completes the proof. □

Remark 1. If l=m=1, then the result of Theorem 4 can be used to construct the static policy of airline seat inventory control.

Theorem 5. Let $X_1 \le \ldots \le X_k$ be the first k ordered early observations from a sample of size m from the two-parameter Weibull distribution (82). Then a lower one-sided conditional γ prediction limit h on the lth order statistic X_l $(l>k)$ in the same sample is given by

$$h=w_h^{1/\hat{\delta}}X_k, \tag{97}$$

where w_h satisfies the equation

$$\Pr\{X_l>h\,|\,z^{(k)}\}=\Pr\left\{(X_l/X_k)^{\hat{\delta}}>(h/X_k)^{\hat{\delta}}\,\middle|\,z^{(k)}\right\}=\Pr\{W>w_h\,|\,z^{(k)}\}$$

$$= \left[\int_0^\infty v_2^{k-2} \prod_{i=1}^k z_i^{v_2} \sum_{j=0}^{l-k-1} \binom{l-k-1}{j} \frac{(-1)^{l-k-1-j}}{m-k-j} \left((m-k-j)(w_h z_k)^{v_2} + j z_k^{v_2} + \sum_{i=1}^k z_i^{v_2} \right)^{-k} dv_2 \right]$$

$$\times \left[\int_0^\infty v_2^{k-2} \prod_{i=1}^k z_i^{v_2} \sum_{j=0}^{l-k-1} \binom{l-k-1}{j} \frac{(-1)^{l-k-1-j}}{m-k-j} \left((m-k) z_k^{v_2} + \sum_{i=1}^k z_i^{v_2} \right)^{-k} dv_2 \right]^{-1} = \gamma, \tag{98}$$

$$\mathbf{z}^{(k)} = (z_1, \ldots, z_k), \tag{99}$$

$$Z_i = \left(\frac{Y_i}{\hat{\beta}} \right)^{\hat{\delta}}, \quad i = 1, \ldots, k, \tag{100}$$

$$w_h = \left(\frac{h}{Y_k} \right)^{\hat{\delta}}, \tag{101}$$

where $\hat{\beta}$ and $\hat{\delta}$ are the maximum likelihood estimates of β and δ based on the first k ordered past observations $X_1 \le \ldots \le X_k$ from a sample of size m from the two-parameter Weibull distribution (82), which can be found from solution of

$$\hat{\beta} = \left(\left[\sum_{i=1}^k x_i^{\hat{\delta}} + (m-k) x_k^{\hat{\delta}} \right] \Big/ k \right)^{1/\hat{\delta}}, \tag{102}$$

and

$$\hat{\delta} = \left[\left(\sum_{i=1}^k x_i^{\hat{\delta}} \ln x_i + (m-k) x_k^{\hat{\delta}} \ln x_k \right) \left(\sum_{i=1}^k x_i^{\hat{\delta}} + (m-k) x_k^{\hat{\delta}} \right)^{-1} - \frac{1}{k} \sum_{i=1}^k \ln x_i \right]^{-1}, \tag{103}$$

(Observe that an upper one-sided conditional α prediction limit h on the lth order statistic X_l based on the first k ordered early-failure observations $X_1 \le \ldots \le X_k$, where $l > k$, from the same sample may be obtained from a lower one-sided conditional γ prediction limit by replacing γ by $1-\gamma$)

Proof. The joint density of $X_1 \le \ldots \le X_k$ and X_l is given by

$$f(x_1, \ldots, x_k, x_l \mid \beta, \delta) = \frac{m!}{(l-k-1)!(m-l)!} \prod_{i=1}^k \frac{\delta}{\beta} \left(\frac{x_i}{\beta} \right)^{\delta-1} \exp\left(-\left(\frac{x_i}{\beta} \right)^{\delta} \right)$$

$$\times \left[\exp\left(-\left(\frac{x_k}{\beta} \right)^{\delta} \right) - \exp\left(-\left(\frac{x_l}{\beta} \right)^{\delta} \right) \right]^{l-k-1} \frac{\delta}{\beta} \left(\frac{x_l}{\beta} \right)^{\delta-1} \exp\left(-\left(\frac{x_l}{\beta} \right)^{\delta} \right)$$

$$\times \exp\left(-(m-1)\left(\frac{x_1}{\beta}\right)^{\delta}\right). \tag{104}$$

Let $\widehat{\beta}$, $\widehat{\delta}$ be the maximum likelihood estimates of β, δ, respectively, based on $X_1 \le \ldots \le X_k$ from a complete sample of size m, and let

$$V_1 = \left(\frac{\widehat{\beta}}{\beta}\right)^{\delta}, \quad V_2 = \frac{\delta}{\widehat{\delta}}, \quad W = \left(\frac{Y_l}{Y_k}\right)^{\widehat{\delta}}, \tag{105}$$

and $Z_i = (Y_i / \widehat{\beta})^{\widehat{\delta}}$, i=1(1)k. Using the invariant embedding technique (Nechval et al., 1999; 2004; 2008; 2010a; 2010b; 2010c; 2010d; 2010e; 2011a; 2011b), we then find in a straightforward manner, that the joint density of V_1, V_2, W, conditional on fixed $z^{(k)} = (z_1, \ldots, z_k)$, is

$$f(v_1, v_2, w \mid z^{(k)}) = \vartheta(z^{(k)}) v_2^{k-1} \prod_{i=1}^{k} z_i^{v_2} (wz_k)^{v_2} v_1^{k} \sum_{j=0}^{l-k-1} \binom{l-k-1}{j} (-1)^{l-k-1-j}$$

$$\times \exp\left[-v_1\left((m-k-j)(wz_k)^{v_2} + jz_k^{v_2} + \sum_{i=1}^{k} z_i^{v_2}\right)\right], \quad v_1 \in (0,\infty),\ v_2 \in (0,\infty),\ w \in (1,\infty), \tag{106}$$

where

$$\vartheta(z^{(k)}) = \left[\int_0^{\infty} v_2^{k-2} \prod_{i=1}^{k} z_i^{v_2} \sum_{j=0}^{l-k-1} \binom{l-k-1}{j} \frac{(-1)^{l-k-1-j}\Gamma(k)}{m-k-j} \left((m-k)z_k^{v_2} + \sum_{i=1}^{k} z_i^{v_2}\right)^{-k} dv_2\right]^{-1} \tag{107}$$

is the normalizing constant. Using (106), we have that

$$\Pr\{Y_l > h \mid z^{(k)}\} = \Pr\left\{\left(\frac{Y_l}{Y_k}\right)^{\widehat{\delta}} > \left(\frac{h}{Y_k}\right)^{\widehat{\delta}} \middle| z^{(k)}\right\}$$

$$= \Pr\{W > w_h \mid z^{(k)}\} = \int_0^{\infty} \int_{w_h}^{\infty} \int_0^{\infty} f(v_1, v_2, w \mid z^{(k)}) dv_1 dw dv_2$$

$$= \left[\int_0^{\infty} v_2^{k-2} \prod_{i=1}^{k} z_i^{v_2} \sum_{j=0}^{l-k-1} \binom{l-k-1}{j} \frac{(-1)^{l-k-1-j}}{m-k-j} \left((m-k-j)(w_h z_k)^{v_2} + jz_k^{v_2} + \sum_{i=1}^{k} z_i^{v_2}\right)^{-k} dv_2\right]$$

$$\times \left[\int_0^{\infty} v_2^{k-2} \prod_{i=1}^{k} z_i^{v_2} \sum_{j=0}^{l-k-1} \binom{l-k-1}{j} \frac{(-1)^{l-k-1-j}}{m-k-j} \left((m-k)z_k^{v_2} + \sum_{i=1}^{k} z_i^{v_2}\right)^{-k} dv_2\right]^{-1}, \tag{108}$$

and the proof is complete. □

Remark 2. If l=m, then the result of Theorem 5 can be used to construct the dynamic policy of airline seat inventory control.

5. Conclusions and directions for future research

In this paper, we develop a new frequentist approach to improve predictive statistical decisions for revenue optimization problems under parametric uncertainty of the underlying distributions for the customer demand. Frequentist probability interpretations of the methods considered are clear. Bayesian methods are not considered here. We note, however, that, although subjective Bayesian prediction has a clear personal probability interpretation, it is not generally clear how this should be applied to non-personal prediction or decisions. Objective Bayesian methods, on the other hand, do not have clear probability interpretations in finite samples.

For constructing the improved statistical decisions, a new technique of invariant embedding of sample statistics in a performance index is proposed. This technique represents a simple and computationally attractive statistical method based on the constructive use of the invariance principle in mathematical statistics. The method used is that of the invariant embedding of sample statistics in a performance index in order to form pivotal quantities, which make it possible to eliminate unknown parameters (i.e., parametric uncertainty) from the problem. It is especially efficient when we deal with asymmetric performance indexes and small data samples

More work is needed, however, to obtain improved or optimal decision rules for the problems of unconstrained and constrained optimization under parameter uncertainty when: (i) the observations are from general continuous exponential families of distributions, (ii) the observations are from discrete exponential families of distributions, (iii) some of the observations are from continuous exponential families of distributions and some from discrete exponential families of distributions, (iv) the observations are from multiparametric or multidimensional distributions, (v) the observations are from truncated distributions, (vi) the observations are censored, (vii) the censored observations are from truncated distributions.

Author details

Nicholas A. Nechval and Maris Purgailis
University of Latvia, Latvia

Acknowledgement

This research was supported in part by Grant No. 06.1936, Grant No. 07.2036, Grant No. 09.1014, and Grant No. 09.1544 from the Latvian Council of Science and the National Institute of Mathematics and Informatics of Latvia.

6. References

Agrawal, N. & Smith, S.A. (1996). Estimating Negative Binomial Demand for Retail Inventory Management with Unobservable Lost Sales. *Naval Res. Logist.*, Vol. 43, pp. 839 –861

Azoury, K.S. (1985). Bayes Solution to Dynamic Inventory Models under Unknown Demand Distribution. *Management Sci.*, Vol. 31, pp. 1150 –1160

Belobaba, P.P. (1987). Survey Paper: Airline Yield Management – an Overview of Seat Inventory Control. *Transportation Science*, Vol. 21, pp. 63 –73

Brumelle, S.L. & J.I. McGill (1993). Airline Seat Allocation with Multiple Nested Fare Classes. *Operations Research*, Vol. 41, pp. 127–137

Conrad, S.A. (1976). Sales Data and the Estimation of Demand. *Oper. Res. Quart.*, Vol. 27, pp. 123 –127

Curry, R.E. (1990). Optimal Airline Seat Allocation with Fare Classes Nested by Origins and Destination. *Transportation Science*, Vol. 24, pp. 193–204

Ding, X.; Puterman, M.L. & Bisi, A. (2002). The Censored Newsvendor and the Optimal Acquisition of Information. *Oper. Res.*, Vol. 50, pp. 517–527

Karlin, S. (1960). Dynamic Inventory Policy with Varying Stochastic Demands. *Management Sci.*, Vol. 6, pp. 231 –258

Lariviere, M.A. & Porteus, E.L. (1999). Stalking Information: Bayesian Inventory Management with Unobserved Lost Sales. *Management Sci.*, Vol. 45, pp. 346 –363

Littlewood, K. (1972). Forecasting and Control of Passenger Bookings. In: *Proceedings of the Airline Group International Federation of Operational Research Societies*, Vol. 12, pp. 95–117

Liyanage, L.H. & Shanthikumar, J.G. (2005). A Practical Inventory Control Policy Using Operational Statistics. *Oper. Res. Lett.*, Vol. 33, pp. 341–348

Nahmias, S. (1994). Demand Estimation in Lost Sales Inventory Systems. *Naval Res. Logist.*, Vol. 41, 739 –757

Nechval N. A. & Purgailis, M. (2010e). Improved State Estimation of Stochastic Systems via a New Technique of Invariant Embedding. In: *Stochastic Control*, Chris Myers (ed.), pp. 167–193. Publisher: Sciyo, Croatia, India

Nechval, N. A. & Vasermanis, E. K. (2004). *Improved Decisions in Statistics*. SIA "Izglitibas soli", Riga

Nechval, N. A. & Nechval, K. N. (1999). Invariant Embedding Technique and Its Statistical Applications. In: *Conference Volume of Contributed Papers of the 52nd Session of the International Statistical Institute*, Finland, Helsinki: ISI–International Statistical Institute,Availqable from http://www.stat.fi/isi99/procee-dings/arkisto/varasto/nech0902.pdf

Nechval, N. A.; Berzins, G.; Purgailis, M. & Nechval, K.N. (2008). Improved Estimation of State of Stochastic Systems via Invariant Embedding Technique. *WSEAS Transactions on Mathematics*, Vol. 7, pp. 141 –159

Nechval, N. A.; Nechval, K. N. & Purgailis, M. (2011b). Statistical Inferences for Future Outcomes with Applications to Maintenance and Reliability. In: *Lecture Notes in*

Engineering and Computer Science: Proceedings of the World Congress on Engineering (WCE 2011), pp. 865--871. London, U.K.

Nechval, N. A.; Purgailis, M.; Berzins, G.; Cikste, K.; Krasts, J. & Nechval, K. N. (2010a). Invariant Embedding Technique and Its Applications for Improvement or Optimization of Statistical Decisions. In: *Analytical and Stochastic Modeling Techniques and Applications*, Al-Begain, K., Fiems, D., Knottenbelt, W. (eds.),. LNCS, Vol. 6148, pp. 306–320. Springer, Heidelberg

Nechval, N. A.; Purgailis, M.; Cikste, K. & Nechval, K. N. (2010d). Planning Inspections of Fatigued Aircraft Structures via Damage Tolerance Approach. In: *Lecture Notes in Engineering and Computer Science: Proceedings of the World Congress on Engineering (WCE 2010)*, pp. 2470 –2475. London, U.K.

Nechval, N. A.; Purgailis, M.; Cikste, K.; Berzins, G. & Nechval, K. N. (2010c). Optimization of Statistical Decisions via an Invariant Embedding Technique. In: *Lecture Notes in Engineering and Computer Science: Proceedings of the World Congress on Engineering (WCE 2010)*, pp. 1776 –1782. London, U.K.

Nechval, N. A.; Purgailis, M.; Cikste, K.; Berzins, G.; Rozevskis, U. & Nechval, K. N. (2010b). Prediction Model Selection and Spare Parts Ordering Policy for Efficient Support of Maintenance and Repair of Equipment. In: *Analytical and Stochastic Modeling Techniques and Applications*, Al-Begain, K., Fiems, D., Knottenbelt, W. (eds.). LNCS, Vol. 6148, pp. 321–338. Springer, Heidelberg

Nechval, N. A.; Purgailis, M.; Nechval, K. N. & Rozevskis, U. (2011a) Optimization of Prediction Intervals for Order Statistics Based on Censored Data. In: *Lecture Notes in Engineering and Computer Science: Proceedings of the World Congress on Engineering (WCE 2011)*, pp. 63 –69. London, U.K.

Nechval, N.A.; Rozite, K. & Strelchonok, V.F. (2006). Optimal Airline Multi-Leg Flight Seat Inventory Control. In: *Computing Anticipatory Systems*, Daniel M. Dubois (ed.), Melville, New York: AIP (American Institute of Physics) Proceedings, Vol. 839, pp. 591-600

Richter, H. (1982). The Differential Revenue Method to Determine Optimal Seat Allotments by Fare Type. In: *Proceedings of the XXII AGIFORS Symposium*, American Airlines, New York, pp. 339 –362

Scarf, H. (1959). Bayes Solutions of Statistical Inventory Problem. *Ann. Math. Statist*. Vol. 30, pp. 490 –508

Weatherford, L.R.; Bodily, S.E. & P.E. Pferifer (1993). Modeling the Customer Arrival Process and Comparing Decision Rules in Perishable Asset Revenue Management Situations. *Transportation Science*, Vol. 27, pp. 239 –251

Wollmer, R.D. (1992). An Airline Seat Management Model for a Single Leg Route when Lower Fare Classes Book First. *Operations Research*, Vol. 40, pp. 26 –37

Chapter 12

Stochastic Based Simulations and Measurements of Some Objective Parameters of Acoustic Quality: Subjective Evaluation of Room Acoustic Quality with Acoustics Optimization in Multimedia Classroom (Analysis with Application)

Vladimir Šimović, Siniša Fajt and Miljenko Krhen

Additional information is available at the end of the chapter

1. Introduction

This chapter offers an original scientific research about stochastic based simulations and measurements of some objective parameters of acoustic quality and subjective evaluation of room acoustic quality, with acoustics optimization in multimedia classroom, complete scientific analysis, and application. The organisation of this chapter is very simple. After a short introduction that is an overview of completed research there are two main chapter parts with some definitions and explanations on the main subjects and research done. The conclusion to this research is contained in the final chapter segment, before the cited references.

According to http://www.answers.com/topic/convolution (2011/12/31), stochastic control methodology (SCM) is applied in a variety of fields including the computing, communications and acoustics optimization in multimedia environment. Also, SCT as a branch of control theory (CT) that deals with systems which involve random variables/signals and which occurrence can only be described in probabilistic terms attempts to predict and minimize the effect of these random signals through the optimization of controller design for acoustics optimization in multimedia room. Such deviations occur when random noise and disturbance processes are present in a control system, so that the system does not follow its prescribed course but deviates from the latter by a randomly varying amount. In contrast to deterministic signals, random signals cannot be described as given functions of time such as a step, a ramp, or a sine wave. The exact function is unknown to the system designer; only some of its average properties are known. A random signal may be generated by one of nature's processes, for instance, wind or multimedia wave-induced forces and moments on a antenna or a multimedia room. Alternatively, it may be generated by human intelligence, for instance, the contour to be followed by a duplicating machine. Outstanding experimental fact about nature's random processes is that these signals are very closely "Gaussian" that is a mathematical concept which describes one or more signals, i1, i2, ..., in having the following properties: The amplitude of each signal is normally distributed and the joint distribution function of any number of signals at the same or different times taken from the set is a multivariate normal distribution. This experimental fact is not surprising in view of the fact that a random process of nature is usually the sum total of the effects of a large number of independent contributing factors. For instance, an ocean-wave height at any particular time and place is the sum of wind-generated waves at previous times over a large area. The underlying mechanism that generates a random process can usually be described in physical or mathematical terms. For instance, the underlying mechanism that generates shot-effect noise is thermionic emission. If the generating mechanism does not change with time, any measured average property of the random process is independent of the time of measurement aside from statistical fluctuations, and the random process is called stationary. If the generating mechanism does change, the random process is called nonstationary. Random noise is a type of noise comprised of transient disturbances which occur at random times; its instantaneous magnitudes are specified only by probability distribution functions which give the fraction of the total time that the magnitude lies within a specified range. Random noise represents: in mathematics a form of random stochastic process arising in control theory, or in physics noise characterized by a large number of overlapping transient disturbances occurring at random, such as thermal noise and shot noise, also known as fluctuation noise.

The room acoustics for multimedia room is complex problem which has to be take in consideration while designing the one, because it consists of several parts, like: analysing and calculating of acoustical parameters, measuring the real condition, simulating on computer, finding the cause and giving the solution for possible problems. Calculating and measuring methods, as well as their results, have to be analised and shown. At the end some proposals to obtain better conditions have to be given, and the simulation with proposed actions have to be made, to find out if those proposals solved the problem found out in the real room (Domitrović, Fajt & Krhen, 2009).

The room acoustic quality rating is at the end always subjective and influenced by stochastic (Simovic & Jr. Simovic, 2010) convolution (Uludag, 1998; Sobolev, 2001). The only question is how sharp is the criteria which should be met, what is the purpose of space and how important is the quality of listening for the listeners. With regard to this, judgment process is extremely complex (Fajt, 2000). On the other hand, it is possible to do objective measurement relatively easily and quickly. Following the above stated, the main idea of the first part of this chapter is that it offers an original scientific discussion with a conclusion concerning the relevance of making judgments about the subjective assessment of quality of space based on objective measurements of the room acoustic quality rating which is at the end always subjective and influenced by convolution, where information about the statistical correlation of subjective assessments and objective measurements of room acoustic quality is shown. In the first part of this work the results of statistical analysis of two acoustically differently treated spaces subjective testing are shown. The number of test participants was 33, of which 8 had musical education. The musical sample duration was 16.5 min. The loudness level has been calibrated, i.e. it was identical in both rooms. Eighteen parameters were evaluated in two ways: with and without the impact of musical memory. These results were compared with hypothetical results based on the experience of the authors. The comparison of subjective judgment and objective measuring results is shown. The possibility of making judgments about the subjective assessment of quality of space based on objective measurements is also given. At the end, information about the statistical correlation (Žužul, Šimović & Leinert-Novosel, 2008) of subjective assessments and objective measurements (Kosić, 2008) of room acoustic quality is shown (Krhen, 1994). Room acoustics is a characteristic of a room in which sound is well transferred and well received (heard), i.e., sound depending on type (speech or music, that is, type of music) and purpose (listening or recording). The development of acoustics – acoustic science, and especially its subcategories, architectural acoustics, is founded on defining objective parameters of spatial room acoustic quality and finding new methods (von zur Gathen & Gerhard, 2003) of their measuring and identifying their relationship with subjective parameters of spatial acoustic quality obtained through subjective testing (Everest & Pohlmann, 2009). That is why the task of this simulation research was to make an estimate, a simulation and measurement of the acoustics of a representative multimedia classroom (classroom at the Department of Electroacoustics, Faculty of Electrical Engineering and Computing, FER). The main idea of the second part of this chapter is that it offers an original scientific discussion with a conclusion concerning the relevance of making a prediction, stochastic based simulations and measurements of (some objective parameters of) acoustic (quality) in a sample multimedia classroom (classroom at the Department of Electroacoustics, Faculty of Electrical Engineering and Computing, or FER). Measurements and simulations of some objective parameters of acoustic quality were conducted in order to determine whether the representative multimedia classroom was acoustic or not, and if not, what measures (with stochastic optimization) should be taken in order for the mentioned room to meet optimal conditions for being acoustic.

2. Introduction to analysis

Subjective rating of acoustic quality is always the most important result which must be as high as possible, and it depends on several factors, such as: training of the listener, familiarity with the topic, musical memory and even the expectations of the listener. These ratings, however, always affect the same parameters, which were assessed in our studies. The statistical based analysis of these results has been made. On the other hand, objective parameters are much easier to measure. Measurements of objective parameters of acoustic quality were made, and at the end their mutual correlation is shown. Why? Because possible convolution problems. See visual explanation of convolution and its applications.

Some of convolution applications are cited bellow. Citation from: http://www.answers.com/topic/convolution (2012/12/31):

"Convolution and related operations are found in many applications of engineering and mathematics:

- In linear acoustics, an echo is the convolution of the original sound with a function representing the various objects that are reflecting it.
- In artificial reverberation (digital signal processing, pro audio), convolution is used to map the impulse response of a real room on a digital audio signal (Krhen, 1994).
- In electrical engineering, the convolution of one function (the input signal) with a second function (the impulse response) gives the output of a linear time-invariant system (LTI). At any given moment, the output is an accumulated effect of all the prior values of the input function, with the most recent values typically having the most influence (expressed as a multiplicative factor). The impulse response function provides that factor as a function of the elapsed time since each input value occurred.
- In digital signal processing and image processing applications (Pap, Kosić & Fajt, 2006), the entire input function is often available for computing every sample of the output function. In that case, the constraint that each output is the effect of only prior inputs can be relaxed.
- Convolution amplifies or attenuates each frequency component of the input independently of the other components.
- In statistics, as noted above, a weighted moving average is a convolution.
- In probability theory, the probability distribution of the sum of two independent random variables is the convolution of their individual distributions."

3. Subjective evaluation of room acoustic quality parameters

For the purposes of subjective acoustic quality evaluation, we examined the following parameters:

1. Noise volume;
2. Intimacy, Presence;
3. Loudness;
4. Reverberance;
5. Tonal Reproduction, Timbre;
6. Sound Definition, Clarity;
7. Echo Disturbance;
8. Speech Intelligibility;
9. Spectral Uniformity, Balance;
10. Sound stage imaging;
11. Dynamics;

12. Distortion;
13. Stability of performance;
14. Brilliance;
15. Bass reproduction;
16. Resonance;
17. Ambience Reproduction, Diffusion;
18. Overall Acoustic Impression.

Express each function in terms of a dummy variable τ. Reflect one of the functions: $g(\tau) \rightarrow g(-\tau)$. Add a time-offset, t, which allows $g(t-\tau)$ to slide along the τ-axis. Start t at -∞ and slide it all the way to +∞. Wherever the two functions intersect, find the integral of their product. In other words, compute a sliding, weighted-average of function $f(\tau)$, where the weighting function is $g(-\tau)$. The resulting waveform (not shown here) is the convolution of functions f and g. If $f(t)$ is a unit impulse, the result of this process is simply $g(t)$, which is therefore called the impulse response.

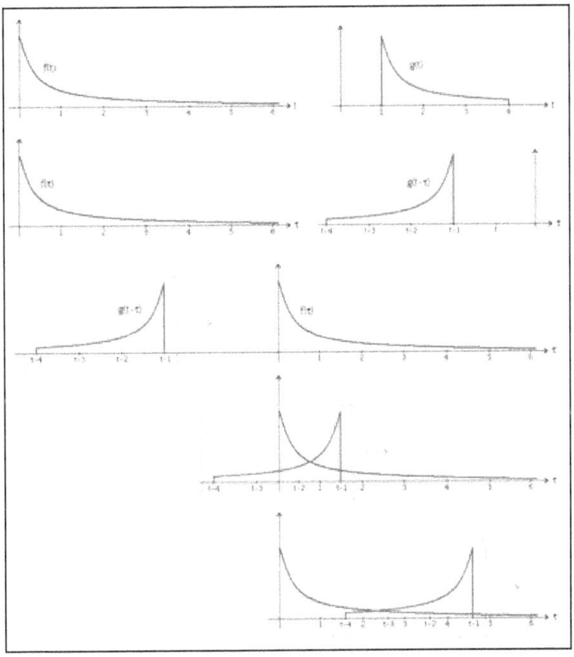

Figure 1. Visual explanation of convolution - according to: http://www.answers.com/topic/convolution (2011/12/31)

One room was acoustically untreated (Room 1), while the other was acoustically treated as a listening room (Room 2). All subjects evaluated parameter values as numeric values in the range from 1 to 5. In each room two tests were made. The goal was to determine the effect of listener's music memory on test results.

For this purpose, a group of subjects first heard a music test pattern, and the evaluation process began 1 minute after the end of the test pattern – Measurement Type A.

After that a 15 min pause was done, and subsequently after the pause the following test started in which subjects entered their ratings during the test sample listening – Measurement Type B.

It is interesting to note that for Room 1, for all parameters except for (3) at least one subject had given the worst score when they did test with the impact of musical memory, while in Type B test, only 4 parameters are not given the worst score.

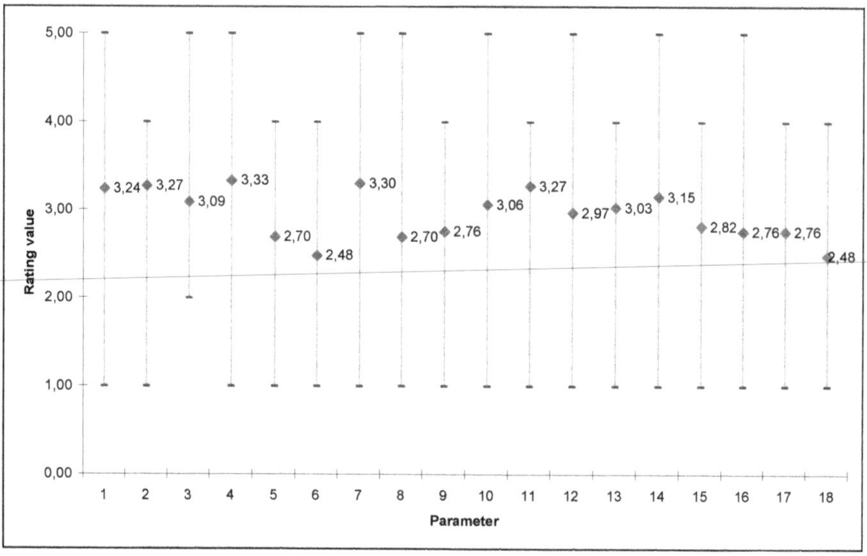

Figure 2. Feedback results for Room 1 / Measurement Type A – Minimum, Maximum and Average Value

Parameter	*1*	*2*	*3*	*4*	*5*	*6*	*7*	*8*	*9*
σRoom 1, Type A	1.07	0.83	0.93	1.03	0.83	0.93	1.22	0.80	0.74
σRoom 1, Type B	0.99	0.61	0.83	1.02	0.81	0.78	0.99	0.86	0.57
σRoom 2, Type A	0.98	0.58	0.55	0.88	0.59	0.58	0.36	0.69	0.65
σRoom 2, Type B	0.64	0.72	0.55	0.85	0.46	0.59	0.17	0.55	0.46
Parameter	*10*	*11*	*12*	*13*	*14*	*15*	*16*	*17*	*18*
σRoom 1, Type A	1.13	0.75	1.29	0.80	1.02	0.87	1.02	0.85	0.78
σRoom 1, Type B	1.02	0.71	1.03	0.72	0.91	0.95	1.15	0.77	0.82
σRoom 2, Type A	0.70	0.64	0.73	0.72	0.74	0.90	0.59	0.86	0.67
σRoom 2, Type B	0.90	0.67	0.53	0.66	0.65	0.89	0.52	0.80	0.67

Table 1. Standard deviation for all feedback results

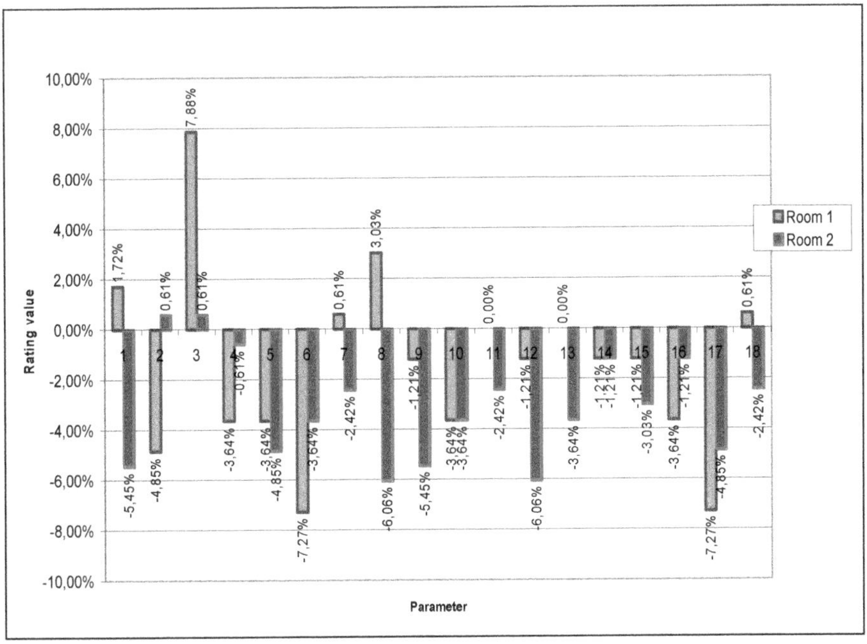

Figure 3. The feedback difference depending on musical memory (Meas.Type A – Meas.Type B) – Positive value: better result without memory effect; Negative value: better result with memory effect

In this evaluation, 8/18 (44.4%) of the parameter (Meas.Type A) and 10/18 (55.5%) parameters (Meas.Type B) also received the best value. This clearly points to a very large dispersion of results, and points (and confirms) the problem that always exists in the subjective assessment of acoustical quality parameters. Standard deviation shows us also the relatively large dispersion of results (Table 1).

In the most cases, individual results are significantly below average: Room 1 – Meas.Type A: 26/33 (78.8%); Meas.Type B: 23/33 (69.7%) and Room 2 – Meas.Type A: 18/33 (54.5%) ; Meas.Type B: 23/33 (69.7%).

Musical memory definitely has an impact on making assessments about the acoustical quality, and we can see (Fig. 2) that the results are in the most cases better with memory effect. This is especially outlined in the Room 2 (16/18 – 88.8%), compared with Room 1 (11/18 – 61%).

When we compare the subjective assessment of the average acoustic impression (18) with an average value of all parameters, we can see that these values are in Room 1 - acoustically untreated room: Meas.Type A: 2.48 / 2.98; Meas.Type B: 2.45 / 3.06 much more different than in Room 2 - acoustically treated room: Meas.Type A: 3.97 / 4.03; Meas.Type B: 4.09 / 4.18.

4. Objective measurements of room acoustic quality parameters

Objective measurements for Room 1 and Room 2 have been made with acoustical SW "ARTA". The measurements were done with 4 kinds of signal generator: sweep, pink noise, white noise and MLS. The sampling frequency was 48 kHz except for the speech intelligibility measurements, where the sampling frequency was 16 kHz for all signals. The correlation between the results is very good.

f [Hz]	T30 [s]	T20 [s]	T10 [s]	EDT [s]	C80 [dB]	C50 [dB]	D50 [%]
63	1.268	1.344	0.877	1.608	1.79	1.12	56.40
125	1.197	1.326	1.182	1.491	0.80	-3.17	32.50
250	1.027	1.165	1.305	1.400	0.58	-1.73	40.20
500	1.164	1.237	1.260	1.429	-0.51	-3.06	33.09
1000	1.139	1.240	1.100	1.315	1.61	-0.61	46.52
2000	1.059	1.101	1.114	1.170	2.30	0.00	50.01
4000	0.873	0.852	0.799	0.904	4.50	1.55	58.86
8000	0.710	0.694	0.697	0.699	6.82	3.35	68.38

Table 2. Objective measurement results for Room 1, sweep signal, sampling f = 48 kHz

Measured speech intelligibility (sweep signal, sampling f = 16 kHz)

for Room 1 is:

STI = 0.5044 (male),
0.4893 (female),
Rating:
FAIR,
%ALcons= 8.8092;
RASTI = 0.5146
Rating: FAIR,
%ALcons = 10.4877,

and for Room 2:

STI = 0.6668 (male),
0.6362 (female)
Rating: GOOD,
%ALcons= 3.2315;
RASTI = 0.7269
Rating: GOOD,
%ALcons = 3.3203.

5. Statistical based analysis - first conclusion

Since it largely depends on the audience and their real possibilities of making assessments, subjective evaluation of room acoustic quality parameters is always difficult to make, as well as to accurately assess the subjective room quality. The resulting dispersion of the results shows that even within a group of people from similar socio-economic groups, education levels and technical knowledge, there are differences. However, since there is still a very good correlation between subjective and objective evaluation of the measured parameters, we can say that there is a very high probability that the subjective evaluation of room acoustic quality will be consistent with objectively measured parameters, and that the majority of people agree with that assessment.

6. Introduction to application: Acoustics based optimization in multimedia classroom

Acoustics is a characteristic of a room where within that space sound is well transferred and well received (heard) according to the type of sound (speech or music, that is type of music) and purpose (listening or recording). The development of acoustics – acoustic science, and especially its sub disciplines, architectural acoustics, is based on defining objective parameters of acoustic quality of a space and in finding methods for their measurement and establishment of their interdependency with subjective parameters of spatial acoustic quality which were obtained through subjective testing. The traditional task of spatial acoustics was to ensure and form conditions which will enable better acoustic transfer from the sound source to the listener. Objective parameters of spatial acoustic quality are determined by the possibility of finding objective methods of measurement. The parameters that determine the acoustic quality of a room are: time of echo, time of early drop of sound energy, room constant, room radius, clarity, mean time, definition, relationship between the reflected and direct energy, support, relative level or power index, sound color, occasional diffusion, the assessed relationship signal/murmur, loss of consonant articulation, speech transfer index, the ratio of useful and harmful sound, fraction of lateral energy, inter-aural coefficient of cross correlation, direction index, and time of initial delay. All measurable parameters of acoustic quality of a room which are interrelated in some way due to the fact that they have basic characteristics of sound waves are mentioned here. It is not necessary to measure and use all parameters in spatial acoustics. Those which can be measured and those that give us reliable information even with subjective parameters on the acoustics of the room being tested should be selected. The results obtained will help take appropriate actions which will improve the acoustic quality of a particular area. Subjective parameters for acoustic quality of a room depend on whether we are referring to speech or music. When determining the acoustic quality of a room with respect to speech the subjective parameter is intelligibility. Subjective parameters for the acoustic quality of a room are: loudness, definition or clarity, reproduction of high frequencies or brilliance, reproduction of low frequencies or fullness, sparkle, sound imagery (stage), localization or locating the sound source, intimacy or closeness, echo, noise level, reverberation, tone reproduction, color of voice, tone or sound, intelligibility of speech, spectral equality, dynamics, area of soundness and transient response, deformities, equal distribution of sound pressure, reproduction of ambiance, spatiality, impression of space, diffusion, resonance audibility and entire acoustic impression.

7. Predicting measures for checking acquired acoustic quality in a classroom

The basic measurement of acoustic quality of a room is the measurement of reverberation time as the most important objective parameter of the acoustic quality of a room since it contains characteristics of a room such as dimension and volume and shape and absorption. Measuring reverberation time can be carried out in various ways and different types of measuring signals and sources of signal measurement are used. Basically, all types of measurement of reverberation time differ according to the principle of big difficulties created by the level of sound pressure of 60 dB above the noise level of the area which is needed according to definition for establishing reverberation time. The oldest way of measuring reverberation time is using the burst (pistol or similar) or noise (noise generator) which through frequency analyzer and logarithmic printer give a curve-like lowering of the sound pressure in a space. The Figure 4 graph shows the reverberation time measuring a protractor – a transparent board with scales for reading the reverberation time. In this type of measurement it is advisable that the level of actuation signal is at least 40-50 dB above environmental noise level.

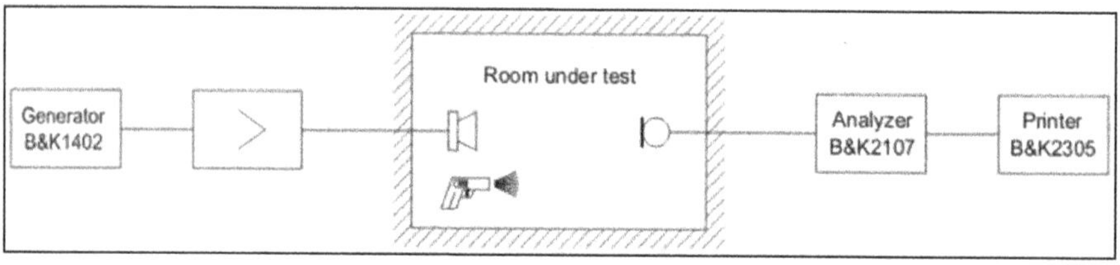

Figure 4. Graphical layout for direct reading of reverberation time using a protractor

Based on the Schroeder integration, a special way of measuring reverberation time from the impulse response possible to implement into a digital sound measuring device was developed (e.g. B&K 2231 using a module for measuring reverberation time BZ 7108). The results are directly read from the instrument according to the octal and terca (tertiary bands) within the set frequency area. Reverberation time is calculated (extrapolation) based on 10, 20 and 30 dB of lowering of the signal level and thus avoiding the need for a 60 dB signal dynamics. The digital sound measurer or SPL meter (B&K 2231 using a module for measuring reverberation time BZ 7108), depending on the environmental noise level, itself generates the required signal level sufficient for measuring, based on the signal received with appropriate indicators of a low level or pre–actuation, independently carries out level correction. Due to the lowered influences of possible mistakes in measurement, four subsequent measurements within the entire frequency area, whose results are stored in digital sound measuring device memory and through statistical analysis can yield results of proper accuracy.

8. Predictions, stochastic based simulations and measurements in the sample multimedia room

The task is to make a prediction, stochastic based simulations and measurements of (some objective parameters of) acoustic (quality) in a sample multimedia classroom (classroom at the Department of Electroacoustics, Faculty of Electrical Engineering and Computing, or FER). Measurements and simulations of some objective parameters of acoustic quality in order to establish opinion whether the sample multimedia classroom is acoustic (according to some rules and conditions) or not, and if not, what measures should be taken in order for that classroom to meet the condition of acoustics. If the same measuring conditions are ensured, objective parameters of must be repeatable and be direct indicators of acoustic quality. The first measurable parameter is the reverberation time (time necessary for level of sound pressure or sound strength is lowered for 60 dB as sound emission is stopped).

Figure 5 shows the layout of the sample multimedia classroom. The numbers refer to positions in the room where reverberation time was measured, and the sound source is also marked with a symbol. Table 3 shows the results of the measurement of reverberation time depending on various frequencies of sound source in the real system.

Frequency	Position 1	Position 2	Position 3	Position 2	Position 4	Position 5
62,5 Hz	1.14 s	1.44 s	1.24 s	1.44 s	1.22 s	1.33 s
125 Hz	1.32 s	1.50 s	1.37 s	1.50 s	1.22 s	1.54 s
250 Hz	1.67 s	1.55 s	1.62 s	1.55 s	1.59 s	1.67 s
500 Hz	1.50 s	1.55 s	1.48 s	1.55 s	1.57 s	1.56 s
1 kHz	1.40 s	1.38 s	1.38 s	1.38 s	1.35 s	1.36 s
2 kHz	0.99 s	1.18 s	1.14 s	1.18 s	1.14 s	1.17 s
4 kHz	0.89 s	0.82 s	0.85 s	0.82 s	0.83 s	0.83 s
8 kHZ	0.67 s	0.63 s	0.99 s	0.63 s	0.64 s	0.65 s

Table 3. Results of reverberation time measurement

The second measurable parameter is the distribution (division of the sound pressure). Sound pressure at a point in the room area is the result of the interference of direct and reflected sound and therefore in the spatial allocation of the sound pressure there are areas of minimum and maximum of sound pressure which are especially expressed in resonant frequencies. This parameter was measured at various frequencies in 30 points defined with introduced coordinates, shown in Figure 6. The results are shown in Table 4.

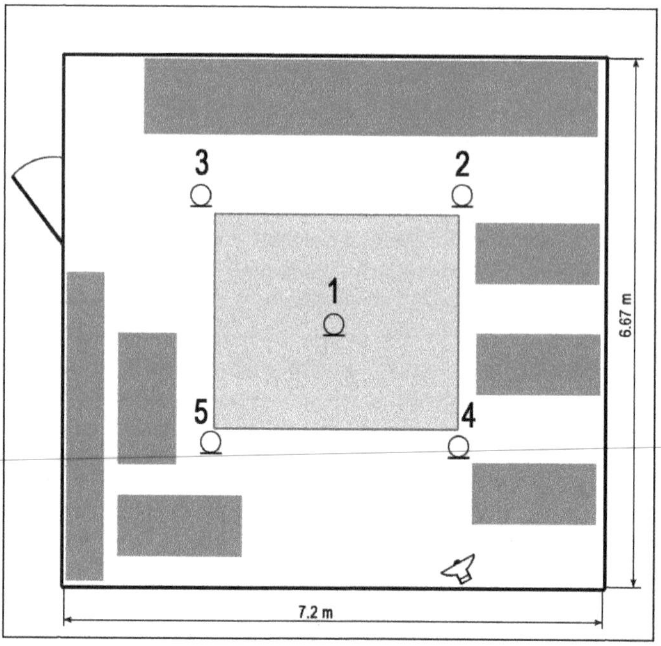

Figure 5. Room layout CX-15 (sample multimedia room)

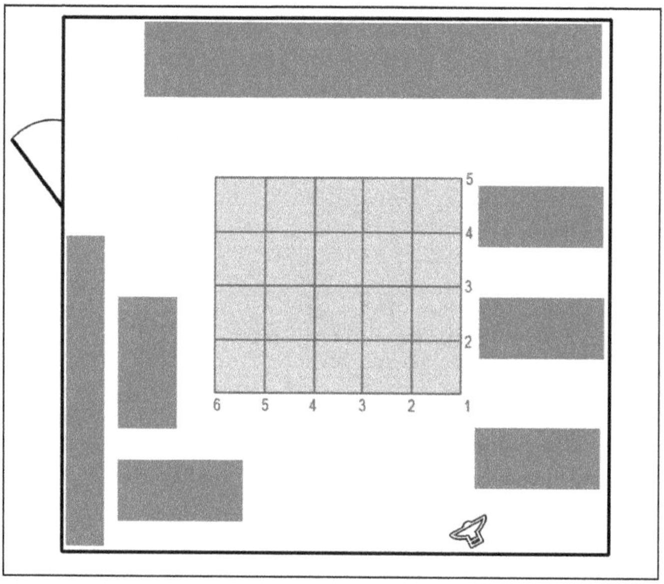

Figure 6. Measurement of sound pressure distribution

Position	31.5 Hz	63 Hz	125 Hz
(1,1)	63.5	80.0	77.5
(1,2)	64.5	82.0	81.0
(1,3)	64.5	82.5	66.0
(1,4)	64.0	80.5	79.0
(1,5)	64.0	78.0	83.0
(2,1)	63.0	71.0	78.0
(2,2)	63.0	74.0	74.0
(2,3)	62.0	78.0	73.0
(2,4)	61.0	79.0	75.0
(2,5)	58.0	77.5	79.0
(3,1)	63.0	79.0	74.5
(3,2)	61.0	74.0	72.0
(3,3)	58.0	65.0	71.0
(3,4)	54.0	74.5	74.0
(3,5)	51.0	76.0	73.0
(4,1)	63.5	83.5	70.5
(4,2)	53.0	78.5	74.5
(4,3)	49.0	74.0	76.0
(4,4)	52.0	75.0	73.0
(4,5)	61.5	76.5	70.0
(5,1)	67.0	84.5	70.0
(5,2)	62.0	79.5	73.0
(5,3)	64.0	75.5	74.0
(5,4)	68.0	77.5	73.0
(5,5)	67.0	77.0	73.0
(6,1)	65.5	80.0	71.0
(6,2)	58.5	79.5	62.0
(6,3)	56.5	77.5	70.0
(6,4)	65.0	79.0	75.5
(6,5)	68.5	83.0	77.0

Table 4. Level of sound pressure in dB for particular frequencies

The obtained results for sound level in decibel (dB) are relative, considering that they are only relevant for differences in levels for particular frequencies. Separate measurement was conducted using Bruel and Kjaer sound measuring device (SPL meter - B&K 2231 using a module for measuring reverberation time BZ 7108) which was attached to a spectral analyzer.

Measurement was conducted in the spot which is marked by number 1 on Figure 7 and the shaded area around that place. The speakers emitted a white murmur and the spectral analyzer recorded the situation for several spots in the area. The results were transferred

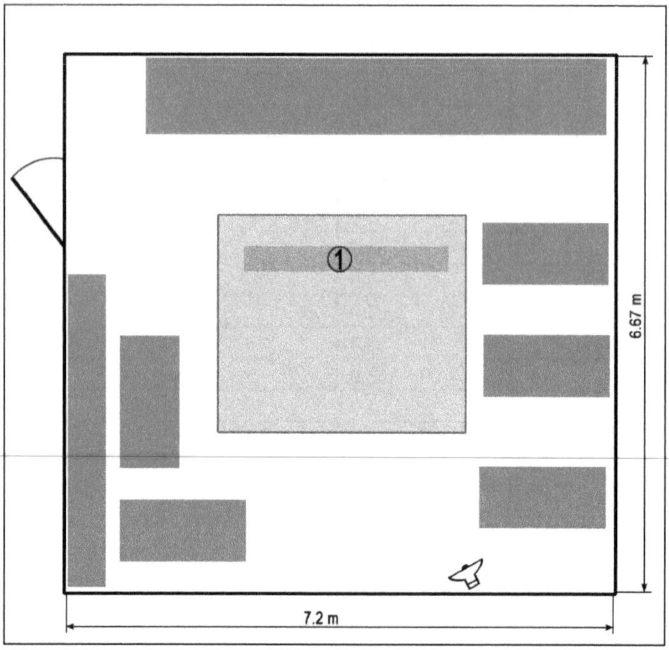

Figure 7. The measurement was carried out in the area marked with number 1 and the shaded area around that spot

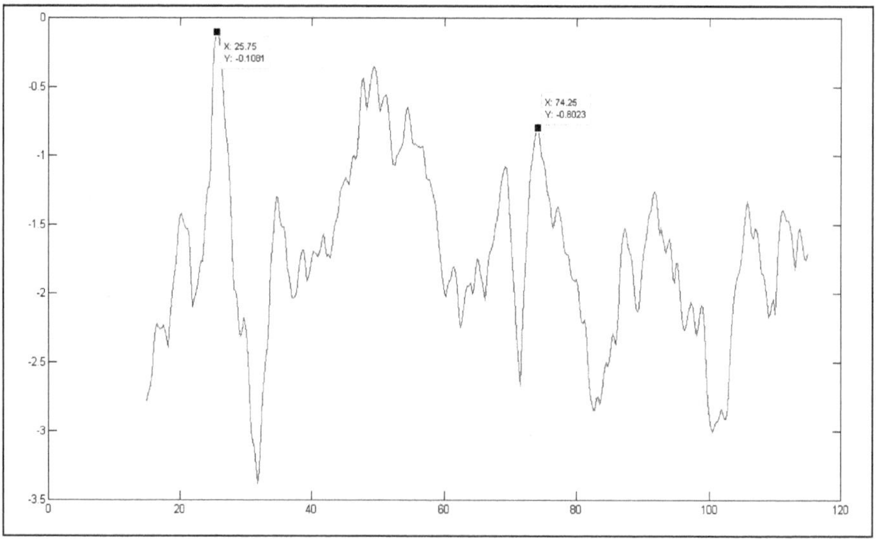

Figure 8. Measured resonances (points which represent maximum standing waves)

into the program Matlab which created a graph shown in Figure 8. Figure 8 shows the dependency of the sound pressure on frequency in spot 1. The results of the measurement in the area are rather similar to the results of measurement in that spot.

Figure 8 shows resonances, that is, points which represent standing waves maximum. The most prominent maxima are at the frequency of about 26 Hz, followed by 50 Hz and 75 Hz. Such maxima show up at resonant frequencies obtained by the formula:

$$f = \frac{c}{2}\sqrt{\frac{p^2}{A^2} + \frac{q^2}{B^2} + \frac{r^2}{C^2}}$$

The measure represents the speed of sound in the air which is c= 340 m/s, A, B and C are room dimensions, and p, q and r are natural, whole numbers. Room dimensions: A=6.7 m; B=7.2 m; C=3.2 m.

Table 5 gives the results of the calculation of resonant frequencies and it is evident that the calculated values coincide with the measured values. The most prominent frequencies are 26 and 74 Hz.

p	q	r	Mod (Hz)
0	1	0	23.60
1	0	0	25.37
1	1	0	34.60
0	2	0	47.20
2	0	0	50.70
0	0	1	53.00
1	2	0	53.60
2	1	0	56.00
0	1	1	58.10
1	0	1	58.90
1	1	1	63.40
2	2	0	69.30
0	2	1	71.00
2	0	1	73.46
0	0	2	106.25
0	1	2	108.80
1	0	2	109.20

Table 5. Resonant frequency calculation results

The following step was the acoustic simulation on the computer. For that purpose the EASE program (version 4.1) was used. The aim of the simulation was to show the reverberation time at various frequencies depending on the construction materials used. For calculating reverberation time, the Eyring formula was used and set with the following expression:

$$T_r = \frac{0.161V}{-S\ln(1-\alpha)}$$

Where V is room volume, S is the sum of all areas in the room, and α average absorption coefficient. The simulation of the current situation of the room was conducted first. The simulation results are presented in Figure 9, which shows that the situation is not satisfactory for frequencies lower than 2 kHz, considering that there is a large deviation in the reverberation time from normal values. Because of that some changes were necessary. In the area (virtual, used in the simulation) absorbers, materials used for "absorbing" sound waves, were used. To be more exact, they were placed on the only available wall in the room.

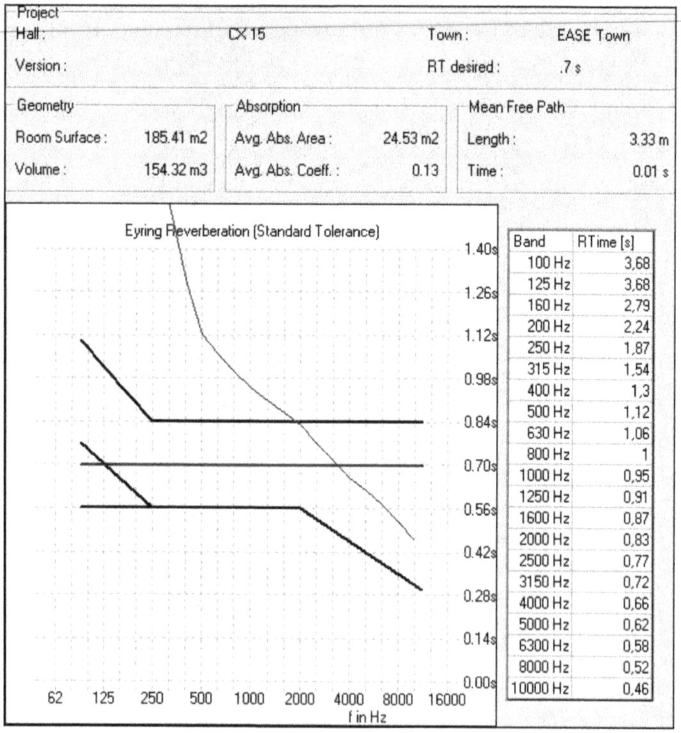

Figure 9. Results of the first simulation

After that, another simulation took place, the results of which are shown in Figure 10. As it can be seen on Figure 10, the results are rather satisfactory considering that there are no great deviations in the reverberation time from optimal values.

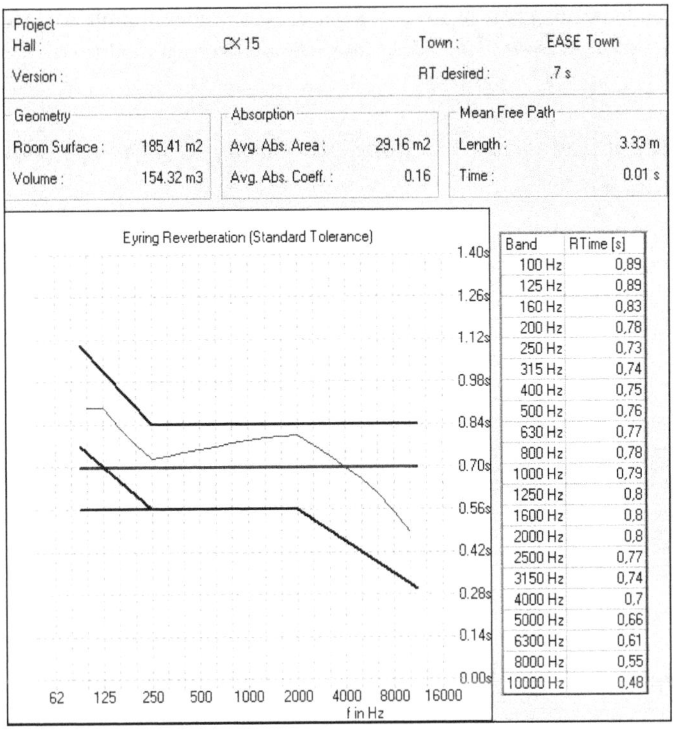

Figure 10. Second simulation results

Therefore, the back wall should be entirely covered with absorbing diffuser whose dependency of absorbency coefficient on frequency is shown in Figure 11.

100Hz	0.8200
125Hz	0.8200
160Hz	0.8467
200Hz	0.8733
250Hz	0.9000
315Hz	0.9300
400Hz	0.9600
500Hz	0.9900
630Hz	0.9900
800Hz	0.9900
1000Hz	0.9900
1250Hz	0.9900
1600Hz	0.9900
2000Hz	0.9900
2500Hz	0.9900
3150Hz	0.9900
4000Hz	0.9900
5000Hz	0.9900
6300Hz	0.9900
8000Hz	0.9900
10000Hz	0.9900

Figure 11. Overview of the dependency of absorbency coefficient on frequency

It is important to mention how in acoustic room simulation, some additional objects such as cupboards, board, etc. were not taken into consideration since they have a marginal influence on the end results constraint.

9. Acoustics optimization in multimedia classroom as an application - second conclusion

The presented procedure of the analysis, simulation and measurement of the sample multimedia classroom, indicates a need for professional design and supervision of the performance of multimedia classrooms considering their ever-growing use from the point of view of long life learning (LL) that is lengthy and frequent stay in them. Such a statement is very important in the case of adult education since they tire more quickly (especially if they are strained in order to hear better and understand speech) than the younger population involved in lifelong learning, that is, in-service teacher training. This is also very important for handicapped groups (partially sighted people and people with partial hearing abilities) involved in long life learning. Furthermore, all classrooms used for foreign language learning should have strictly defined adequate acoustic conditions, also confirmed by this simulation (optimization).

10. Conclusion

Since it largely depends on the audience and their real possibilities of making assessments, subjective evaluation of room acoustic quality parameters is always difficult to make, as well as to accurately assess the subjective room quality. The resulting dispersion of the results shows that even within a group of people from similar socio-economic groups, education levels and technical knowledge, there are differences. However, since there is still a very good correlation between subjective and objective evaluation of the measured parameters, we can say that there is a very high probability that the subjective evaluation of room acoustic quality will be consistent with objectively measured parameters, and that the majority of people agree with that assessment. The presented procedure of the analysis, stochastic simulation and measurement of the sample multimedia classroom, indicates a need for professional design and supervision of the performance of multimedia classrooms considering their ever-growing use from the point of view of long LL that is lengthy and frequent stay in them. Such a statement is very important in the case of adult education since they tire more quickly (especially if they are strained in order to hear better and understand speech) than the younger population involved in lifelong learning, that is, in-service teacher training. This is also very important for handicapped groups (partially sighted people and people with partial hearing abilities) involved in long life learning. Furthermore, all classrooms used for foreign language learning should have strictly defined adequate acoustic conditions, also confirmed by this simulation (optimization).

The suggestion for further development opens up a new area for improving the system of real simulation according to the effort in the organization of complex research. The system of analysis, comparison and monitoring has to be based on set and realistic stochastic procedures, control and values in both main chapter parts, and the results obtained are expected to open new areas for research and modelling. This research was a part of main Scientific research named "Analytical Model for Monitoring of New Education Technologies for Long Life Learning"conducted by Ministry of Science, Education and Sports of the Republic of Croatia (Registered Number 227-2271694-1699).

Author details

Vladimir Šimović, Siniša Fajt and Miljenko Krhen
University of Zagreb, Croatia

11. References

Answers.com, Convolution at answers.com, http://www.answers.com/topic/convolution (2011/12/31)

Domitrović, H., Fajt, S., Krhen, M. (2009), "Multimedia Room Acoustical Design", Proceedings Elmar 2009, Zadar, 221-224 (ISSN: 1334-2630; ISBN: 978-953-7044-10-7)

Everest, F.A., Pohlmann, K.C. (2009), "Master Handbook of Acoustics", 5th ed. Tab Electronics

Fajt, S. (2000), "Vrednovanje kvalitete akustički obrađenih prostora", PhD Thesis, University of Zagreb, FER, Zagreb

Kosić T. (2008), „Razvoj analitičkog modela korištenja multimedije i tiskovina u cjeloživotnom učenju" – doctoral disertation, FFZG, Zagreb

Krhen, M. (1994), "Ocjena akustičke kvalitete prostora", Master Thesis, University of Zagreb, FER, Zagreb

Pap, K., Kosić, T., Fajt, S. (2006), "Research of Workflows in Graphic Production and Creating Digital Workflow Knowledge Databases"; 1st Special Focus Symposium - "CISKS: Communication and Information Sciences in the Knowledge Society", (ed. by Ljubica Bakić-Tomić, Vladimir Šimović and George E. Lasker), InterSymp-2006, Baden-Baden, pp. 34-40

Simovic, V. and Simovic, V. Jr. (2010). Pertinence and Information Needs of Different Subjects on Markets and Appropriate Operative (Tactical or Strategic) Stochastic Control Approaches, Stochastic Control, Chris Myers (Ed.), ISBN: 978-953-307-121-3, InTech, Available from: http://www.intechopen.com/articles/show/title/pertinence-and-information-needs-of-different-subjects-on-markets-and-appropriate-operative-tactical

Sobolev, V.I. (2001), "Convolution of functions", in Hazewinkel, Michiel, Encyclopedia of Mathematics, Springer, ISBN 978-1556080104, http://www.encyclopediaofmath.org/index.php?title=C/c026430

Uludag, A. M. (1998), "On possible deterioration of smoothness under the operation of convolution", J. Math. Anal. Appl. 227 no. 2, 335–358

von zur Gathen, J.; Gerhard, J. (2003), Modern Computer Algebra, Cambridge University Press, ISBN 0-521-82646-2

Žužul, J., Šimović, V., Leinert-Novosel, S. (2008), "Statistika u informacijskom društvu", Europski centar za napredna i sustavna istraživanja ECNSI, Zagreb

Chapter 13

Identifiability of Quantized Linear Systems

Ying Shen and Hui Zhang

Additional information is available at the end of the chapter

1. Introduction

On-line parameter identification is a key problem of adaptive control and also an important part of self-tuning regulator (STR) (Astrom & Wittenmark, 1994). In the widely equipped large-scale systems, distributed systems and remote systems, the plants, controllers, actuators and sensors are connected by communication channels which possess only finite communication capability due to, e.g., data loss, bandwidth constraint, and access constraint. From a heuristic analysis perspective, the existence of communication constraints has the effect of complicating what are otherwise well-understood control problems, including the traditional methods, such as the H_∞ control (Fu & Xie, 2005), and even the basic theoretic notions, such as the stability (De Persis & Mazenc, 2010).

Due to the constraints of the communication channels, it is difficult to transmit data with infinite precision. Quantization is an effective way of reducing the use of transmission resource, and then meeting the bandwidth constraint of the communication channels. However, quantization is a lossy compression method, and hence the performance of parameter identification, even the validity or effectiveness of identification may be changed by quantization, along with which the performance of adaptive control may deteriorate. This has attracted plenty of works. The problem of system identification with quantized observation was investigated in (Wang, et al, 2003, 2008, 2010), where issues of optimal identification errors, time complexity, optimal input design, and the impact of disturbances and unmodeled dynamics on identification accuracy and complexity are included.

In the light of the fundamental effect of quantization on system identification, it is necessary to pay attention to the parameter identifiability property of quantized systems. The concept of identifiability has been defined by maximal information criterion in (Durgaryan & Pashchenko, 2001): the system is parameter identifiable by maximal information criterion, if the mutual information between actual output and model output is greater than zero. However, the concept of identifiability in (Durgaryan & Pashchenko, 2001) is defined in principle, based on which there is no practical result. Reference (Zhang, 2003; Zhang & Sun, 1996; Baram & Kailath, 1988) have discussed the problem of states estimability, which is related closely with parameter identifiability, for that input-output description of linear systems with Gauss-Markov parameters can be transformed to state space model, and then the problem of parameter identifiability can be treated as state estimability. Reference (Zhang, 2003) has proposed the definition of parameter identifiability under the criterion of minimum maximum error entropy (MMSE) estimation referring to the definition of states estimability, and also obtained some useful conclusions. Reference (Wang & Zhang, 2011) has studied the parameter identifiability of linear systems under access constraints. However, there is few work on that for quantized systems.

This paper mainly analyzes the parameter identifiability of quantized linear systems with Gauss-Markov parameters from information theoretic point of view. The definition of parameter identifiability proposed in (Zhang, 2003) is reviewed: the linear system with Gauss-Markov parameters is identifiable, if and only if the mutual information between the actual value and estimates of parameters is greater than zero, which is extended to quantized systems by considering the intrinsic property of the system. Then the parameter identifiability of linear systems with quantized outputs is analyzed and the criterion of parameter identifiability is obtained based on the measure of mutual information. Furthermore, the convergence property of the quantized parameter identifiability Gramian is analyzed.

The rest of the paper is organized as follows: In section 2, we introduce the model that we are interested in; Section 3 discusses the existing definition of parameter identifiability, proposes our new definition, and gives analytic conclusion focusing on quantized linear systems with Gauss-Markov parameters; The convergence property of Gramian matrix of parameter identifiability for quantized systems is discussed in section 4; Section 5 and 6 are illustrative simulation and conclusion, respectively.

2. System description

Consider the following SISO linear system expressed by Auto-Regressive and Moving Average Model (ARMA) (Astrom & Wittenmark, 1994):

$$\begin{aligned} y(k) + a_1(k)y(k-1) + \cdots + a_n(k)y(k-n) \\ = b_1(k)u(k-1) + \cdots + b_m(k)u(k-m) + e(k) \end{aligned} \tag{1}$$

where $u(k)$ and $y(k)$ are input and output of the system respectively, stochastic noise sequence $\{e(k)\}$ is Gaussian and white with zero-mean and covariance R, and uncorrelated with $y(k)$, $y(k-1)\cdots$ and $u(k)$, $u(k-1)\cdots$. Let

$$\theta(k) = [b_1(k) \cdots b_m(k)\, a_1(k) \cdots a_n(k)]^{\mathrm{T}} \tag{2}$$

be the parameter vector, where $(\cdot)^{\mathrm{T}}$ denotes the operation of transposition, and let

$$F(k) = [u(k-1) \cdots u(k-m) - y(k-1) \cdots - y(k-n)]^{\mathrm{T}} \tag{3}$$

then system (1) can be described as

$$y(k) = F^{\mathrm{T}}(k)\theta(k) + e(k) \tag{4}$$

Suppose that the parameter $\theta(k)$ can be modeled by a Gauss-Markov process, i.e.,

$$\theta(k+1) = A\theta(k) + Bw(k) \tag{5}$$

where A, B are known matrices with appropriate dimensions; noise sequence $\{w(k)\}$ is Gaussian and white with zero-mean and covariance Q; initial value of the parameter $\theta(0)$ is Gaussian with mean $\bar{\theta}$ and covariance $\Pi(0)$. Suppose that $e(k)$, $w(k)$ and $\theta(0)$ are mutually uncorrelated. Hence, linear system (1) with Gauss-Markov parameters can be described by (4) and (5), i.e.,

$$\begin{cases} \theta(k+1) = A\theta(k) + Bw(k) \\ y(k) = F^{\mathrm{T}}(k)\theta(k) + e(k) \end{cases} \tag{6}$$

Due to bandwidth constraint of the channel, quantization is required. The discussion in the present paper does not focus on a special quantizer, but on general N-level quantization (Curry 1970; Gray and Neuhoff 1998) which can be described as:

$$y_{\mathrm{q}}(k) = Q(y(k)) = z_l, \text{ for } y(k) \in \delta_l, l = 1,2,\cdots,N \tag{7}$$

where $Q(\cdot)$ is the general quantizer, $y_{\mathrm{q}}(k) \in \{z_1, z_2, \cdots, z_N\}$ denote the quantizer outputs with $z_l, l = 1,\cdots,N$ as the reproduction values; $\delta_l = (a_l, a_{l+1}]$, $l = 1,2,\cdots,N$ denote the quantization intervals, where $\{a_i\}_{i=1}^{N+1}$, $-\infty = a_1 < a_2 < \cdots < a_{N+1} = +\infty$ are the thresholds of the quantizer.

The channel is assumed to be lossless. $y_{\mathrm{q}}(k) \in \{z_1, z_2, \cdots, z_N\}$ is transmitted and then received at the channel receiver. z_i, $i = 1,2,\cdots,N$ are symbols denoting the ith quantization interval and not necessarily real numbers, hence, further decoding is required, as follows

$$y_{\mathrm{q}}^{*}(k) = D_k(y_{\mathrm{q}}(k)) \tag{8}$$

where $D_k(\cdot)$ is assumed to be a one to one mapping. A common decoding method (Curry, 1970) is

$$D_k(y_{\mathrm{q}}(k)) = \mathrm{E}\{y(k) \mid y_{\mathrm{q}}(k)=z_l\},\ l = 1,2,\cdots,N$$

where E{.} is the operation of expectation.

3. Parameter identifiability

3.1. Definition of parameter identifiability

Reference (Zhang, 2003) proposed the definition of parameter identifiability for system (6) referring to the definition of state estimability under MMEE.

Let $\hat{\theta}(k)$ be the MMEE estimation of $\theta(k)$ based on $F(k)$, and $\tilde{\theta}(k)=\theta(k)-\hat{\theta}(k)$ be the estimation error. Define the prior and posterior mean-square estimation error matrices respectively as

$$\Pi(k) = \mathrm{E}\{(\theta(k)-\bar{\theta}(k))(\theta(k)-\bar{\theta}(k))^{\mathrm{T}}\}$$

$$P(k) = \mathrm{E}\{\tilde{\theta}(k)\tilde{\theta}^{\mathrm{T}}(k)\}$$

where $\bar{\theta}(k)$ is the mean of $\theta(k)$, i.e. $\bar{\theta}(k)=\mathrm{E}\{\theta(k)\}$.

Definition 1(Zhang, 2003): The linear system (1) with Gauss-Markov parameters (i.e. system (6)) is identifiable, if and only if

$$I(\theta(k);\hat{\theta}(k)) > 0, \forall k \geq n+m-1 \tag{9}$$

where $I(\cdot;\cdot)$ denotes mutual information.

Based on Definition 1, the following conclusion was obtained in (Zhang, 2003).

Lemma 1(Zhang, 2003): The linear system (1) with Gauss-Markov parameters (i.e. system (6)) is identifiable, if and only if, the identifiability Gramian

$$W_k^{\mathrm{id}} = \sum_{j=k}^{0} A^{k-j}\Pi(j)F(j)F^{\mathrm{T}}(j)\Pi(j)(A^{k-j})^{\mathrm{T}}, \ \forall k \geq n+m-1 \tag{10}$$

has full rank, i.e. $rank(W_k^{\mathrm{id}}) = n+m$, $\forall k \geq n+m-1$.

In the present paper, we propose an alternative definition of parameter identifiability for the quantized system (6)(7)(8) from information theoretic point of view, as follows.

Definition 2: The quantized linear system with Gauss-Markov parameters (6)(7)(8) is identifiable, if and only if

$$I(\theta(k);Y_{\mathrm{q}}^{*k}) > 0, \ \forall \mathrm{k} \geq n+m-1 \tag{11}$$

where $Y_{\mathrm{q}}^{*k} = [y_{\mathrm{q}}^{*}(0)\, y_{\mathrm{q}}^{*}(1) \cdots y_{\mathrm{q}}^{*}(k)]^{\mathrm{T}}$.

Remark 1:

1. From information theory (Cover & Thomas, 2006), condition (11) is equivalent to that

$$H(\theta(k)) > H(\hat{\theta}(k)), \forall k \geq n+m-1 \tag{12}$$

where $H(\cdot)$ denotes entropy, i.e. the prior error entropy $H(\theta(k))$ is strictly greater than posterior error entropy $H(\hat{\theta}(k))$;

2. Definition 1 considers the mutual information between the actual value and estimate of parameters, so it relies on the estimation principle, while Definition 2 takes into account the intrinsic property of the system independent of the estimator used. If Definition 2 is adopted to analyze unquantized system (6), the identifiability condition (11) turns into

$$I(\theta(k); Y^k) > 0, \forall \mathrm{k} \geq n+m-1 \tag{13}$$

where $Y^k = [y(0)\, y(1) \cdots y(k)]^{\mathrm{T}}$, and condition (9) is equivalent to (13) for linear Gaussian system (6) (Zhang, 2003). Hence, in some sense, Definition 2 is a more general one than Definition 1.

3.2. Identifiability analysis of quantized linear systems

Mutual information $I(\cdot;\cdot)$ (Cover & Thomas, 2006) is a measure of the information amount commonly contained in, and the statistic dependence between two random variables. $I(\cdot;\cdot) \geq 0$ with equality if and only if these two random variables are independent. Therefore (11) based on the information theoretic Definition 2 indicates that system is parameter identifiable if and only if any direction of parameter space is not orthogonal to all the past (quantized) measurements (Baram & Kailath, 1988), i.e. $\forall g \in \mathbf{R}^{n+m}, g \neq 0$,

$$g^{\mathrm{T}} \mathrm{E}\{(\theta(k) - \bar{\theta}(k))(y_{\mathrm{q}}^*(j) - \bar{y}_{\mathrm{q}}^*(j))\} \neq 0, \exists j \leq k, \forall k \geq n+m-1 \tag{14}$$

where $\bar{y}_{\mathrm{q}}^*(j)$ is the mean of $y_{\mathrm{q}}^*(j)$, i.e. $\bar{y}_{\mathrm{q}}^*(j) = \mathrm{E}\{y_{\mathrm{q}}^*(j)\}$.

From the Gauss-Markov property of the parameters, we have

$$\theta(k) = A^{k-\mathrm{j}}\theta(j) + \sum_{i=0}^{k-j-1} A^{k-j-i-1} Bw(j+i) \tag{15}$$

$$\bar{\theta}(k) = A^{k-j}\bar{\theta}(j) \tag{16}$$

where $\sum_{i=0}^{-1} A^{-i-1} Bw(k+i) = 0$ when $j = k$, then

$$\begin{aligned} &\mathrm{E}\{(\theta(k) - \bar{\theta}(k))(y_{\mathrm{q}}^*(j) - \bar{y}_{\mathrm{q}}^*(j))\} \\ &= \mathrm{E}\{(\theta(k) - \bar{\theta}(k)) \cdot y_{\mathrm{q}}^*(j)\} \\ &= \mathrm{E}\{[A^{k-j}(\theta(j) - \bar{\theta}(j)) + \sum_{i=0}^{k-j-1} A^{k-j-i-1} Bw(j+i)] \cdot y_{\mathrm{q}}^*(j)\} \\ &= A^{k-j} \mathrm{E}\{(\theta(j) - \bar{\theta}(j)) \cdot y_{\mathrm{q}}^*(j)\} \\ &= A^{k-j} \mathrm{E}\{\theta'(j) \cdot y_{\mathrm{q}}^*(j)\} \end{aligned} \tag{17}$$

where $\theta'(j) = \theta(j) - \bar{\theta}(j)$. Let $\theta' \triangleq \theta'(j)$, $y_{\mathrm{q}}^{*} \triangleq y_{\mathrm{q}}^{*}(j)$, and the time variable j of other relevant time-variant variables (e.g. $F(j)$, $\Pi(j)$) is omitted for notational simplicity, then

$$\begin{aligned} \mathrm{E}\{\theta'(j)y_{\mathrm{q}}^{*}(j)\} \triangleq \mathrm{E}\{\theta' \cdot y_{\mathrm{q}}^{*}\} &= \sum_{l=1}^{N} \int_{\theta' \in \mathbf{R}^{n+m}} \theta' \cdot D_j(z_l) \cdot p(\theta', y_{\mathrm{q}}^{*} = D_j(z_l)) \mathrm{d}\theta' \\ &= \sum_{l=1}^{N} D_j(z_l) \int_{\theta' \in \mathbf{R}^{n+m}} \theta' p(\theta', y_{\mathrm{q}}^{*} = D_j(z_l)) \mathrm{d}\theta' \end{aligned} \tag{18}$$

Based on the Bayesian law,

$$\begin{aligned} & p(\theta', y_{\mathrm{q}}^{*} = D_j(z_l)) \\ &= p(y_{\mathrm{q}}^{*} = D_j(z_l) \mid \theta') p(\theta') \\ &= p(a_l < y \le a_{l+1} \mid \theta') p(\theta') \\ &= p(a_l < F^{\mathrm{T}}\theta + e \le a_{l+1} \mid \theta') \mathrm{p}(\theta') \\ &= p(a_l < F^{\mathrm{T}}(\theta' + \bar{\theta}) + e \le a_{l+1} \mid \theta') p(\theta') \\ &= p(a_l - F^{\mathrm{T}}\bar{\theta} - F^{\mathrm{T}}\theta' < e \le a_{l+1} - F^{\mathrm{T}}\bar{\theta} - F^{\mathrm{T}}\theta' \mid \theta') p(\theta') \\ &= p(a_l - F^{\mathrm{T}}\bar{\theta} - F^{\mathrm{T}}\theta' < e \le a_{l+1} - F^{\mathrm{T}}\bar{\theta} - F^{\mathrm{T}}\theta') p(\theta') \end{aligned} \tag{19}$$

where the last equality is based on the fact that $\theta'(j)$ and $e(j)$ are stochastically independent, then

$$\begin{aligned} & \int_{\theta' \in \mathbf{R}^{n+m}} \theta' p(\theta', y_{\mathrm{q}}^{*} = D_j(z_l)) \mathrm{d}\theta' \\ = & \int_{\theta' \in \mathbf{R}^{n+m}} \theta' p(a_l - F^{\mathrm{T}}\bar{\theta} - F^{\mathrm{T}}\theta' < e \le a_{l+1} - F^{\mathrm{T}}\bar{\theta} - F^{\mathrm{T}}\theta') p(\theta') \mathrm{d}\theta' \\ = & \int_{\theta' \in \mathbf{R}^{n+m}} \theta' (T(\frac{a_{l+1} - F^{\mathrm{T}}\bar{\theta} - F^{\mathrm{T}}\theta'}{\sqrt{R}}) - T(\frac{a_l - F^{\mathrm{T}}\bar{\theta} - F^{\mathrm{T}}\theta'}{\sqrt{R}})) G(\theta', 0, \Pi) \mathrm{d}\theta' \end{aligned} \tag{20}$$

where $T(.)$ is the probability distribution function of standardized normalized distribution (note that, $T(.)$ is different from the tail function defined in (Ribeiro et al., 2006; You et al., 2011)), $G(\theta', 0, \Pi)$ denotes the probability density function which means that stochastic vector θ' is Gaussian with zero-mean and covariance Π . Following the similar line of argument as in (Ribeiro et al., 2006; You et al., 2011), we calculate the integral (20) and then obtain

$$\int_{\theta' \in \mathbf{R}^{n+m}} \theta' p(\theta', y_{\mathrm{q}}^{*} = D_j(z_l)) \mathrm{d}\theta' = (\phi(\frac{a_l - F^{\mathrm{T}}\bar{\theta}}{\sqrt{F^{\mathrm{T}}\Pi F + R}}) - \phi(\frac{a_{l+1} - F^{\mathrm{T}}\bar{\theta}}{\sqrt{F^{\mathrm{T}}\Pi F + R}})) \frac{\Pi F}{\sqrt{F^{\mathrm{T}}\Pi F + R}} \tag{21}$$

where $\phi(\cdot)$ is the probability density function of standardized normalized distribution. Substitute (21) into (18), we have

$$\mathrm{E}\{\theta'(j)y_{\mathrm{q}}^{*}(j)\} = \sum_{l=1}^{N} \frac{D_j(z_l)}{\sqrt{F^{\mathrm{T}}(j)\Pi(j)F(j) + R}} (\phi(\frac{a_l - F^{\mathrm{T}}(j)\bar{\theta}(j)}{\sqrt{F^{\mathrm{T}}(j)\Pi(j)F(j) + R}}) - \phi(\frac{a_{l+1} - F^{\mathrm{T}}(j)\bar{\theta}(j)}{\sqrt{F^{\mathrm{T}}(j)\Pi(j)F(j) + R}})) \Pi(j)F(j). \tag{22}$$

Combining (22) with (17) and (14), we get that $g^{\mathrm{T}}\mathrm{E}\{(\theta(k)-\bar{\theta}(k))(y_{\mathrm{q}}^{*}(j)-\bar{y}_{\mathrm{q}}^{*}(j))\}\neq 0$ is equivalent to

$$g^{\mathrm{T}}\sum_{l=1}^{N}\frac{D_j(z_l)}{\sqrt{F^{\mathrm{T}}(j)\Pi(j)F(j)+R}}\left(\phi\left(\frac{a_l-F^{\mathrm{T}}(j)\bar{\theta}(j)}{\sqrt{F^{\mathrm{T}}(j)\Pi(j)F(j)+R}}\right)-\phi\left(\frac{a_{l+1}-F^{\mathrm{T}}(j)\bar{\theta}(j)}{\sqrt{F^{\mathrm{T}}(j)\Pi(j)F(j)+R}}\right)\right)A^{k-j}\Pi(j)F(j)\neq 0, \quad \exists j\leq k, \forall k\geq n+m-1. \tag{23}$$

Let

$$\psi(j)\triangleq\sum_{l=1}^{N}\frac{D_j(z_l)}{\sqrt{F^{\mathrm{T}}(j)\Pi(j)F(j)+R}}\left(\phi\left(\frac{a_l-F^{\mathrm{T}}(j)\bar{\theta}(j)}{\sqrt{F^{\mathrm{T}}(j)\Pi(j)F(j)+R}}\right)-\phi\left(\frac{a_{l+1}-F^{\mathrm{T}}(j)\bar{\theta}(j)}{\sqrt{F^{\mathrm{T}}(j)\Pi(j)F(j)+R}}\right)\right) \tag{24}$$

Then (23) is equivalent to that

$$W_k^{\mathrm{idq}}=\sum_{j=k}^{0}\psi^2(j)A^{k-j}\Pi(j)F(j)F^{\mathrm{T}}(j)\Pi(j)(A^{k-j})^{\mathrm{T}}, \forall k\geq n+m-1 \tag{25}$$

has full rank. We conclude the above analysis as follows.

Theorem 1: The quantized linear system with Gauss-Markov parameters (6)(7)(8) is parameter identifiable, if and only if

$$rank\, W_k^{\mathrm{idq}}=n+m, \forall k\geq n+m-1 \tag{26}$$

Remark 2:

1. In Theorem 1, $\psi(j), j=0,1,2,\cdots,k$ is defined by the quantizer, while $A^{k-j}\Pi(j)F(j)F^{\mathrm{T}}(j)\Pi(j)(A^{k-j})^{\mathrm{T}}, j=0,1,2,\cdots,k$, which is the same part as in the unquantized system, reflects the intrinsic properties of the system. Hence, the full rank requirement of W_k^{idq} shows that the parameter identifiability of the quantized system is defined by quantizer and intrinsic properties of the system jointly;
2. When quantization level $N=1$, i.e. $a_1=-\infty$, $a_2=+\infty$, $\psi(j)\equiv 0$, then $W_k^{\mathrm{idq}}\equiv 0$, condition (26) is not satisfied and the system is not identifiable. This is consistent with the intuition. From (10) and (25), it can be observed that the difference between unquantized estimability Gramian W_k^{id} and quantized estimability Gramian W_k^{idq} is that the later includes additional weights $\psi^2(j), j=0,1,2,\cdots,k$. As a result, it can be seen that besides the situation of quantization level $N=1$, the matrix W_k^{idq} may become singular due to the property of $\psi^2(j)$, though W_k^{id} has full rank;
3. The quantizer in Theorem 1 is time-invariant. However, by using the above analysis method, a conclusion similar to Theorem 1 can be derived for time-variant quantizer, except that the weights in the estimability Gramian reflects the time-variant property of the quantizer, i.e.,

$$\psi(j)\triangleq\sum_{l=1}^{N}\frac{D_j(z_l(j))}{\sqrt{F^{\mathrm{T}}(j)\Pi(j)F(j)+R}}\left(\phi\left(\frac{a_l(j)-F^{\mathrm{T}}(j)\bar{\theta}(j)}{\sqrt{F^{\mathrm{T}}(j)\Pi(j)F(j)+R}}\right)-\phi\left(\frac{a_{l+1}(j)-F^{\mathrm{T}}(j)\bar{\theta}(j)}{\sqrt{F^{\mathrm{T}}(j)\Pi(j)F(j)+R}}\right)\right) \tag{27}$$

4. Condition (26) is equivalent to that the matrix sequence $\{\psi(j)A^{k-j}\Pi(j)F(j),\ j\le k\}$ has column rank $n+m$. $\{\psi(j)A^{k-j}\Pi(j)F(j),\ j\le k\}$ can be decomposed as $\{\psi(j)A^{k-j}\Pi(j)F(j),\ j\le k\}diag\{\psi(0),\psi(1),\cdots,\psi(k)\}$, where $diag\{\cdot\}$ denotes diagonal matrix. Hence the parameter identifiability of the original system can be preserved if $\psi(j)\ne 0, j=0,1,2,\cdots$. Especially, condition (26) is equivalent to Lemma 1 for (6). Suppose we can design a time-variant quantizer as follows

$$\begin{aligned} &D_j(z_l(j)) = c_l\sqrt{F^{\mathrm{T}}(j)\Pi(j)F(j)+R},\ l=1,2,\cdots,N,\\ &a_l(j) = d_l\sqrt{F^{\mathrm{T}}(j)\Pi(j)F(j)+R}+F^{\mathrm{T}}(j)\bar{\theta}(j),\ l=1,2,\cdots,N+1,\ j=0,1,2,\cdots,k, \end{aligned} \tag{28}$$

where c_l and d_l are constants which make

$$\psi(j)=\sum_{l=1}^{N}c_l(\phi(d_l)-\phi(d_{l+1}))\ne 0 \tag{29}$$

be the same for every j, thus,

$$W_k^{\mathrm{idq}} = (\sum_{l=1}^{N}c_l(\phi(d_l)-\phi(d_{l+1})))^2\sum_{j=k}^{0}\psi^2(j)A^{k-j}\Pi(j)F(j)F^{\mathrm{T}}(j)\Pi(j)(A^{k-j})^{\mathrm{T}}. \tag{30}$$

By comparing (30) with (10), we can find that such a time-variant quantizer does not change the parameter identifiability of the original system if and only if (29) is satisfied;

5. The parameter identifiability of the system can be preserved even if the quantization level is low as long as the quantizer is designed reasonably. Especially, when quantization level $N=2$, set $c_1=-1$, $c_2=1$, $d_1=-\infty$, $d_2=0$, $d_3=+\infty$ in the formula (28), then $\psi(j)\equiv\sqrt{2/\pi}, j=0,1,2,\cdots$, namely a coarse quantizer of 1 bit can preserve the parameter identifiability of the original system.

4. Convergence analysis

In this section, we discuss the convergence property of the Gramian in Theorem 1, i.e., the convergence property of $\psi(j), j=0,1,2,\cdots$.

We know that

$$\sum_{l=1}^{N}\frac{F^{\mathrm{T}}(j)\bar{\theta}(j)}{\sqrt{F^{\mathrm{T}}(j)\Pi(j)F(j)+R}}(\phi(\frac{a_l-F^{\mathrm{T}}(j)\bar{\theta}(j)}{\sqrt{F^{\mathrm{T}}(j)\Pi(j)F(j)+R}})-\phi(\frac{a_{l+1}-F^{\mathrm{T}}(j)\bar{\theta}(j)}{\sqrt{F^{\mathrm{T}}(j)\Pi(j)F(j)+R}}))=0 \tag{31}$$

by the property of $\phi(\cdot)$, then $\psi(j)$ can be re-expressed as

$$\psi(j)=\sum_{l=1}^{N}\frac{D_j(z_l)-F^{\mathrm{T}}(j)\bar{\theta}(j)}{\sqrt{F^{\mathrm{T}}(j)\Pi(j)F(j)+R}}(\phi(\frac{a_l-F^{\mathrm{T}}(j)\bar{\theta}(j)}{\sqrt{F^{\mathrm{T}}(j)\Pi(j)F(j)+R}})-\phi(\frac{a_{l+1}-F^{\mathrm{T}}(j)\bar{\theta}(j)}{\sqrt{F^{\mathrm{T}}(j)\Pi(j)F(j)+R}})) \tag{32}$$

Let $\Delta = \sup_{1\le k\le N} \Delta_l$,where $\Delta_l = a_{l+1} - a_l$, $l = 1,2,\cdots,N$, let $\tilde{a}_l(j) \triangleq a_l - F^{\mathrm{T}}(j)\overline{\theta}(j)$, then $\Delta_l = \tilde{a}_{l+1}(j) - \tilde{a}_l(j)$, $\sqrt{\cdot} \triangleq \sqrt{F^{\mathrm{T}}(j)\Pi(j)F(j)+R}$, $\overset{*}{z}_l(j) \triangleq D_j(z_l) - F^{\mathrm{T}}(j)\overline{\theta}(j)$. Consider the convergence property of the Gramian W_k^{idq} when $\Delta \to 0$. Note that $N \to \infty$ when $\Delta \to 0$, and then

$$
\begin{aligned}
\lim_{N\to\infty} \psi(j) &= \lim_{N\to\infty} \sum_{l=1}^{N} \frac{\overset{*}{z}_l(j)}{\sqrt{\cdot}}\left(\phi\left(\frac{\tilde{a}_l(j)}{\sqrt{\cdot}}\right) - \phi\left(\frac{\tilde{a}_{l+1}(j)}{\sqrt{\cdot}}\right)\right) \\
&= \sum_{l=1}^{\infty} \lim_{\Delta_l\to 0} \frac{\overset{*}{z}_l(j)}{\sqrt{\cdot}}\left(\phi\left(\frac{\tilde{a}_l(j)}{\sqrt{\cdot}}\right) - \phi\left(\frac{\tilde{a}_l(j)+\Delta_l}{\sqrt{\cdot}}\right)\right) \\
&= \sum_{l=1}^{\infty} \lim_{\Delta_l\to 0} \frac{\overset{*}{z}_l(j)}{\sqrt{\cdot}} \frac{\phi\left(\frac{\tilde{a}_l(j)}{\sqrt{\cdot}}\right) - \phi\left(\frac{\tilde{a}_l(j)+\Delta_l}{\sqrt{\cdot}}\right)}{-\frac{\Delta_l}{\sqrt{\cdot}}} . \left(-\frac{\Delta_l}{\sqrt{\cdot}}\right) \\
&= \sum_{l=1}^{\infty} \lim_{\Delta_l\to 0} \frac{\overset{*}{z}_l(j)}{\sqrt{\cdot}} \frac{\mathrm{d}}{\mathrm{d}s}\phi(s)\Big|_{s=\frac{\tilde{a}_l(j)}{\sqrt{\cdot}}} \left(-\frac{\Delta_l}{\sqrt{\cdot}}\right) \\
&= \sum_{l=1}^{\infty} \lim_{\Delta_l\to 0} \left(\frac{\overset{*}{z}_l(j)}{\sqrt{\cdot}}\right)^2 \frac{\exp\{-\frac{1}{2}(\frac{\overset{*}{z}_l(j)}{\sqrt{\cdot}})^2\}}{\sqrt{2\pi}} \frac{\Delta_l}{\sqrt{\cdot}} \\
&= \int_{\mathbf{R}} \left(\frac{\overset{*}{z}_l(j)}{\sqrt{\cdot}}\right)^2 \frac{\exp\{-\frac{1}{2}(\frac{\overset{*}{z}_l(j)}{\sqrt{\cdot}})^2\}}{\sqrt{2\pi}} \frac{\mathrm{d}\overset{*}{z}_l(j)}{\sqrt{\cdot}}
\end{aligned}
\tag{33}
$$

Let $r = \frac{\overset{*}{z}_l(j)}{\sqrt{\cdot}}$, then $\mathrm{d}r = \frac{\mathrm{d}\overset{*}{z}_l(j)}{\sqrt{\cdot}}$,

$$
\lim_{N\to\infty} \psi(j) = \int_{\mathbf{R}} r^2 \frac{e^{-\frac{r^2}{2}}}{\sqrt{2\pi}} \mathrm{d}r = 1. \tag{34}
$$

Remark 3: Equation (34) implies the convergence of Theorem 1 to Lemma 1 when $\Delta \to 0$, i.e. the quantized identifiability Gramian W_k^{idq} converges to the unquantized identifiability Gramian W_k^{id}.

5. Simulation

In order to illustrate our main conclusion, the following system is simulated with the tool of Matlab:

$$
y(k) = b(k)u(k-1) + e(k)
$$

where $b(k)$ is the parameter to be identified, and can be modeled as a Gauss-Markov process. Then the system model can be transformed to

$$\begin{cases} \theta(k+1) = a\theta(k) + w(k) \\ y(k) = F^{\mathrm{T}}(k)\theta(k) + e(k) \end{cases}$$

where $\theta(k) = b(k)$, a=0.5. $e(k)$, $w(k)$ and $\theta(0)$ are mutually statistically uncorrelated, their covariance are Q=1, R=0.1, $\Pi(0) = 1$ respectively, and the mean of $\theta(0)$ is $\bar{\theta} = 1$. Here we set $F(k) = u(k-1) = 2\sin(k) + 3$ as the assumed system input (i.e., the control signal, which can be considered to be generated, for example, by the adaptive controller), where the additive term "3" plays the role of avoiding the problem of "turn-off" (Astrom & Wittenmark, 1994).

To do the illustrative simulation, an optimal filter is required though the analysis about parameter identifiability is independent of the estimator used. The discussed linear system with Gauss-Markov parameter is transformed to state space model, and then the problem of parameter identification can be treated as states estimation. A number of quantized state estimators have been proposed by scholars in various areas, and we choose the Gaussian fit algorithm (Curry, 1970) as the filter in this section for that this filter which bases on the Gaussian assumption is near optimal and convenient to be implemented. Note that, in this simulated model, $F(k)$ is defined by $u(k-1)$ completely, so it is known at the channel receiver; however, in general model (1) $a_i(k), i = 1,2,\cdots,n$ and $b_i(k), i = 1,2,\cdots,m$ are to be identified, then $F(k)$ is defined by $u(k-1)\cdots u(k-m)$ and $-y(k-1)\cdots - y(k-n)$ jointly. So in general, the quantized signals $y_{\mathrm{q}}(k-1),\cdots,y_{\mathrm{q}}(k-n)$, instead of the actual outputs $y(k-1),\cdots,y(k-n)$ are received at the channel receiver, thus $y(k-1),\cdots,y(k-n)$ in $F(k)$ should be replaced by their estimates.

The analysis about parameter identifiability of quantized systems is suitable for any rational quantizer. Here the Max-Lloyd quantizer (Proakis, 2001) generally adopted in areas of communication and signal processing is employed. In the following statement, cases of quantization level $N = 4$ and $N = 2$ in (7) are simulated, respectively. The thresholds of the 4 level Max-Lloyd quantizer are {–∞, –0.9816, 0, 0.9816, +∞} and the outputs of the quantizer are {–1.51, –0.4528, 0.4528, 1.51} when the signal to be quantized is standardized normally distributed. In the case of 2 level quantizer, the thresholds are {–∞, 0, +∞} and the outputs of the quantizer are {–0.7979, 0.7979}. Hence the thresholds of the time-variant quantizers with 4 and 2 levels are respectively $\sigma_{y(k)} \times$ {–∞, –0.9816, 0, 0.9816, +∞}+ $E_{y(k)} \times$ {1, 1, 1, 1, 1} and $\sigma_{y(k)} \times$ {–∞, 0, +∞}+ $E_{y(k)} \times$ {1, 1, 1}, the outputs of the quantizers are $\sigma_{y(k)} \times$ {–1.51, –0.4528, 0.4528, 1.51}+ $E_{y(k)} \times$ {1, 1, 1, 1} and $\sigma_{y(k)} \times$ {–0.7979, 0.7979}+ $E_{y(k)} \times$ {1, 1}, where $E_{y(k)}$ and $\sigma_{y(k)}$ are the mean and standard deviation of the output $y(k)$ respectively.

It is obvious that the above model is parameter identifiable by Lemma 1 when it is unquantized. We get $\psi(j) \equiv 0.8823$ by calculating the weight $\psi(j)$ in equation (27) when the 4 level time-variant Max-Lloyd quantizer is used and $\psi(j) \equiv 0.6366$ when quantization level $N = 2$. Hence the parameter identifiability will not be changed by the quantization according to Remark 2, i.e. the quantized system is still parameter identifiable, theoretically. The simulation results of the quantized system shown in Fig. 1 ($N = 4$) and Fig. 2 ($N = 2$) illustrate the above conclusion.

In Fig. 1(a) and Fig. 2(a), actual values of parameter are denoted by solid lines and the estimates are denoted by dotted lines. Estimation errors are shown in Fig. 1(b) and Fig. 2(b). Fig. 1 and Fig. 2 show that the estimate can track the real value of the parameter when the outputs are quantized coarsely. The curves of prior error entropy and posterior error entropy are shown in Fig. 1(c) and Fig. 2(c). The entropy is calculated by

$$H(x) = \frac{n}{2}\ln 2\pi e + \frac{1}{2}\ln|C| \tag{35}$$

where $x \in \mathbf{R}^n$ is a Gaussian vector with covariance C, $|\cdot|$ denotes determinant. For quantized systems, the probability distribution of estimation error $\tilde{\theta}(k)$ is unknown, but is supposed to make the entropy of $\tilde{\theta}(k)$ maximal according to "maximal entropy principle" of Jaynes (Jaynes, 1957), namely, the uncertainty of $\tilde{\theta}(k)$ is supposed to be maximal in the situation of lack of prior information, hence $\tilde{\theta}(k)$ is assumed to be Gaussian, and thus (35) can be adopted to calculate the entropy of $\tilde{\theta}(k)$ in this simulation. We can observe from Fig. 1(c) and Fig. 2(c) that the posterior error entropy is strictly smaller than prior error entropy from the initial time instant. This indicates that this quantized system is parameter identifiable, and these observations consist with our analysis mentioned above perfectly. Besides, we can observe that the estimation error when quantization level $N = 2$ is greater than that in the case of quantization level $N = 4$ though the system is parameter identifiable in both of the two quantization cases. This shows that systems with different quantizers lead to different estimation precision, though all of them are parameter identifiable when rational quantizers are used.

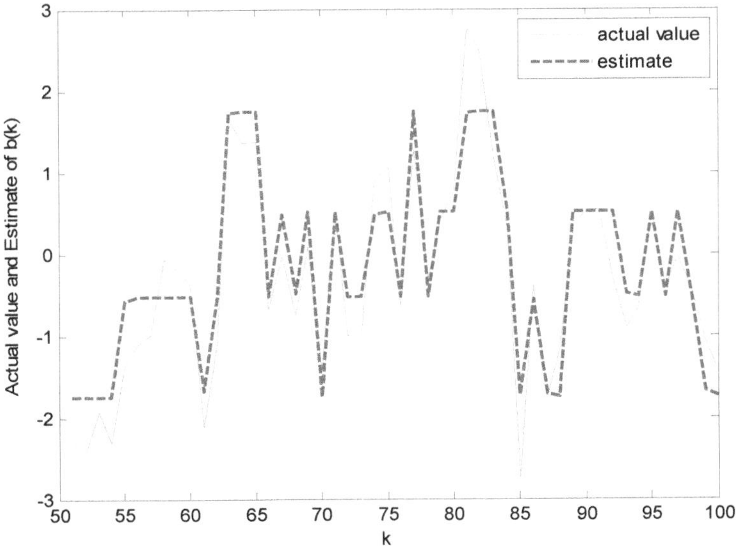

(a)

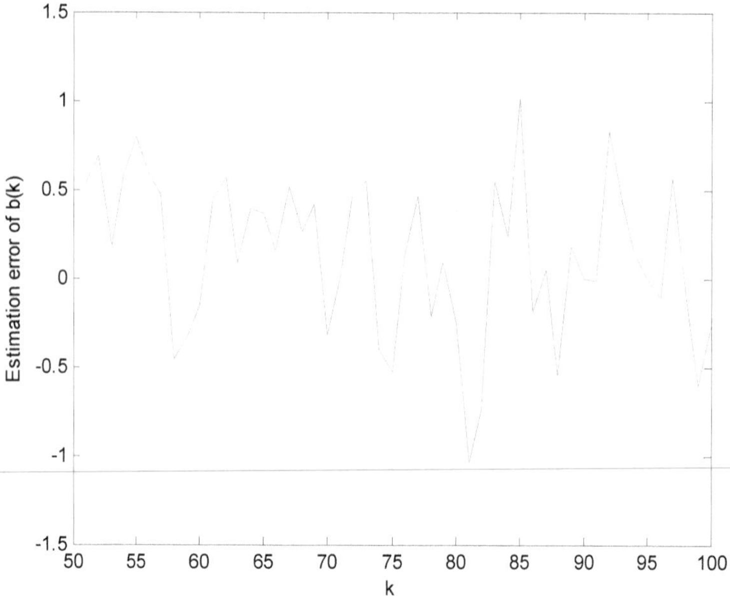

(b)

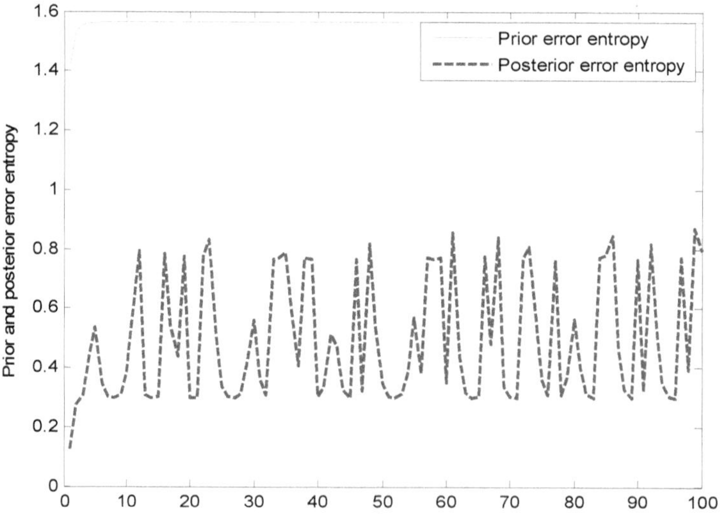

(c)

Figure 1. (a) Actual value and estimate of b(k), N=4, (b) Estimation error of b(k), N=4, (c) Prior and posterior error entropy, N=4

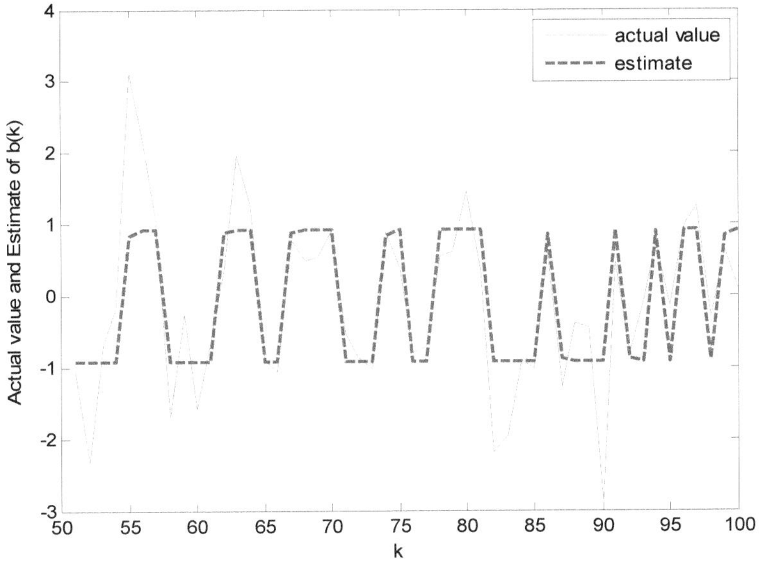
4
3
2
1
0
-1
-2
-3
50
55
60
65
70
75
80
85
90
95
100
k
Actual value and Estimate of b(k)
actual value
estimate

(a)

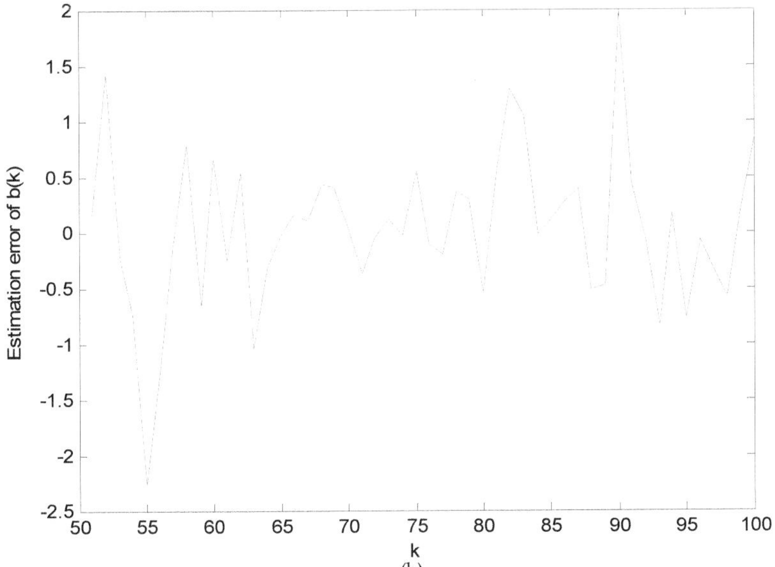
2
1.5
1
0.5
0
-0.5
-1
-1.5
-2
-2.5
50
55
60
65
70
75
80
85
90
95
100
k
Estimation error of b(k)

(b)

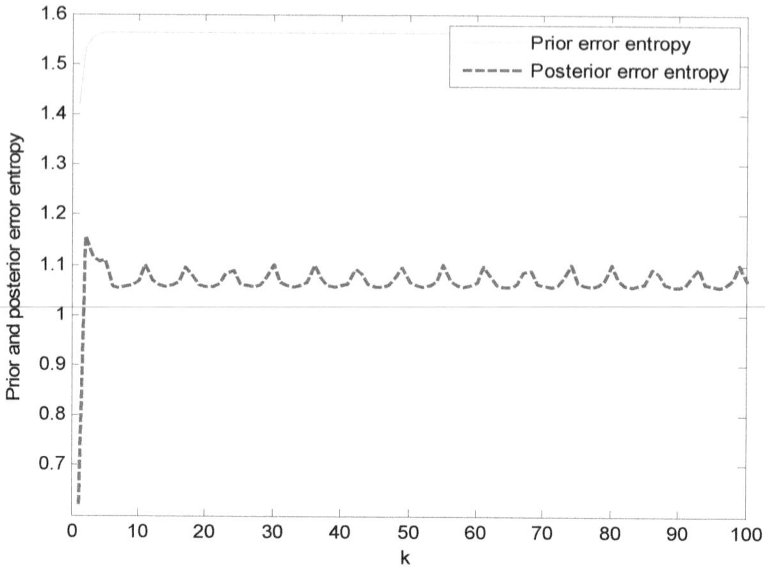

(c)

Figure 2. (a) Actual state and estimate of b(k), N=2, (b) Estimation error of b(k), N=2, (c) Prior and posterior error entropy, N=2

6. Conclusion

This paper discusses the parameter identifiability of quantized linear systems with Gauss-Markov parameters from information theoretic point of view. The existing definition concerning this property is reviewed and new definition is proposed for quantized systems. Criterion function, the Gramian of parameter identifiability for quantized systems is analyzed based on the quantity of mutual information. The derived conclusions consist with our intuition very well and also provide us with intrinsic perspective for the quantizer design. The analysis shows that the Gramian of quantized systems converge to that of unquantized systems when the quantization intervals turn to zero, and a well designed quantizer can preserve the identifiability of the original system even if the quantizer is as coarse as one bit. The analytical analysis is verified by the illustrative simulation.

Author details

Ying Shen
China Mobile Group, Zhejiang Co., Ltd. Huzhou Branch, Huzhou, China

Hui Zhang*
State Key Laboratory of Industrial Control Technology, Institute of Industrial Process Control, Department of Control Science and Engineering, Zhejiang University, Hangzhou, China

Acknowledgement

This work was supported by the Natural Science Foundation of China under Grand No. 61174063 & 60736021.

7. References

Astrom, K. J. & Wittenmark, B. (1994). *Adaptive Control* (2nd edition), Addison-Wesley Publishing Company, ISBN 0201558661, Boston, MA, USA.

Baram, Y. & Kailath T. (1988). Estimability and Regulability of Linear Systems. *IEEE Transactions on Automatic Control*, vol. 33, no. 12, (December 1988), pp. 1116-1121, ISSN 00189286.

Cover, T. M. & Thomas, J. A. (2006). *Elements of Information Theory* (2nd edition). John Wiley & Sons, Inc., ISBN 10-0-471-24195-4, New Jersey.

Curry, R. E. (1970). *Estimation and Control with Quantized Measurements* (1st edition). The M.I.T Press, ISBN 0-262-53216-6, London.

De Persis, C. & Mazenc, F. (2010). Stability of Quantized Time-Delay Nonlinear Systems: A Lyapunov-Krasowskii-Functional Approach. *Mathematics of Control, Signals, & Systems*, vol. 21, no. 4, (August 2010), pp. 337-370, ISSN 09324194.

Durgaryan, I. S. & Pashchenko, F. F. (2001). Identification of Objects by the Maximal Information Criterion. *Automation & Remote Control*, vol. 62, no. 7, (July 2001), pp. 1104-1114, ISSN 00051179.

Fu, M.-Y. & Xie, L.-H. (2005). The Sector Bound Approach to Quantized Feedback Control. *IEEE Transactions on Automatic Control*, vol. 50, no. 11, (November 2005), pp. 1698-1711, ISSN 0018-9286.

Gray, R. M. & Neuhoff, D. L. (1998). Quantization. *IEEE Transactions on Information Theory*, vol. 44, no. 6, (October 1998), pp. 2325-2383, ISSN 00189448.

Jaynes, E. T. (1957). Information Theory and Statistical Mechanics, I & II. *Physical Review*, vol. 106, no. 4, (May 1957), pp. 620-630, ISSN 0031899X, vol. 108, no. 2, (October 1957), pp. 171-190, ISSN 0031899X.

Proakis, J. G. (2001). *Digital Communications* (4th edition). The McGraw-Hill Companies, ISBN 0-07-232111-3, New York, NY.

* Corresponding Author

Ribeiro, A. ; Giannakis, G. B. & Roumeliotis, S. I. (2006). SOI-KF: Distributed Kalman Filtering with Low-Cost Communications Using The Sign of Innovations. *IEEE Transactions on Signal Processing*, vol. 54, no. 12, (December 2006), pp. 4782-4795, ISSN 1053587X.

Wang, L.-J. & Zhang, H. (2011). On Parameter Identifiability of Linear Multivariable Systems with Communication Access Constraints. *Journal of Zhejiang University (Engineering Science)*, vol. 45, no. 12, (December 2011), pp. 102-107, ISSN 1008-973X, in Chinese.

Wang, L.-Y. ; Zhang, J.-F. & Yin, G. G. (2003). System Identification Using Binary Sensors. *IEEE Transactions on Automatic Control*, vol. 48, no. 11, (November 2003), pp. 1892-1907, ISSN 00189286.

Wang, L.-Y. ; Yin, G. G. ; Zhao, Y.-L. & Zhang, J.-F. (2008). Identification Input Design for Consistent Parameter Estimation of Linear Systems With Binary-Valued Output Observations. *IEEE Transactions on Automatic Control*, vol. 53, no. 4, (May 2008), pp. 867-880, ISSN 00189286.

Wang, L.-Y. ; Yin, G. G. & Xu, C.-Z. (2010). Identification of Cascaded Systems with Linear and Quantized Observations. *Asian Journal of Control*, vol. 12, no. 1, (January 2010), pp. 1-14, ISSN 15618625.

You, K.-Y. ; Xie, L.-H. ; Sun, S.-L. & Xiao, W.-D. (2011). Quantized Fltering of Linear Stochastic Systems. *Transactions of the Institute of Measurement and Control*, vol. 33, no. 6, (August 2011), pp. 683-698, ISSN 01423312.

Zhang, H. (2003). Information Descriptions and Approaches in Control Systems, Ph.D Dissertation, Department of Control Science & Engineering, Zhejiang University, in Chinese.

Zhang, H. & Sun, Y.-X. (1996). Estimability of Stochastic Systems – An Information Theory Approach. *Control Theory and Applications*, vol. 13, no. 5, (October 1996), pp. 567-572, ISSN 10008152, in Chinese.

Permissions

The contributors of this book come from diverse backgrounds, making this book a truly international effort. This book will bring forth new frontiers with its revolutionizing research information and detailed analysis of the nascent developments around the world.

We would like to thank Prof. Ivan Ganchev Ivanov, for lending his expertise to make the book truly unique. He has played a crucial role in the development of this book. Without his invaluable contribution this book wouldn't have been possible. He has made vital efforts to compile up to date information on the varied aspects of this subject to make this book a valuable addition to the collection of many professionals and students.

This book was conceptualized with the vision of imparting up-to-date information and advanced data in this field. To ensure the same, a matchless editorial board was set up. Every individual on the board went through rigorous rounds of assessment to prove their worth. After which they invested a large part of their time researching and compiling the most relevant data for our readers. Conferences and sessions were held from time to time between the editorial board and the contributing authors to present the data in the most comprehensible form. The editorial team has worked tirelessly to provide valuable and valid information to help people across the globe.

Every chapter published in this book has been scrutinized by our experts. Their significance has been extensively debated. The topics covered herein carry significant findings which will fuel the growth of the discipline. They may even be implemented as practical applications or may be referred to as a beginning point for another development. Chapters in this book were first published by InTech; hereby published with permission under the Creative Commons Attribution License or equivalent.

The editorial board has been involved in producing this book since its inception. They have spent rigorous hours researching and exploring the diverse topics which have resulted in the successful publishing of this book. They have passed on their knowledge of decades through this book. To expedite this challenging task, the publisher supported the team at every step. A small team of assistant editors was also appointed to further simplify the editing procedure and attain best results for the readers.

Our editorial team has been hand-picked from every corner of the world. Their multi-ethnicity adds dynamic inputs to the discussions which result in innovative outcomes. These outcomes are then further discussed with the researchers and contributors who give their valuable feedback and opinion regarding the same. The feedback is then collaborated with the researches and they are edited in a comprehensive manner to aid the understanding of the subject.

Apart from the editorial board, the designing team has also invested a significant amount of their time in understanding the subject and creating the most relevant covers. They scrutinized every image to scout for the most suitable representation of the subject and create an appropriate cover for the book.

The publishing team has been involved in this book since its early stages. They were actively engaged in every process, be it collecting the data, connecting with the contributors or procuring relevant information. The team has been an ardent support to the editorial, designing and production team. Their endless efforts to recruit the best for this project, has resulted in the accomplishment of this book. They are a veteran in the field of academics and their pool of knowledge is as vast as their experience in printing. Their expertise and guidance has proved useful at every step. Their uncompromising quality standards have made this book an exceptional effort. Their encouragement from time to time has been an inspiration for everyone.

The publisher and the editorial board hope that this book will prove to be a valuable piece of knowledge for researchers, students, practitioners and scholars across the globe.

List of Contributors

J. Linares-Pérez
Dpto. de Estadística e I.O. Universidad de Granada. Avda. Fuentenueva,18071. Granada, Spain

R. Caballero-Águila
Dpto. de Estadística e I.O. Universidad de Jaén. Paraje Las Lagunillas, 23071. Jaén, Spain

I. García-Garrido
Dpto. de Estadística e I.O. Universidad de Granada. Avda. Fuentenueva,18071. Granada, Spain

David Opeyemi
Rufus Giwa Polytechnic, Owo, Nigeria

Serena Doria
Department of Engineering and Geology, University G. d'Annunzio, Chieti-Pescara, Italy

Uwe Küchler
Humboldt University Berlin, Germany

Vyacheslav A. Vasiliev
Tomsk State University, Russia

López-Ruiz Ricardo
Department of Computer Science, Faculty of Science, Universidad de Zaragoza, Zaragoza, Spain
BIFI, Institute for Biocomputation and Physics of Complex Systems, Universidad de Zaragoza, Zaragoza, Spain

Sañudo Jaime
Department of Physics, Faculty of Science, Universidad de Extremadura, Badajoz, Spain
BIFI, Institute for Biocomputation and Physics of Complex Systems, Universidad de Zaragoza, Zaragoza, Spain

Jingtao Shi
Shandong University, P. R. China

Tze Leung Lai
Department of Statistics, Stanford University, U.S.A.

Tiong Wee Lim
Department of Statistics and Applied Probability, National University of Singapore, Republic of Singapore

Sergey V. Sokolov
Rostov State University of Means of Communication, Russia

Ivan Ivanov
Sofia University "St.Kl.Ohridski", Bulgaria

Raúl Fierro
Instituto de Matemática, Pontificia Universidad Católica de Valparaíso, Centro de Investigación y Modelamiento de Fenómenos Aleatorios, Universidad de Valparaíso, Chile

Nicholas A. Nechval and Maris Purgailis
University of Latvia, Latvia

Vladimir Šimović, Siniša Fajt and Miljenko Krhen
University of Zagreb, Croatia

Ying Shen
China Mobile Group, Zhejiang Co., Ltd., Huzhou Branch, Huzhou, China

Hui Zhang
State Key Laboratory of Industrial Control Technology, Institute of Industrial Process Control, Department of Control Science and Engineering, Zhejiang University, Hangzhou, China

Printed in the USA
CPSIA information can be obtained
at www.ICGtesting.com
JSHW011453221024
72173JS00005B/1052

9 781632 403544